D0146127

UNITED STATES
WEST COAST

Other Titles in
ABC-CLIO'S
NATURE AND HUMAN SOCIETIES SERIES

NATURE AND HUMAN SOCIETIES

UNITED STATES
WEST COAST
An Environmental History

Adam M. Sowards
Mark Stoll, Series Editor

Property of
St. John Fisher College
Lavery Library
Rochester, N.Y. 14618

A B C  C L I O

Santa Barbara, California • Denver, Colorado • Oxford, England

Copyright © 2007 by ABC-CLIO, Inc.

All rights reserved. No part of this publication may be reproduced, stored in a retrieval system, or transmitted, in any form or by any means, electronic, mechanical, photocopying, recording, or otherwise, except for the inclusion of brief quotations in a review, without prior permission in writing from the publishers.

Library of Congress Cataloging-in-Publication Data
Sowards, Adam M.
  United States West Coast : an environmental history / Adam M. Sowards.
    p. cm. — (ABC-CLIO's nature and human societies series)
  Includes bibliographical references and index.
    ISBN 978-1-85109-909-2 (hard copy : alk. paper) — ISBN 978-1-85109-910-8 (ebook)
    1. Human ecology—Pacific Coast (U.S.)—History.   2. Nature—Effect of human beings on—Pacific Coast (U.S.)   3. Pacific Coast (U.S.)—Environmental conditions.
I. Title.
GF504.P33S69 2007
304.20979—dc22

2007014541

11 10 09 08 07   1 2 3 4 5 6 7 8 9 10

Production Editor: Anna A. Moore
Editorial Assistant: Sara Springer
Production Manager: Don Schmidt
Media Editor: Jed DeOrsay
Media Resources Coordinator: Ellen Brenna Dougherty
Media Resources Manager: Caroline Price
File Manager: Paula Gerard

ABC-CLIO, Inc.
130 Cremona Drive, P.O. Box 1911
Santa Barbara, California 93116–1911

This book is also available on the World Wide Web as an ebook.
Visit http://www.abc-clio.com for details.

This book is printed on acid-free paper. ∞

Manufactured in the United States of America

To the memory of my grandparents, who rooted me in the West's past:
Vernon Sowards (1912–1982)
Lillian Maxine Wills Sowards (1913–1996)
Elwin Paul Matthews (1916–1989)
Ruth Virlea Burns Matthews (1917–2004)

CONTENTS

SERIES FOREWORD

Long ago, only time and the elements shaped the face of the earth, the black abysses of the oceans, and the winds and blue welkin of heaven. As continents floated on the mantle, they collided and threw up mountains or drifted apart and made seas. Volcanoes built mountains out of fiery material from deep within the earth. Mountains and rivers of ice ground and gorged. Winds and waters sculpted and razed. Erosion buffered and salted the seas. The concert of living things created and balanced the gases of the air and moderated the earth's temperature.

The world is very different now. From the moment our ancestors emerged from the southern forests and grasslands to follow the melting glaciers or to cross the seas, all has changed. Today the universal force transforming the earth, the seas, and the air is for the first time a single form of life: we humans. We shape the world, sometimes for our purposes and often by accident. Where forests once towered, fertile fields or barren deserts or crowded cities now lie. Where the sun once warmed the heather, forests now shade the land. We exterminate one creature only to bring another from across the globe to take its place. We pull down mountains and excavate craters and caverns; drain swamps and make lakes; divert, straighten, and stop rivers. From the highest winds to the deepest currents, the world teems with chemical concoctions that only we can brew. Even the very climate warms from our activity.

And as we work our will upon the land, as we grasp the things around us to fashion them into instruments of our survival, our social relations, and our creativity, we find in turn our lives and even our individual and collective destinies shaped and given direction by natural forces, some controlled, some uncontrolled, and some unleashed. What is more, uniquely among the creatures, we come to know and love the places where we live. For us, the world has always abounded with unseen life and manifest meaning. Invisible beings have hidden in springs, in mountains, in groves, in the quiet sky and the thunder of the clouds, in the deep waters. Places of beauty from magnificent mountains to small, winding brooks have captured our imaginations and our affection. We have perceived a mind like our own, but greater, designing, creating, and guiding the universe around us.

The authors of the books in this series endeavor to tell the remarkable epic of the intertwined fates of humanity and the natural world. It is a story only now coming to be fully known. Although traditional historians have told the drama of men and women of the past, for more than three decades now, many historians have added the natural world as a third actor. Environmental history by that name emerged in the 1970s in the United States. Historians quickly took an interest and created a professional society, the American Society for Environmental History, and a professional journal, now called *Environmental History*. U.S. environmental history flourished and attracted foreign scholars. By 1990 the international dimensions were clear; European scholars joined together to create the European Society for Environmental History in 2001, with its journal, *Environment and History*. A Latin American and Caribbean Society for Environmental History should not be far behind. With an abundant and growing literature of world environmental history now available, a true world environmental history can appear.

This series is organized geographically into regions determined as much as possible by environmental and ecological factors, and secondarily by historical and historiographical boundaries. Befitting the vast environmental historical literature on the United States, four volumes tell the stories of the North, the South, the Plains and Mountain West, and the Pacific Coast. Other volumes trace the environmental histories of Canada and Alaska, Latin America and the Caribbean, Northern Europe, the Mediterranean region, sub-Saharan Africa, South Asia, Southeast Asia, East Asia, and Australia and Oceania. Authors from around the globe, experts in the various regions, have written these volumes, almost all of which are the first to convey the complete environmental history of their subjects. Each author has, as much as possible, written the twin stories of the human influence on the land and of the land's manifold influences on its human occupants. Every volume contains a narrative analysis of a region along with a body of reference material. This series constitutes the most complete environmental history of the globe ever assembled, chronicling the astonishing tragedies and triumphs of the human transformation of the earth.

The process of creating the series, recruiting the authors from around the world, and editing their manuscripts has been an immensely rewarding experience for me. I cannot thank the authors enough for all of their effort in realizing these volumes. I owe a great debt to Kevin Downing, who first approached me about the series, and Steven Danver at ABC-CLIO, who has shepherded the volumes through delays and crises all the way to publication. Their unfaltering support for and belief in the series were essential to its successful completion.

Mark Stoll
Department of History, Texas Tech University
Lubbock, Texas

PREFACE

I grew up on a small farm at the edge of the Tulalip Indian Reservation in Marysville, Washington, about forty miles north of Seattle. It was not a large farm; indeed, by the time I arrived in the early 1970s, it did not provide our family's main income. Nevertheless, I could not help but be imprinted by the place. I have come to believe it is why I became an environmental historian. Although I certainly did not recognize it at the time, that small farm and its neighborhood exemplified many elements of the region's environmental history.

Perhaps a century or more ago, the land remained covered by forest. It is conceivable that the local Native population stripped cedar bark from trees on the land and fished salmon from the small creek that coursed through the trees. Then came loggers and railroads, a combination that meant forests turned to lumber for local markets. As they did in so many places in North America, farmers set about converting the cutover land into productive farms. Indeed, early in the twentieth century that patch of land became the first dairy farm owned and operated by Tulalips. This transformation to farming marked the first significant change in land use since the last ice age.

For the better part of the twentieth century, successive farmers planted some crops but mostly raised domestic animals. My own family raised chickens for a commercial laying flock and replacement heifers for nearby dairies. Neighboring land had cattle and horses, too. In mid-century, however, the federal government invested millions of dollars in building a superhighway system, a portion of which I would be able to see from my living room window. Interstate 5, which runs the length of the West Coast, demanded more changes in our neighborhood. A developer bought the farm next to ours and dug two enormous holes, using the soil to shape the bed on which the road engineers built the highway. Groundwater filled those holes, and so I grew up across the street from two large, human-made lakes. In addition, the highway changed drainage patterns, which affected our farm making some places wetter and some drier. Another significant shift had occurred.

With population pressures and profit to be made, that developer decided to turn the erstwhile farm into a lakefront development. In the late 1980s, then, that land converted to a suburban subdivision, complete with chemical-drenched lawns and too-large, close-together homes. The ever-expanding Seattle metropolitan area fueled this suburban growth. My father, who had worked our land as a farmhand before he and my mother bought it in the 1960s, took a job in town in the late 1970s at a store that primarily sold farm supplies. As the transformation of farm-to-suburb swept across the county, his store's products changed. Now he serves hobby farmers and soccer moms buying expensive dog food far more than he sells to working farmers. The suburban landscape now dominates the neighborhood, although my parents' land continues to keep suburban development at bay.

Not all is in ecological decline, however. In recent years, salmon have been seen in the creek and ditches. Some of the remaining farmers do not wish the authorities to know, for fear endangered species laws may force them to change their agricultural practices. Trees have returned so much so that the once-unobstructed view of the Cascade Mountains that I enjoyed from my living room is gone. Only the peaks of the mountains are visible. The gouging clear-cuts that were once visible appear from a distance to be healing.

Many of the themes that appear in this book were ones that, unbeknownst to me, I lived in the shadow of. These ecological transformations—forest to farm to suburb—characterize much of the region's environmental history.

BOUNDARIES

Mark Stoll offered me the freedom to define the boundaries of this study as I saw fit. Although originally conceived as being a volume on the West Coast of North America, certain choices and inclinations on my part have both enlarged and shrunk those borders. In an attempt to make the political and economic contexts more manageable, I have committed a grand sin of environmental history and delimited the study by political borders, instead of ecological ones. Although there are good arguments to include British Columbia (and Alaska, as well) in the north and Mexico in the south, this book remains centered on what became by 1848 United States territory. However, in an attempt to use some ecological criteria to establish the edges of the study area, I elected to follow watersheds inland from the coast. Since the Columbia-Snake river systems gather water from much of the Northwest interior and push through the Cascade Mountains, I have extended the original boundary eastward, over the mountains. In practical effect, then, this region I call interchangeably the Pacific West or Far West constitutes most or all of the modern states of California, Oregon, Washington, and Idaho. Although these boundaries may not be entirely satisfactory (even to me), they allow the study to encompass a generous concept of the United States West Coast.

Adam M. Sowards

ACKNOWLEDGMENTS

At the 2004 Western History Association meeting, I sat eating lunch with friends and some new acquaintances. One of the individuals I met, Steve Danver, suggested that I write this book. I thought he was joking. Within the week, however, Steve, an editor for ABC-CLIO, and Mark Stoll, the editor of the Nature and Human Societies Series, were hard at work successfully convincing me to undertake this project. Although I am now more careful with whom I eat at conferences, I am grateful to them both. Throughout this process Steve has provided sage advice and has remained a cheerful editor, answering my inane questions with a promptness that puts most any editor to shame. The rest of the editorial staff at ABC-CLIO has been similarly professional and helpful. Mark, too, has been incredibly helpful, patient, and positive—all qualities I needed as I sometimes struggled through difficult material and circumstances. He provided confidence when I couldn't muster any, and even more important, he gave me freedom to pursue this project as I saw it. A writer could hardly ask for more.

The University of Idaho provided a supportive environment conducive to completing this project. Kathy Aiken, my erstwhile department chair and now dean, encouraged me to write this book and wisely suggested that I use the university's John Calhoun Smith Memorial Fund to pay for successive research assistants: Valerie Park, Amy Canfield, Rob Bobier, and Dan Karalus. Amy and Dan were especially helpful at critical times with many of the reference materials. Rodney Frey read and discussed Chapter Two and provided a perspective that I needed. I called on several historians for counsel while puzzling through the region's environmental history and for suggestions for primary sources. I thank them all, especially Andrew Duffin, who served as a constant sounding board and good friend.

My family sustains me. On a daily basis, Elizabeth and Ella lived with this book as long as I did. I gained from their support, patience, and companionship far more than I probably deserved or can express. My family in Washington, Idaho, Colorado, and Connecticut have encouraged and supported me in countless ways for years and kindly asked about the book's progress with an interest I deeply appreciated.

As I was finalizing the book's contract, my maternal grandmother died. Because she was my last living grandparent and because those grandparents root me firmly in the Western past, I dedicate this book to them.

Adam M. Sowards
University of Idaho

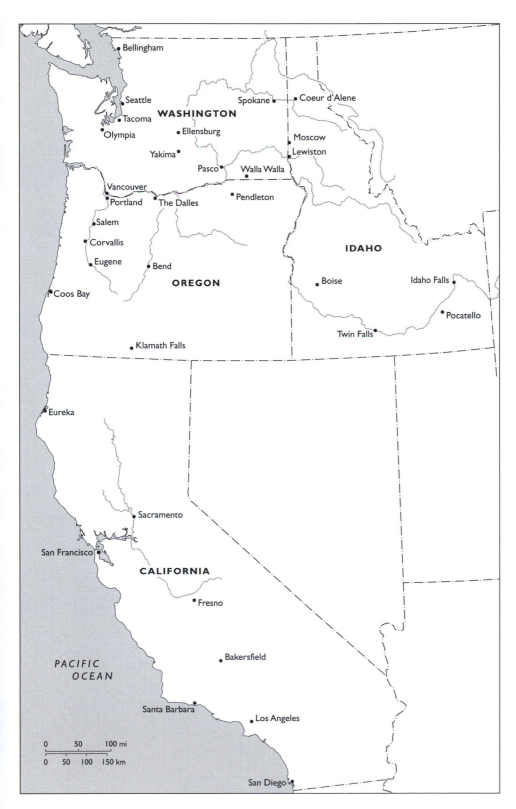

Washington, Oregon, Idaho, and California: Major Cities

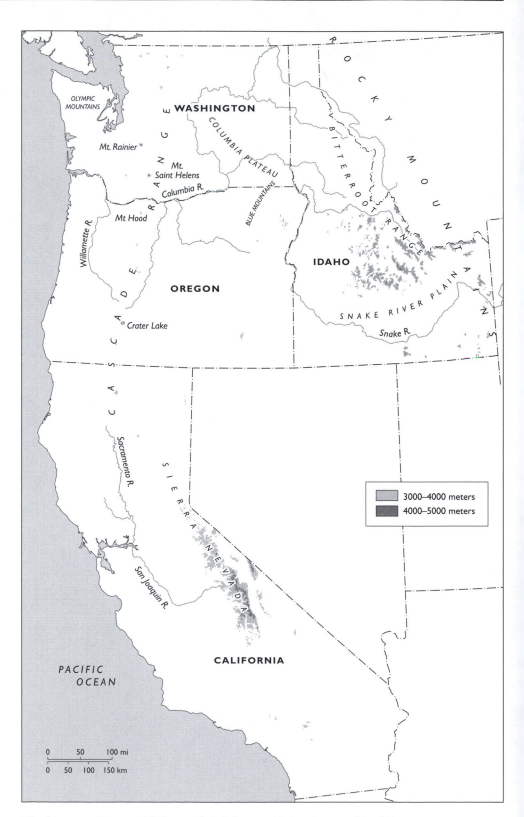

Washington, Oregon, Idaho, and California: Major Geographical Features

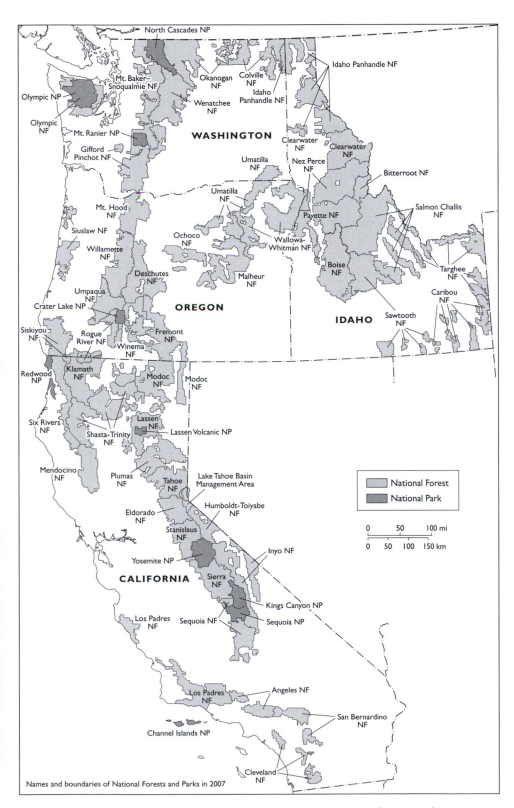

North Cascades NP

Mt. Baker–Snoqualmie NF

Okanogan NF

Colville NF

Idaho Panhandle NF

Idaho Panhandle NF

Olympic NP

Olympic NF

Mt. Ranier NP

Wenatchee NF

Gifford Pinchot NF

WASHINGTON

Umatilla NF

Clearwater NF

Clearwater NF

Nez Perce NF

Bitterroot NF

Umatilla NF

Mt. Hood NF

Salmon Challis NF

Siuslaw NF

Payette NF

Willamette NF

Ochoco NF

Wallowa-Whitman NF

Targhee NF

Deschutes NF

Boise NF

Caribou NF

Umpaqua NF

Malheur NF

Crater Lake NP

OREGON

IDAHO

Siskiyou NF

Rogue River NF

Fremont NF

Sawtooth NF

Winema NF

Redwood NP

Klamath NF

Modoc NF

Modoc NF

Six Rivers NF

Lassen NF

Lassen Volcanic NP

Shasta-Trinity NF

Mendocino NF

Plumas NF

Tahoe NF

Lake Tahoe Basin Management Area

Humboldt-Toiyabe NF

Eldorado NF

Stanislaus NF

Inyo NF

Yosemite NP

Sierra NF

CALIFORNIA

Kings Canyon NP

Los Padres NF

Sequoia NF

Sequoia NP

| | National Forest |
| | National Park |

0   50   100 mi

0   50   100   150 km

Los Padres NF

Angeles NF

San Bernardino NF

Channel Islands NP

Cleveland NF

Names and boundaries of National Forests and Parks in 2007

Washington, Oregon, Idaho, and California: National Parks and National Forests

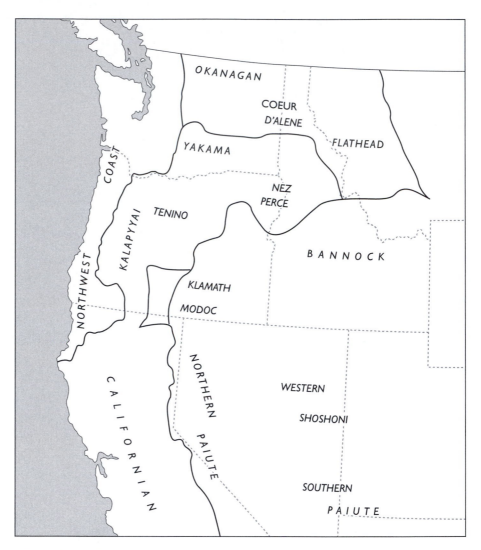

Major Native American Groups and Their Locations

A LAND OF UPHEAVAL
AND DIVERSITY

On 18 May 1980, Mount Saint Helens dramatically erupted. Volcanic ash—hundreds of thousands of tons—blanketed the region and brought darkness to daytime skies; floods unleashed great power and carried tons of debris down through once-quiet watersheds; millions of trees blew over as if they were toothpicks; and sixty people died (Ewert 1999, 4). Almost fourteen years later, in the early morning of 17 January 1994, residents of Southern California were shaken awake by the powerful magnitude 6.7 Northridge earthquake. The aftermath left 25,000 people homeless and killed 72, 16 of whom were crushed when an apartment building collapsed on itself. Damage estimates were placed at $42 billion (Davis 1998, 7, 31–32). These natural disasters reminded westerners of nature's power. Although most considered them catastrophic—and it is hard to argue with that characterization—they were quite small compared with the volcanic and seismic events that shaped the region over a thousand millennia.

The Pacific West has a deep history of dramatic upheavals that produced a land of ecological contrasts and frequent landscape transformations. The environment's autonomy and dynamism are rooted in this deep past, and the nature the first humans encountered was a product of its accumulated history.

MAJOR FORCES

Like all regions, the Far West is a result of major geologic forces. For millions of years, constant—sometimes cataclysmic, usually subtle—change shaped the region. Plates collided. Mountains rose. Mountains eroded. Volcanoes erupted. Floods scoured. Glaciers scraped. Flora and fauna evolved and died. During deep time, ecological changes developed entirely without human interference. Although seemingly remote, these periods influence the region's environmental history still.

Plate Tectonics

Continents float on plates that make up the outer 60 miles of the earth called the lithosphere. These plates constantly move. They may grate against one another, move away from each other, or crash directly into one another. In the oceans, ridges develop where plates move apart and basalt magma pours through the opening to form the oceanic crust. Oceanic trenches develop where plates collide. About 200 million years ago, the North American continent met with the Pacific Ocean's floor. The collision of the oceanic and continental lithospheres resulted in an oceanic trench where the Pacific Ocean floor submerged beneath the North American continent. The process moved slowly, and continents traveled only about two inches a year. Nevertheless, these collisions millions of years ago brought forth many of the region's landmarks, such as many of its mountain ranges (Alt and Hyndman 2000, 1–2; Alt and Hyndman 1989, 1–9; Alt and Hyndman 1984, 1–9; Alt and Hyndman 1978, 1–5).

In addition to the subversion of the ocean floor, microcontinental islands in the eastern Pacific rammed into and merged with the North American continent. For example, to form much of the interior Northwest, the Okanogan microcontinent connected with North America about 100 million years ago. Fifty million years later, the North Cascade microcontinent joined the west coast, adding more landmass to the Northwest. Meanwhile, the subversion of the ocean floor created coastal mountain ranges all along the coast. These mountains formed out of the lighter rocks from the seafloor that were scraped off as the rest of the seafloor dove beneath the North American continent. The Sierra Nevada of California and the Blue and Wallowa ranges in eastern Oregon, now hundreds of miles to the interior, were originally coastal ranges created when spreading ocean floors met moving continents. The movement of plates, however, cannot explain all of the region's geography (Alt and Hyndman 1978, 3–5; Alt and Hyndman 1984, 1–9,13–15; Alt and Hyndman 1989, 21–22; Alt and Hyndman 2000, 2–4).

Volcanism

Volcanic activity along the western edge of North America contributed dramatic and important episodes to the Pacific West's environmental history. Much of the northern interior, for example, formed from basalt lava flows. The Columbia Plateau, which constitutes the bulk of the inland Northwest, grew from basalt lava flows 17 million and 11 million years ago. The Snake River

Plain in southern Idaho also formed from volcanic activity that left a 70-mile-wide stretch of ash and lava during the Miocene and more recent Pleistocene basalt flows. In addition, the basalt lava flows that came from the oceanic ridges and spread over the ocean floor melt once they reach about 60 miles below the surface and return to the surface as volcanoes. Moreover, wherever a coastal range built from subverting plates, an inland series of volcanoes developed 50 to 100 miles away. Collectively, the volcanoes or volcanic plateaus that spewed their molten basalt or ash have gone extinct, but a few remnants are evident in the inland mountains, such as those in northeastern Oregon (Mueller and Mueller 1997, 17; Orr and Orr 1996, 221, 288; Alt and Hyndman 1978, 3–5; Alt and Hyndman 1984, 1–9; Alt and Hyndman 1989, 25–28; Alt and Hyndman 2000, 2–4).

However, much younger volcanoes still dot the Cascade Mountain range. They rise among the mountains, comprising several of the most distinctive peaks of the range, including Mount Shasta, Lassen Peak, Mount Hood, and Mount Rainer. They are part of a Pacific Rim of Fire, a string of volcanoes that largely encircle the Pacific Ocean. These prominent peaks tend to rise approximately a mile above their neighboring mountains reaching as high as 14,410 feet (Mount Rainier) and as low as 7,985 (Mount Newberry). These volcanoes appeared during the Pliocene some 7 million years ago. The youngest volcano in the Cascades, Mount Saint Helens, is also the youngest in the continental United States. The continued presence of volcanic activity in the Far West demonstrates that the processes of shaping the land of the Pacific West continue (Harris 1980, 15–23; McKee 1972, 193–217).

The eruptions of Mount Mazama are instructive of volcanoes' power. Scientists estimate Mount Mazama rose around 12,000 feet above sea level. A series of eruptions beginning 7,900 years before present (BP) covered with Mazama Ash a massive 350,000 square miles of northern Nevada, Idaho, Oregon, and Washington, along with southern British Columbia, Alberta, perhaps Saskatchewan, and western Montana and portions of Wyoming. At the volcano's base, the ash from the initial eruptions is 20 feet deep, while 70 miles distant it remains a foot deep. Meanwhile, the intense heat of the eruption made charcoal out of the nearby trees. Fifteen to seventeen cubic miles of Mount Mazama's bulk disappeared. The eruptions collapsed the center of the mountain 2,500 feet. In place of Mount Mazama is now Crater Lake, a beautiful lake 4,000 feet deep and approximately 5 miles across, managed by the National Park Service. Together with plate tectonics, volcanism produced most of the terrestrial shape of the Far West (Harmon 2002, 10–12; Harris 1980, 85–105; McKee 1972, 212–217; Orr and Orr 1996, 108–10).

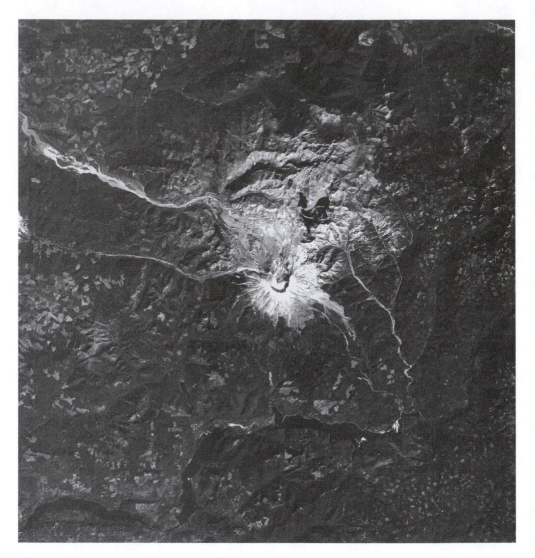

This photograph, taken from a satellite, shows Mount Saint Helens on September 28, 2004, only a few days after small earthquakes shook the region and a dome of molten rock began to rise in the volcano's crater. In 1980, the volcano erupted with a blast that wreaked devastation across 200 square miles. (U.S. Geological Survey)

Glaciation

The last major geologic force is glaciation. As it does everywhere, global climate patterns influence the region's glacial history. Alternately cooling and heating, the climate continually changes. Dramatic climate change can be explained partially by the placement of continental landmasses. Continents and oceans have been situated so that warm tropical ocean currents easily moved into polar regions. Now, Antarctica blocks currents in the south and land-masses almost entirely surround the Arctic Ocean, preventing those warm currents from reaching the poles. These factors have produced the current glacial age, creating cool summers in high latitudes that cannot generally melt the snow and ice that form during winter. The most severe ice ages, called glacial ages, are infrequent in the earth's history and include many periods of glaciations lasting sixty to ninety percent of the age. During these periods, glaciers work as erosive agents, carrying rocks along with them as they move up and down mountains and valleys. This movement carved trenches as glaciers receded, significantly changing the terrain. The subsequent mild interludes are called interglacials; we currently live in an interglacial. About 18,000 years ago, the climate cycle shifted and summers warmed, marking the maximum reach of the last glaciation. After that, the earth transitioned into an interglacial, one that sparked important ecological and ultimately cultural changes in the Pacific West. Ice sheets retreated, and plants and animals colonized western North America, forming most of the region's contemporary biota (Harris 1980, 34–39; Pielou 1991, 5–9).

Like volcanism, glaciation could occasionally produce significant effects in a short period. Glacial Lake Missoula rested just south of the Cordilleran ice sheet west of the Rocky Mountains; it was about 300 kilometers long with a volume of 2,000 cubic kilometers, comparable to today's Lake Ontario. It periodically emptied and filled about forty times over 1,500 years. The northeast boundary of the lake was the ice sheet; the southwest was bounded by rising land. A narrow ice dam in the Purcell Trench lobe in the present-day Lake Pend Oreille valley governed the cycle. As the climate warmed and the ice sheet melted, meltwater filled the lake. As it became too full, the ice dam would begin to float, no longer functioning as a dam. Then, Lake Missoula would flood the land in the Columbia River valley. The entire lake could drain away in less than two weeks. The ice dam would then re-form, and the lake would refill. The cycle occurred every twenty to sixty years. Among its most visible effects were the channeled scablands, an area scraped largely clean of life and soil by these floods. These spectacular floods were violent and ecologically deadly. As soon as a new aquatic ecosystem would begin to redevelop, the next flood would largely reduce it to little life. Most of the eastern Washington lakes and rivers, then, lacked much aquatic life. The floods also took much vegetation and soil away. Rich loess would have initially covered the channeled scablands, creating grasslands and woodlands. These massive floods demonstrate how major forces like glaciation interacted with the biota to assemble the Far West (Pielou 1991).

REGIONS

The Far West is magnificently diverse. Any attempt at providing a comprehensive description of the countless bioregions in the Pacific West is a fool's errand. Nevertheless, representative environmental types are highlighted in the following sections to furnish generalized patterns of several typical bioregions.

Valleys

The Pacific West contains three large valleys that have figured prominently in its environmental history. Beginning in the north is the Puget Sound Lowland, a basin surrounding Puget Sound and surrounded by Olympic Mountains on the west and the Cascade Range on the east. During the last glaciation, this area lay 3,000 to 5,000 feet beneath the southernmost reaches of the Cordilleran ice sheet. The glacier's retreat between 14,000 BP and 11,000 BP carved many landforms like the foothills surrounding the sound. The Strait of Juan de Fuca, channels in Puget Sound, and other regional waterways formed from the subsidence of rocks and multiple glaciations with their abundant meltwater. Grasses and pines colonized the basin, followed by other vegetation. Today, the basin rests in the western hemlock zone with Douglas fir largely dominating in historic times with the southern reaches of the Puget lowland in open grasslands, "liberally interspersed with conifer forest" (Kruckeberg 1999, 66). An array of animals has populated the basin. They tend not to be conspicuous, but as naturalist Arthur R. Kruckeberg has explained, "the animals are there: bear, cougar, elk, and deer command places in the forest ecosystem, as do a host of rodents, birds, and insects. Without many of them, forests would be impenetrable. Tunneling, burrowing, and browsing animals, especially herbivores, modify and shape the green landscapes" (Kruckeberg 1999, 69). By the time humans arrived in the Puget Sound Lowlands, the region had already undergone many geographic evolutions (Kruckeberg 1999, 52–59, 66–69; Orr and Orr 1996, 332–34).

Further south, western Oregon's Willamette Valley rests between mountains, the Cascades on the east and the Coast Range on the west. The Willamette River flows south to north throughout the valley, draining the vast watershed between the two mountain ranges. As with the Puget Sound Lowlands, glaciation profoundly influenced the valley during the Pleistocene, as it repeatedly filled with flood waters and silt deposits, a process that helped create the excellent soil that attracted human settlement. These floods from the Columbia River could inundate the valley with 400 feet of water and much debris. Following the last ice age, forests colonized the mountains surrounding the valley in species suitable to the new warmer climate. By the time the first humans occupied the valley, a variety of tree species flourished, including ash, cottonwood, willow, alder, and maple. However, most prominent along the valley floor were the tall grasses, such as tufted hairgrass, meadow barley, and bluegrass. Some of these grasses reached several feet high. Abundant grasslands meant abundant animals, too. Deer lived throughout the valley during the year, and elk fed on the grasses along their migratory paths. Because of this abundance, the Willamette Valley has long been a center for natural productivity and biodiversity (Boag 1992, 1–10, 15, 61; Orr and Orr 1996, 333–35; Wells and Anzinger 2001, 19–26).

California's Central Valley is perhaps the most impressive of the Far West's large valleys. It is nearly 500 miles long, an average of 40 miles wide, quite flat, and low (from just below sea level to 400 feet above). Precipitation is a scant 5 to 20 inches on average, but wide variations are common. Two major rivers— the San Joaquin in the south and the Sacramento in the north—flow through the Central Valley and make it well watered. Before intensive agricultural development in the nineteenth and twentieth centuries, many lakes, marshes, and wetlands and their associated flora and fauna dotted the valley. For instance, nearly a million acres of riparian mixed deciduous forests had evolved by the time Europeans arrived. Two million acres of tule marsh in the valley furnished excellent habitat for migrating birds. And 13 million acres of open grasslands and another 10 million with an oak overstory created a most impressive biome. Bunchgrasses fed substantial herds of deer, elk, and antelope (Barbour et al. 1993, 70–97; Norris and Webb 1990, 55–56, 412–35).

By the time humans arrived, these valleys were the results of a long evolutionary history. Their geography (e.g., climate, soils, water sources) made them arguably the most productive environments in the region, ultimately encouraging the most intensive human changes.

Mountains

An impressive array of mountain ranges hemmed in these large lowlands. Crushing plate tectonics and exploding volcanoes created these mountains, which influenced the region in myriad ways. They shaped precipitation patterns, provided habitat to animals and plants, furnished abundant mineral and timber resources, formed icons of regional identity, and became site to some of the most intense environmental conflicts in the region's history.

From southern British Columbia to northern California, the Cascade Range dominates much of the landscape. A number of volcanoes dot the range. In the

Holocene, the Cascades have experienced on average at least one significant eruption per century. Moreover, many of the region's iconic peaks and national parks (e.g., Mount Adams and Mount Rainier) are dormant volcanoes. The largest peak in the range is Mount Rainier, but the Cascades contain dozens of peaks higher than 8,000 feet. The Cascades separate the wet west side of the Pacific Northwest from the arid interior. The mountains mark this division of physiographic provinces sharply. The west side of this 700-mile-long range is dominated by densely packed Douglas-fir forests, while the east side of the 30- to 80-mile-wide mountain range is covered more sparsely by ponderosa pine forests and open grasslands. Much of this biotic contrast is attributable to the dramatic precipitation patterns—the west side along the coast can reach 150 inches annually, while the east side in central Oregon and Washington can have fewer than 10 inches. This climatic shift comes as moisture-laden clouds from the Pacific Ocean dump their heavy water as they rise over the peaks. The mountains are, arguably, the defining geographic feature of the Pacific Northwest and do much to shape the region's geography and its subsequent politics and culture (Harris 1980, 15–23).

Perhaps the most storied mountains in North America, however, are the Sierra Nevada range. They range about 430 miles long from the Cascades in the north to Tehachapi Mountains in the south and are 60 to 80 miles wide. More than 500 peaks exceed 12,000 feet; Mount Whitney is the highest in the continental United States at 14,495 feet. This enormous range holds precipitation generated from Pacific storms in its many lakes, streams, and rivers. Unlike the Cascades, where the biota basically divides at the crest, a much larger variety of floral communities exists within the Sierras, including conifer forests, meadows, chaparral, and others. Of course, among the mountains, too, were the minerals that spawned one of the greatest gold rushes in human history (Beesley 2004, 1–9; Hutchinson 1969, 22; Norris and Webb 1990, 63; Storer and Usinger 1963, 3–35).

Although none are as important as the Cascades or Sierras, other ranges matter. The Coast Ranges in California and Oregon and the Olympic Mountains in Washington near the Pacific Coast are nowhere near as dramatic as the parallel ranges 150 miles inland. These coastal mountains are blanketed by well-watered forests, including the only rain forest in the continental United States and the magnificent California redwood stands. Further in the interior, the Blue Mountains of northeastern Oregon cover 50,000 square miles and hold a variety of forest and range ecosystems that support similarly diverse wildlife populations. The western edge of the Northern Rockies that makes up the boundary of this region includes the heavily forested Bitterroot Mountains. Besides ample stands of white pines and other valuable timber, this area of northern Idaho contains commercially valuable mineral deposits, the results of

At 14,491 feet, California's Mount Whitney is the highest point in the 48 contiguous states. It is located along the Sierra Nevada Mountain Range, which provides much of Southern California with water. (Karl Sterne)

geologic history. All of central Idaho is covered by the Idaho batholith, an area of 16,000 square miles covered by a rough and craggy mountainous landscape. This vast region also contains canyons drained by the Clearwater, Salmon, Payette, and Boise rivers, all tributaries of the Snake River. Taken together, the "minor" mountain ranges of the Pacific West cover much of the region, furnish major timber and mineral resources later used for economic growth, provide important ecological functions such as watershed protection, and offer significant recreational opportunities (Hudson 2002, 313–18; Langston 1995, 16–29; Orr and Orr 1996, 166,194, 314–16).

Plateaus

Between these mountain ranges, the inland Far West, especially in its northern reaches, is dominated by two large plateaus, the Columbia Plateau and the Snake River Plain (Hudson 2002, 324).

The Columbia Plateau dominates most of eastern Washington and parts of eastern Oregon and northern Idaho. Numerous basalt lava flows laid the groundwork for this immense plateau. The Cascade Mountains bound the plateau on the west and block much precipitation so that the plateau receives only 6 to 20 inches of precipitation annually. Before irrigation arrived in the twentieth century, flora was not abundant, except for native bunchgrasses and sagebrush. Two distinct areas are within the plateau. The first, the Palouse, rests largely in southwestern Washington and is a rolling hill landscape covered by rich and fertile loess soil up to 200 feet deep deposited there by volcanic eruptions and shaped into hills by centuries of the west wind. Just north of the Palouse a different, far less fertile, landscape dominates the plateau in the channeled scablands. Although this region from present-day Spokane to Pasco, Washington, initially was covered with volcanic sediments similar to those of the Palouse, the Lake Missoula floods scoured the soil down to the basalt. Consequently, these scablands have barely a dusting of soil. This dry, almost barren landscape typifies the plateau more than the undulating Palouse hills (Meinig 1995, 3–25; Mueller 1997, 17; Orr and Orr 1996, 307).

South and east of the Columbia Plateau is southern Idaho's Snake River Plain. The plain is extremely dry but the Snake runs the 400-mile length of the plain, often through deep canyons. The river's course developed its modern form about 15,000 years ago when Lake Bonneville, which covered most of northern Utah, burst out of its shores and tore through southern Idaho, sculpting the plain's canyons while carving its way westward. Because the volcanic plain is porous, rivers and streams often disappear underground only to reappear downstream or in springs at various distances. Despite the plain's deep aridity, then, water defines it and humans faced significant challenges in living in such a space (Fiege 1999, 13–15; Hudson 2002, 325; Orr and Orr 1996, 220).

River Systems

The region's river systems are complex, and their ecosystem functions are central to various environments and the plant and animal life that depend on them. Probably no other bit of nature does so much, so consistently, for so long. Although they are not the exclusive part of the hydrologic cycle, rivers remain the most visible and create riparian areas that are perhaps the most vital and productive ecosystems. For the many dry areas in the Far West, rivers constitute the region's lifeblood. As river systems carry water long distances, they transfer sediments and carve the edges of the river bank, creating more sediment to bring downstream. These erosive impacts can be quite dramatic, as various canyons demonstrate. Moreover, rivers sometimes flood, releasing nutrients and sediments into nearby floodplains, often replenishing soil on unpredictable schedules. They also change constantly. Channels move; courses alter; they increase and decrease their flows based on climatic fluctuations. Dams—ice or rock—occasionally block their progress to the sea, all this before human engineering could much more effectively impede the great rivers of the West. Certainly, too, they provide essential water for plants and habitat for fish, insects, and other animals. The Pacific West's river systems experienced some of the greatest ecological changes over the course of human occupation and manipulation (Leopold 1974).

Two large river systems, both consisting of two main rivers and their many tributaries, drain vast areas of the Far West. The first is the Columbia–Snake River system. The Columbia originates in Canada's Columbia Lake, which is serenely nestled among the Rocky and Purcell mountains. It flows 1,200 miles generally south and west to empty into the Pacific Ocean. Meanwhile, the Snake originates in Yellowstone National Park and flows more than 1,000 miles. Before joining the Columbia near Pasco, Washington, it runs through North America's deepest canyon, Hells Canyon. The system drains approximately 259,000 square miles, an area two and a half times the size of Great Britain. Besides water, the Columbia carried 7.5 million tons of sediments annually before dams considerably slowed it and trapped many of those deposits. This sediment load created the Columbia Bar at the river's mouth, which hid the river's entrance from many European explorers. The river was arguably the world's greatest salmon stream. It drops about 2,700 feet in elevation, and at The Dalles, where the Columbia punches through the Cascade Mountains, the river typically flowed at a rate of more than 200,000 cubic feet per second before dams reduced that. Its volume and its relatively steep decline made it immensely attractive for hydroelectric development. This powerful and large river system connects diverse ecoregions, and it experienced a significant transformation by human hands in the twentieth century (Dietrich 1995, 100–103; Reisner 1993, 155; White 1995, 6).

The second river system is that of the San Joaquin and Sacramento rivers in central California. These rivers "moved like twisting threads through the vast Central Valley trough" (Hundley 2001, 5). The northern third of the Central Valley is drained by the Sacramento River and its tributaries, for nearly 400 winding river miles. The Sacramento's volume is about three times that of the San Joaquin River because its source is in the wetter Trinity Mountains. In the southern Central Valley, the San Joaquin River travels more than 300 miles from its source in the arid southern Sierra Nevadas. The two rivers join, form a large delta, and flow into the San Francisco Bay. While the Columbia River in

The Snake River originates near the Continental Divide in Yellowstone National Park, then flows roughly 1,000 miles before joining the Columbia River near Pasco, Washington. (PhotoDisc/ Getty Images)

the Northwest demonstrates the power of immense volume, the Sacramento– San Joaquin river system shows the power that rivers possess from flooding. Indeed, these rivers' tendency to leave their banks illustrates well that river systems are important far beyond their banks. The annual flooding in the Central Valley profoundly shaped the valley's geography and ecology. The Sacramento normally carries about 5,000 cubic feet per second, but floods can move as much as 600,000 cubic feet of water per second. The river's channel could not contain such floods, and water routinely spread out through the valley floor, creating what one historian likened to "a vast inland sea a hundred miles long" (Kelley 1989, 5). Thus, agriculture subsequently transformed what was once largely a marshy and lake-filled landscape. Indeed, the first Spanish colonists called it the "Valley of the Lakes," a moniker that no longer reflects the landscape (Dasmann 1965, 14). This relatively wet landscape furnished habitat for abundant fish, birds, and other wildlife, not to mention the tules and prodigious wild grasses that once characterized the Central Valley. As the river systems affected large areas beyond their banks, their alteration contributed to wide-ranging ecological changes (Hundley 2001, 5–8; Kelley 1989, 5–6).

Southern California

As it does culturally, Southern California occupies a unique position geographically and thus merits its own description. Part-desert, part-mountains, part-basin, part-coast, Southern California landscapes are home not only to millions of people but also to a complex natural environment, starting with its geology. Several faults, including the famous San Andreas, run through Southern California, making it subject to significant geological change without a moment's notice (Norris and Webb 1990). A series of mountains in the Transverse and Peninsular ranges bound the Southern California region. These mountains, while contributing to the scenery, arguably gain most of their importance by trapping air in the 1,630-square-mile Los Angeles Basin (Davis 2002, 57). In addition to mountains and the stagnant air they contain, Southern California possesses productive riverine environments. Although today the Los Angeles River is paved almost completely, it has long been an important ecological system. It once "meandered this way and that through a dense forest of willow and sycamore, elderberry and wild grape. Its overflow filled vast marshlands that were home to myriad waterfowl and small animals. . . .So lush was this landscape and so unusual was it in the dry country that the river was a focus of settlement long before the first white man set foot in the areas" (Gumprecht 1999, 2). Its paving suggests just one of the ways Southern California has been radically reconfigured. The geographic combination of an arid climate and pyrophytic vegetation also made Southern California prone to wildfires. Among other things, this complex of potential earthquakes, floods, fires, and surprisingly frequent tornadoes led writer Mike Davis to dub Angelenos' relationship to the natural world an "ecology of fear" (Davis 1998). Southern California's environment, then, enjoyed a unique environmental history even before several million people came to call it home.

ARRIVAL OF HUMANS

The last piece needed to set the scene for this story is the arrival of human populations. Native peoples offer their own explanations of when and how they ar-

Clovis spear points found in Washington State. These have the distinctive narrow profile and longitudinal groove that allowed the stones to be hafted to a wooden shaft. Archaeologists estimate that Clovis culture peoples inhabited the Pacific Northwest between 5,000 and 11,500 years ago. (Warren Morgan/Corbis)

rived (often up from the earth, down from the sky, or from animal ancestors), which complement the traditional perspective of an Asian migration. Questions persist about when the first humans arrived in the Americas, where they came from, and how they dispersed. What does seem beyond debate in scholarly circles is that around 14,000 BP, the human population in the Western Hemisphere increased significantly. Widespread archaeological evidence of a fluted point on various tools, first found near Clovis, New Mexico, and subsequently located at many sites throughout the Americas, has given rise to the name Clovis culture to describe the diverse human populations who shared this technology. Clovis people may not have been the first humans in North America, however. There are archaeological sites that some scholars have suggested contain pre-Clovis artifacts. The archaeological establishment has not confirmed these sites. Perhaps with improved dating techniques in the future a pre-Clovis site might be confirmed. Regardless, any pre-Clovis population would have been small and would have possessed a comparatively unsophisticated technology that made a minimal impact on the surrounding environment (Haynes 2002; Pielou 1991, 113).

Most scholars believe humans traveled from Asia across a land bridge between Siberia and Alaska when the ice age had lowered the sea level by about 300 feet. The bridge, or Beringia, appeared at various times during the last glaciation and was likely open from 25,000 BP and 11,000 BP, after which climate change raised sea levels and covered Beringia under water. In addition, some have posited an ice-free corridor between the two vast ice sheets that covered the North American continent, the Cordilleran and Laurentide ice sheets. That corridor may have allowed humans to migrate south to what became the United States. However, that corridor also may have been too inhospitable to allow the migration. A likely alternative is that humans migrated along the coasts, either by land or sea. Dating the arrival in the Pacific West is imprecise, but archaeological sites on the Northwest Coast indicate that the earliest inhabitants arrived between 10,000 and 5,000 BP. Such timing seems likely to be revised in the future (Fagan 2000, 73–81, 205–213).

CONCLUSION

Thus, the shape of the Far West at about 12,000 BP was in a state of flux: glaciers retreating, new plants and animals arriving, climate warming. The biggest change, though, was the arrival, or the vast increase, of a new species—*Homo sapiens*. So, beginning at the end of the last ice age, culture came for the first time to shape—sometimes quite dramatically—the nature of land, water, animal, and plant life in the Pacific West. Scholars may debate the degree and kind of ecological change, but none deny that a major watershed was crossed.

REFERENCES

Alt, David, and Donald W. Hyndman. *Roadside Geology of Oregon.* Missoula: Mountain Press, 1978.

Alt, David, and Donald W. Hyndman. *Roadside Geology of Washington.* Missoula: Mountain Press, 1984.

Alt, David, and Donald W. Hyndman. *Roadside Geology of Idaho.* Missoula: Mountain Press, 1989.

Alt, David, and Donald W. Hyndman. *Roadside Geology of Northern and Central California.* Missoula: Mountain Press, 2000.

Barbour, Michael, Bruce Pavlik, Frank Drysdale, and Susan Lindstrom. *California's Changing Landscapes: Diversity and Conservation of California Vegetation.* Sacramento: California Native Plant Society, 1993.

Beesley, David. *Crow's Range: An Environmental History of the Sierra Nevada.* Reno: University of Nevada Press, 2004.

Boag, Peter G. *Environment and Experience: Settlement Culture in Nineteenth-Century Oregon.* Berkeley: University of California Press, 1992.

Dasmann, Raymond R. *The Destruction of California.* New York: Macmillan Company, 1965.

Davis, Devra. *When Smoke Ran Like Water: Tales of Environmental Deception and the Battle against Pollution.* New York: Basic Books, 2002.

Davis, Mike. *Ecology of Fear: Los Angeles and the Imagination of Disaster.* New York: Henry Holt and Company, 1998.

Dietrich, William. *Northwest Passage: The Great Columbia River.* New York: Simon and Schuster, 1995.

Ewert, Eric C. "Setting the Pacific Northwest Stage: The Influence of the Natural Environment." In *Northwest Lands, Northwest Peoples: Reading in Environmental History,* edited by Dale D. Goble and Paul W. Hirt, 3–25. Seattle: University of Washington Press, 1999.

Fagan, Brian M. *Ancient North America: The Archaeology of a Continent.* 3rd ed. New York: Thames and Hudson, 2000.

Fiege, Mark. *Irrigated Eden: The Making of an Agricultural Landscape in the American West.* Seattle: University of Washington Press, 1999.

Gumprecht, Blake. *The Los Angeles River: Its Life, Death, and Possible Rebirth.* Baltimore, MD: The Johns Hopkins University Press, 1999.

Harmon, Rick. *Crater Lake National Park: A History.* Corvallis: Oregon State University Press, 2002.

Harris, Stephen L. *Fire and Ice: The Cascade Volcanoes.* Rev. ed. Seattle, WA: The Mountaineers and Pacific Search Press, 1980.

Haynes, Gary. *The Early Settlement of North America: The Clovis Era.* New York: Cambridge University Press, 2002.

Hudson, John C. *Across This Land: A Regional Geography of the United States and Canada.* Baltimore, MD: The Johns Hopkins University Press, 2002.

Hundley, Norris, Jr. *The Great Thirst: Californians and Water: A History.* Rev. ed. Berkeley: University of California Press, 2001.

Hutchinson, W. H. *California: Two Centuries of Man, Land, and Growth in the Golden State.* Palo Alto, CA: American West Publishing Company, 1969.

Kelley, Robert. *Battling the Inland Sea: Floods, Public Policy, and the Sacramento Valley.* Berkeley: University of California Press, 1989.

Kruckeberg, Arthur R. "A Natural History of the Puget Sound Basin." In *Northwest Lands, Northwest Peoples: Readings in Environmental History,* edited by Dale D. Goble and Paul W. Hirt, 51–78. Seattle: University of Washington Press, 1999.

Langston, Nancy. *Forest Dreams, Forest Nightmares: The Paradox of Old Growth in the Inland West.* Seattle: University of Washington, 1995.

Leopold, Luna B. *Water: A Primer.* San Francisco, CA: W. H. Freeman and Company, 1974.

McKee, Bates. *Cascadia: The Geologic Evolution of the Pacific Northwest.* New York: McGraw-Hill, 1972.

Meinig, D. W. *The Great Columbia Plain: A Historical Geography, 1805–1910.* 1968 Reprint, Seattle: University of Washington Press, 1995.

Mueller, Marge, and Ted Mueller. *Fire, Faults, and Floods: A Road and Trail Guide Exploring the Origins of the Columbia River Basin.* Moscow: University of Idaho Press, 1997.

Norris, Robert M., and Robert W. Webb. *Geology of California.* 2nd ed. New York: John Wiley and Sons, 1990.

Orr, Elizabeth L., and William N. Orr. *Geology of the Pacific Northwest.* New York: McGraw-Hill, 1996.

Pielou, E. C. *After the Ice Age: The Return of Life to Glaciated North America.* Chicago: University of Chicago Press, 1991.

Reisner, Marc. *Cadillac Desert: The American West and Its Disappearing Water.* Revised and updated. New York: Penguin, 1993.

Storer, Tracy I., and Robert L. Usinger. *Sierra Nevada Natural History: An Illustrated Handbook.* Berkeley: University of California Press, 1963.

Wells, Gail, and Dawn Anzinger. *Lewis and Clark Meet Oregon's Forests: Lessons from Dynamic Nature.* Portland: Oregon Forest Resources Institute, 2001.

White, Richard. *The Organic Machine: The Remaking of the Columbia River.* New York: Hill and Wang, 1995.

2

RECIPROCITY AND
THE INDIGENOUS LANDSCAPE

In the Inland Northwest, a woman with a digging tool, a *pitse'*, stood in a field searching for the blue flowers that would mark a patch of camas. The woman gently loosened the bulb as her companions gathered around her and the plant. It was the year's first camas. Back into the earth went an offering of tobacco, as the woman gave a prayer for thanksgiving and asked permission to gather the camas. She returned the flowering stalk of the plant before moving on to take more. After gathering camas, families would process the plant to be consumed months later during winter ceremonies (Frey 2001, 159–160). The harvest was ritualized. Those involved were participating intimately in the natural world, giving offerings as they took or borrowed from the land. This reciprocity guided (and guides) Native peoples' relationship to the world that surrounds them.

Competing disciplines, imprecise evidence, and competing interpretations make studying the environmental impact of Native peoples a bewildering experience. A generation or two ago, few scholars even acknowledged that indigenous populations contributed much to the shape of the landscape. Instead, scholars saw Native cultures as too primitive, too few, and too backward. Today, renewed investigations and increased willingness to view Native people as agents of change have created a vastly different picture, and many scholars now credit Native populations with significant effects on plant and animal populations and distribution. Moreover, they are beginning to recognize the deep understanding of Natives' ecological knowledge.

ESTABLISHING HUMAN-ECOLOGICAL REGIMES

Indigenous populations perceived themselves as existing within reciprocal relationships with the natural world. The animals and fish, the forests and rivers were partners and coinhabitants of the world. The link that concerns environmental historians most is the one from Native people toward the natural world, although it is imperative to recognize that the association went both ways. American Indians possessed a great capacity to affect land, water, plant, and animal resources through this partnership with hunting and fishing, as well as through the widespread and nuanced use of fire and extensive cultivation practices. These four types of relationships do not constitute an exhaustive inventory of human influence over the Pacific West, but they offer a wide survey of the principal ways of organizing and reorganizing the environment. More importantly, they are central to the cultures of this place.

Hunting

Hunting is a long-standing subsistence practice, and indigenous hunting was ecologically significant. Although central to environmental history, the nature of pre-Columbian hunting is imprecise owing to conflicting data and interpretations. A scholarly consensus has not emerged and may never be achieved. Nevertheless, diverse scholars—not to mention diverse Native traditions— demonstrate that hunting represented a central place in broader ecological strategies and generated a considerable effect on animals and the lands.

Massive waves of extinction have occurred throughout history, both predating and postdating humans' arrival in particular landscapes. Coincident with the arrival, or vast increase, of humans in North America came an enormous extinction episode that resulted in the decline of most of North America's large mammals, or megafauna. The continent's archaeological record contains evidence of a number of species, between thirty-five and forty, of large mammals rapidly meeting their demise around 13 millennia ago (Pielou 1991, 251; Krech 1999, 33). Earlier extinction episodes may have been even greater. For instance, five million years ago a wave of extinctions eliminated far more species. However, those earlier lost species were much smaller, and thus, their extinction may ultimately have been less consequential ecologically (Pielou 1991, 252). The species that died out around 13,200 years before present (BP) included North American horses and camels, the dire wolf, and giant beavers. The larger mammoth and mastodon lasted until about 12,800 BP, while a greater predator, the short-faced bear, survived another thousand years (Flannery 2001, 188). Although no one doubts the megafaunal extinctions, scholarly consensus on its causes—cultural or climatic—remains elusive.

Paul S. Martin has been the leading scholar arguing that humans, rather than climate change, caused the megafauna extinctions. Globally, Martin has correlated the arrival of humans on an island or continent with die-offs of large mammals following closely. On continents, such as Africa, where the human presence began earlier, few species extinctions occurred during eras of global climate change (Martin 1984, 354–403). According to Martin, megafauna evolved for centuries without human predation, and the arrival of humans skillful in hunting made quick work of these unsuspecting animals. To bolster this argument, twelve species of mammals that originated in North America died out, while a number of other mammal species that migrated from Asia, where there had been a long presence of human hunters, survived the Pleistocene extinctions (Pielou 1991, 256). This explanation suggests that animals with experience of human predation could withstand such pressures, while those without the evolutionary knowledge of the hunt could not.

Other scholars have objected to Martin's overkill hypothesis. The most prominent critics have argued that climatic reasons are the more likely cause, as the North American extinctions corresponded to the end of the last major ice age. The drier climate may have decimated the habitat necessary to sustain large megafaunal grazers. Perhaps water supplies also dried up, jeopardizing animal populations (Fagan 2000, 89). Moreover, all previous episodes of mass extinction had been climate-caused (Flannery 2001, 191). Still another objection to the overkill hypothesis is that there are too few archaeological kill sites (Pielou 1991, 258). With such a mass extinction, archaeological kill sites ought to be more numerous. In the end, the uncertain nature of archaeological evidence and questions over dating likely will prevent absolute certainty in assigning causes.

Ecologist Charles E. Kay has suggested a solution to this perennial conundrum by arguing that scholars have erroneously assumed central elements about megafauna. All who have studied this debate have missed a biological fact, Kay maintains. They have presumed that the megafauna "were food-limited. That is, they assumed that predators had no significant effect on herbivore populations, and thus the herbivores were exceedingly numerous" (Kay 2002, 239). Alternatively, Kay argues that the megafauna were predator-limited, not food-limited. In other words, the megafauna's biggest vulnerability was not the availability of food but the presence of predators. If this is the case, he posits that the megafauna populations may have been much smaller, only 1 to 10 percent of others' estimates. A much smaller population would, of course, have made extinction easier, as well as made the population more vulnerable to both climate-induced environmental change and human or animal predation. The arrival of a "super-predator," humans, meant a "cascading trophic effect—where the addition of one factor, in this case human hunting, causes the entire system to change" (Kay 2002, 240). This model answers many of the objections to the overkill hypothesis: that there were too few kill sites, too few people, and too many extinct species. Simply put, human hunters disrupted the balance, which then allowed other predators to finish off the megafauna. Perhaps Kay's interpretation of few animals pushed over the edge by hunters only to be finished off by other factors can move scholars toward a consensus (Kay 2002, 238–261).

Regardless of ultimate causes, two important consequences of the Pleistocene extinctions stand out. First, the episode deeply impoverished North America's mammal population. Potentially domesticable animals that would surely have changed North American history by allowing agricultural herding did not survive (Crosby 1986; Steinberg 2002, 13). Second, it left an ecological vacuum with few large grazing mammals left compared with the Eurasian and African continents. When European animals arrived in the Pacific West beginning in the eighteenth century, the ramifications of that drained ecological niche would become apparent as new domestic grazing species filled the void.

After the megafaunal extinctions 12 to 13 millennia ago, Native hunters do not seem to have extinguished animal populations, but they certainly affected them. Human hunting and habitat management may have greatly shaped the population and distribution of certain species, attracting, displacing, and sometimes diminishing certain animal populations. Explanations of the more recent impacts of Native hunting have predictably fractured largely along disciplinary lines. Influenced by evolutionary ecology, archaeologists have argued that a species' overriding goal is to maximize reproduction. As such, human populations are likely to behave in ways consistent with natural selection, a type of simple brutish behavior many would dispute. Using optimal foraging models, archaeologists argue that populations will employ forage strategies that will "produce the greatest return in energy relative to time and effort expended" (Fagan 2000, 237). Theoretically, this would mean hunters worked to obtain a higher-ranked food source, often a large mammal such as a deer, while passing over lower-ranked foods, such as smaller animals or plants. In a hypothetically abundant landscape, hunters would be highly selective and concentrate on higher-order foods. As abundance declined, however, they would become less selective. Such behavior seems confirmed at times by the archaeological record, which shows that higher-ranked foods declined as human populations increased. A corresponding rise in lower-ranked foods came with costs in time and energy, as it required more labor to obtain as much food. Significantly, as cultures exploited lower-order foods, they simultaneously innovated with technology, likely in an effort to increase their foraging efficiency. From the theoretical perspective of evolutionary ecology, and supported by some archaeological data, it seems that Native populations did not practice any careful conservation strategy, such as limiting the killing of young or female mammals. Resourceful exploitation of animals seemed to be the order of the day (Fagan 2000, 238–239; Hildebrandt and Jones 2002, 102–109).

Such a perspective is controversial because it conflicts with abundant ethnographic and anthropological evidence that privileges Native oral and written testimony over theoretical models or archaeological evidence. It also neglects to consider the ideal of reciprocity that runs through Native cultures. Perhaps best characterized by Thomas C. Blackburn and Kat Anderson's book, *Before the Wilderness: Environmental Management to Native Californians*, this approach emphasizes Native peoples' agency in managing and shaping the environment to a great degree. This widespread ecological manipulation came with a clear conservation ethos designed to maintain, promote, and conserve productive environments. These scholars argue that Native management proved sustainable over generations because of sociopolitical traditions that governed resource management and consumption. Finally, Native peoples and scholars maintain that it was use itself that ensured the abundance in and utility of the landscape; thus, without reciprocal exchanges with the environment, Native peoples could not have survived (Blackburn and Anderson 1993, 19 and passim; Hildebrandt and Jones 2002, 105–107). Regardless of the view one takes or the evidence one privileges, any notion that Native populations trod lightly and had little influence is no longer tenable. The archaeological and anthropological evidence concludes undoubtedly that human beings constantly adapted with their environment, profoundly shaping their surroundings as they were also shaped by them. And in the end, this reciprocity met cultural and ecological needs.

Indeed, hunting in the Far West exerted a profound influence on populations. Hunters depleted optimal game, so Native communities increased consumption of smaller game increasingly over the past 3,000 years. The smaller species were less vulnerable to humans and because of more prolific reproduction could withstand these hunting pressures. One good example of this trend is pinniped hunting. In localized environments, Northwest Coast Natives displaced pinnipeds, like seals and sea lions, from their mainland rookery sites. Before 1500 BP, there was much evidence of hunting at mainland rookeries, but after that the evidence shows increasing hunting at offshore rookeries. Rising human populations, along with increasingly sophisticated transportation (e.g., oceangoing canoes) and weapon technology, correlated to pinnipeds leaving mainland rookeries for offshore island rookeries. This trend demonstrated the power of culture and hunting technology to displace an animal population through predatory pressure. Moreover, the shift illustrated how resource scarcity gave rise to greater social stratification. With more complex technology and the concomitant capital and labor investments required for producing canoes and weapons came complicated notions of property ownership in oceangoing watercraft. Among a suite of other reasons, the ability to maximize marine

Salishan Indians paddle canoes on Willapa Bay, Washington, around 1913. (Edward S. Curtis/ Library of Congress)

mammal resources and to improve hunting efficiency encouraged the technological innovations that improved oceangoing canoes among the Chumash using the Santa Barbara Channel and the Nootkans using the Strait of Juan de Fuca. Such technological change in turn led to greater social complexity and increasingly hierarchical communities. These examples indicate that when access to environmental resources was not free and open, Native societies increased their social complexity and hierarchy (Arnold 1995; Hildebrandt and Jones 2002, 76–97).

Another complication to reconstructing this coastal hunting regime, however, is the fact that marine mammals decline as ocean surface temperatures rise, such as during periods of El Niño. Thus, during these times, hunting would necessarily decline because of diminishing populations or face the possibility of permanently hurting the species' reproductive capacity. Moreover, with lower populations and perhaps displaced populations, pinnipeds became marginalized as part of the hunting-gathering economy. The record indicates fluctuations in marine mammal hunting, although it cannot determine whether the change was induced by human overhunting, environmental changes, or, more likely, a combination of the two (Colten and Arnold 1998).

Marine mammals were not the only prey for humans. In the Pacific Northwest, Native groups exploited an array of animals and developed several hunting strategies. For example, using a practice called the *battue*, Kalapuya men in the Willamette Valley burned a circle that gradually enclosed animals in an increasingly small area. Eventually, they could kill the animals. The *battue* focused on deer at the end of the summer when grass conditions made burning easier and when deer are their heaviest. The Kalapuya also used fire less directly in hunting. After their annual summer burning, fall rains revitalized the grasses that elk and migrating waterfowl browsed. Thus, the freshest grasses attracted the animals in a predetermined and concentrated area, making them easier targets to Kalapuyas (Boag 1992, 15). Indigenous groups in the Willamette Valley also used bows and arrows and snares in their hunting. They preferred deer and elk, but occasionally hunted larger mammals (e.g., bears), smaller ones (e.g., raccoons, marmots), or birds (e.g., grouse) (Bunting 1997, 9). Diverse strategies derived from deep knowledge of the region's ecology and were part of multiple subsistence practices.

Certain conservation practices developed among tribal groups. For example, among the Salish in the Puget Sound region, tradition held that young animals were never killed, and groups did not kill more than they needed for subsistence and trading purposes. Violating these norms led to shame and other social burdens. However, some tribes that were more dependent on the sea appeared to be less strict about rules concerning waste for land-based hunting. This discrepancy demonstrates that Native peoples were somewhat specialized, so the marine-oriented tribes did not develop the same hunting ethic, knowledge, or dependence as other groups (White 1992, 29–30).

In the interior plateau lands of the Northwest, hunting constituted an important part of Native subsistence, too. The Coeur d'Alene, for instance, used dogs or designated men as drivers to force deer into water where hunters awaited onshore or in canoes from which they would "spear, club, drown, or shoot arrows at the oncoming animals" (Frey 2001, 36). A European in the 1830s or 1840s observed the Coeur d'Alene kill 600 deer in a single communal hunt. As with fish, they dried meat and stored it for consumption in winter months. Before the hunts, men would prepare themselves through fasting or sweat-bathing, praying for a successful hunt. After a kill, men would often thank the animal "for giving himself up" (Frey 2001, 36). Moreover, a spiritual leader, or shaman, would attempt to regulate the entire community's relationship to wildlife and the larger environment, singing *suumesh* songs to ensure a successful hunt. As this example suggests, subsistence practices were not simply about providing physical sustenance; they signified larger connections between the cultural and the natural (Frey 2001, 35–37, 46–48).

The ecological impact of hunting differed from place to place. The smaller Native populations in the interior relied on fishing and cultivating more than hunting, minimizing the overall effects of hunting. However, in parts of California and other areas of high population density, Natives may have caused near-extinctions of some animals. Archaeological sites along the coast at San Francisco Bay, in the Central Valley near Sacramento, and in the Sierra Nevadas show clear declines through the past several thousand years in preferred fish and wildlife (e.g., sturgeon, tule elk, black-tailed deer). Some scholars believe several vertebrate populations were nearly extinct by the late-Holocene and the eve of contact with Europeans, leading to increased violence among tribal groups battling over scarce resources (Broughton 2002, 48–66). One unexpected consequence of this wildlife decline was localized increases in biodiversity. With large game somewhat depleted, smaller animals and some plants flourished where otherwise they would have been unable to. The absence of certain wildlife freed up biota space for other animals and plants (Preston 2002, 131). Thus, Natives' hunting practices profoundly shaped the animal and plant populations, a change that in turn affected culture.

Human hunting reveals much about the Far West's pre-Columbian environmental history. It shows Natives' capacity to affect resources. They contributed to or caused massive extinctions, and they chased seals off coastal rookeries. More typically, hunting functioned as a part of a larger human-ecological regime in which distinguishing between hunting and fire or cultivation practices is impossible. Natives adapted to changes in environmental conditions and animal populations. The hunting regimes evolved to meet a dynamic environment and to satisfy cultural needs. In that respect, it is little different from other subsistence practices. It also illuminates how interconnected Native economic and cultural traditions are with environmental factors.

Salmon

Salmon fishing constituted a key component to many Western tribes' culture and subsistence. Because it has remained so central, it furnishes an ideal example of how ecology, economy, and culture interact interdependently. For several millennia, Indian fishers operated a highly intensive harvest, but they did so sustainably. Most of the Pacific West was within easy reach of salmon except for California south of Point Conception, just north of Santa Barbara, where climatic changes mark the southern reach of the species (McEvoy 1986, 3). Understanding how Native fishers operated within cultural and ecological constraints reveals the important lessons in the region's environmental history.

Pacific salmon (*Oncorhynchus*) are one of nature's greatly complex creatures. These fish, born in cool (40–50 degrees Fahrenheit) streambeds and lakes throughout most of the region, spend between two months and three years maturing in inland waters. Then, they change physiologically as they descend Western rivers ultimately arriving in the Pacific Ocean. The process called "smoltification" allows salmon to live in saltwater. The Columbia River, arguably the greatest salmon river in the world, historically drained three-fourths of its water during the hottest summer months, the time when salmon smolts needed help going downstream. In the open sea, they swim as far north as the Gulf of Alaska, spending from one to six years in the ocean. Different species reach different sizes before returning to their natal streams to spawn. Pink and sockeye reach five to eight pounds before returning after one to three years. Chum, steelhead (an anadromous trout), and coho grow larger, to eighteen to twenty-two pounds, in the same period. The biggest, the chinook, reach eighteen to sixty pounds in their three to six years in the Pacific. Climatic changes in the ocean, such as El Niño, can significantly influence salmon populations, especially in their southern limits in northern California waters. After reaching maturity in the open sea, salmon return upstream through often tortuous rivers, climbing rapids and mountains streams until they reach their original spawning beds. How and why they are able to trace their way back largely remains a scientific mystery. The return journey transforms the salmon physiologically once again, as they stop eating and live off their body fat. Their color changes from silver to brown or red and green; males' backs swell giving them the colloquial name of humpies, even while their jaws become more hooked, their digestive system contracts, and their reproductive organs enlarge. After expending enormous amounts of energy to reach their spawning grounds, they spawn and die (Taylor 1999, 5–6; McEvoy 1986, 7; Harden, 1996, 70).

Throughout most of the region, Native cultures depended on the fish's abundance and reliability. Although ocean currents and climatic changes did affect salmon populations, salmon in the Northwest were typically reliable and thus for many areas contributed a more reliable protein source than hunting (Taylor 1999, 14). Drawing on this dependability, Native groups consumed the fish in prodigious amounts. Some estimates indicate that some Native groups ate or otherwise used about two pounds of fish daily per person (Harden 1996, 63; Boxberger 2000, 13). Salmon might be eaten fresh, but much of it was dried and stored in the fall for winter and spring consumption. Northwest tribes who lived at The Dalles where the Columbia River forces its way through the

Cascade Mountains specialized in pemmican made of dried salmon. Indian women crushed the salmon and dried it on outdoor racks. After the arid air dried the salmon, the pemmican would be packed in ninety-pound baskets. The process allowed the pemmican to remain edible, reportedly up to several years. A household might prepare 3,000 pounds for their own consumption and another 1,000 pounds for trade (Taylor 1999, 24). They often traded this commodity downstream, which had a wetter climate where this type of preparation would have been impracticable (Boyd 1996, 55). Significantly, the ability to store salmon, a skill developed at least 2,500 years ago, allowed consumption to be spread spatially and temporally. That generated greater pressures on salmon populations than if consumption were limited only to when the fish were freshly available (Taylor 1999, 24).

Although certainty is impossible, a number of scholars have attempted estimates at the catch and consumption of salmon among Pacific Western groups. The Lummi in northwestern Washington may have consumed 600 pounds of salmon per capita annually (Boxberger 2000, 13). On the Columbia Plateau, great diversity existed in salmon consumption. Some groups, such as the Coeur d'Alene or Flathead, consumed salmon for only about two percent of their diet. On the other hand, the Thompson relied on salmon for more than 60 percent of their diet, and many other Native groups (e.g., Wishram, Umatilla, Nez Perce) depended on the fish for one-quarter to one-third of their caloric intake (Hunn 1990, 150). Groups in the Central Valley of California may have taken 9 million pounds a year, which approximated the industrial catch at the end of the nineteenth century (McEvoy 1986, 22). A Yurok family in the Klamath River country might possess a ton of dried salmon in their home (Heizer and Elsasser 1980, 85). In the Oregon country, where salmon culture was best elaborated, the catch reached astounding levels. An individual on the Oregon Coast reported catching 300 fish by himself in one night. The Nez Perce, an inland group that relied less on salmon because of its distance from the ocean, still claimed 50 sites from which they could harvest 300 to 700 fish daily (Taylor 1999, 20).

All of this harvesting added up. The annual salmon runs in the Pacific Northwest were between 11 and 16 million fish. The best numbers and theories available now suggest that the indigenous harvest neared 42 million pounds or about 4.5 to 6.3 million fish. This meant approximately 28 to 57 percent of the entire run would be taken annually. Comparatively, only nine times did the industrial fishery catch so much in its heyday at the turn of the twentieth century (Taylor 1999, 23).

How were Native groups able to take so many fish? Native groups employed a variety of methods that were, in the words of one scholar, "frighteningly efficient" in taking salmon from various rivers (Taylor 1999, 20). Along

Native Americans fish using platforms and handheld nets along the Columbia River, ca. 1899. In the twentieth century, dams inundated such Native fishing sites. (Library of Congress)

the Columbia River, especially at the gorge where the river narrowed and rapids ran, Native men stood on wooden platforms overhanging the river and used dipnets or spears to capture salmon fighting their way upstream. This dangerous practice, still used, required enormous skill. Further upstream, the Colville used reed baskets. Occasionally, Native fishers poisoned salmon. In California, scoop nets, harpoons, and weirs were common. Most effective and widespread were seines and gillnets. Indians made these nets of iris, cedar bark, silk grass, Indian hemp, and bear grass. James G. Swan, an American who observed Indian seines in the mid-nineteenth century, reported their immense size: 100 to 600 feet long and 7 to 16 feet deep. All of these techniques required intense investments of labor in both preparation and execution, hinting at the importance of these activities to Native cultures. When the salmon runs began, Natives faced a limited time to complete their task of obtaining enough salmon to last through the leaner months (Taylor 1999, 18–20; McEvoy 1986, 22; Swezey and Heizer, 1993, 299).

With such proficient technology and deep dependence on salmon for year-round subsistence, Native peoples required a management regime to ensure long-term sustainability. Oral traditions and ritual ceremonies created those practices. Examining those activities reveals both cultural and ecological understanding and offers reasons for the limited impact despite such heavy use of salmon. The traditions, collectively, focus on respect and moderation (Taylor 1999, 27).

Traditions varied throughout the Far West, but common elements circumscribed salmon harvesting. Salmon is a central character in many Native oral traditions. Often Salmon serves the role of determining a group's destiny and thus a group must respect it. Regeneration constitutes another common theme, a theme mimicked from nature and salmon's annual return. Ceremonies tended to require disposing some of the first fish, often placing it back into the river, thus ensuring the continuing regeneration of salmon runs. Universally, the traditions demanded respect for salmon. They also demonstrated the responsibility Indians must bear to ensure the return. Indeed, according to Taylor, "fear played as strong a role as reverence in shaping worship" (1999, 31). Failure of the fish to return would cause considerable cultural and economic hardship. Consequently, First Salmon ceremonies and other rituals were designed to satisfy Salmon's requirements. Incidentally, during First Salmon ceremonies when fishing was prohibited, salmon escaped upstream, allowing spawning and limiting the harvest. Thus, these ceremonies represented a conservation practice. Limitations on consumption through cultural taboos also moderated Indian use of salmon. Proscriptions regulated access to scarcer species (e.g., chum and steelhead), and they limited eating to certain times. In addition, nets could only be used a specific number of times. Taken together, these cultural beliefs and practices served important conservation measures (Taylor 1999, 30–36).

Native fishers also constructed elaborate legalistic rituals to govern salmon use. The more dependent a group was on salmon, the greater the complexity. In the Lower Klamath region, for instance, salmon fishing remained a highly regulated practice. The regional inhabitants—the Yurok, Hupa, and Karok—were wealthy, individualistic, and legalistic. They developed ideas of monetary value. Common law developed among the Yurok and neighbors, including a system of arbitration for disputes. Such cultural complexity helped shape and regulate resource consumption. These groups could have exploited the fisheries more extensively and supported higher tribal populations, but they chose not to (McEvoy 1986, 32–34).

Ownership of fishing sites and equipment was a central cultural factor. It has become common to state that American Indians had no concept of private property. While Native groups hardly constructed property in the same manner as subsequent European and American residents did, they certainly possessed notions of property rights. In northern California, individuals could own property and might transfer property rights to other individuals, while some areas stayed open to the entire community. However, for groups living away from the areas where salmon fishing dominated, private property mattered less. Generally speaking, the more important the resource, the more elaborate the legal structure created to regulate its use (McEvoy 1986, 30, 34). For instance, individual Lummi inherited reef net sites and owned the technology and rituals necessary to use them through kinship networks (Boxberger 2000, 13–14). Kinship rights over fishing sites limited access to salmon. Although fathers passed the sites to sons in most circumstances, individuals were expected to share the caught fish, a sharing that promoted community interdependence (Taylor 1999, 37–38).

Religious practices helped cement these relationships among community members and between the community and nature. Religion centered on salmon's reliable return, and it propitiated the fish. Indeed, tribal groups included ritual specialists who presided over first fish rites and the building of weirs (Swezey and Heizer 1993, 300; McEvoy 1986, 30–31). First fish rites were important, and until completed, harvesting was taboo. For instance, in the Yurok village of Kepel, Native fishers annually constructed an extensive weir. Preparatory ceremonies could last two months, and constructing the weir took ten days. It stood for another ten days while fishers harvested abundant salmon, constituting a large share of the ensuing winter's food. Yuroks believed the *woge*, ancestral people, taught them about the world before moving upstream and turning into mountains, trees, and animals. Thus, it remained critical that they dismantle the Kepel weir to allow salmon upstream to their ancestors. Also, during first fish rites, the Yurok would invite neighboring upstream tribes to watch and participate, helping to ensure interdependence and to guard against intertribal conflict. These cultural beliefs and economic practices collectively encouraged reciprocity with upstream tribes, preventing conflict over resource use (McEvoy 1986, 31, 35–37).

Besides cultural prescriptions, two natural phenomena also limited Natives' annual take. First, catching salmon proved no easy task. Not all areas lent themselves to easy harvesting. Areas where rapids slowed upstream progress were the most productive, especially the area on the Columbia River between Celilo Falls and The Dalles and at Priest Rapids, the Cascades, and Kettle Falls. Such concentrated sites limited the annual take. Second, salmon are not available year-round. Several distinct salmon runs occur each year, but relatively few salmon were obtainable in the fall until summer; late June through October were the prime fishing months. During those peak times, the salmon runs crested for only a few days. Thus, constant salmon fishing did not occur, and they did not need to consider the diverse nutritional options available to Native groups (Hunn 1990, 149–151).

The history of the Native salmon fisheries, thus, reveals the intricacies of the interaction among ecology, economy, and culture. The life cycle of anadromous fish and the seasonal flow of the West's rivers set natural parameters, while Native peoples used their technology to take a significant proportion of salmon from annual runs for their own consumption and for complex exchanges among coastal and interior tribes. Meanwhile, cultural values and practices moderated use. As historian Arthur McEvoy has explained concerning California Indians, but equally applicable for all fishing groups in the Far West, "California Indians managed their fisheries and other resources by strategically gearing their productive effort to the ecological realities of the world as they understood them, so as at once to lead comfortable lives, to distribute wealth equitably, and to sustain their resources and their economies over the long run" (McEvoy 1986, 32). Although salmon figured centrally in the culture of many Pacific Western tribes' subsistence practices and cultural traditions, it represented only part of a complex and broad-based economic and ecological regime that was well articulated.

Fire

Unmistakably, fire was the greatest tool indigenous populations possessed to shape the environment. Not all used it. Not all who used it did so in the same way. Not all regions within the Pacific West were particularly conducive to anthropogenic—much less natural—fire. And, of course, not all fires that shaped the landscape were deliberately set. Nevertheless, much of the Far West bears evidence of extensive and long-lasting anthropogenic fire regimes.

Scholars have debated, occasionally rancorously, about the extent to which Native peoples' fires shaped the landscape that Europeans and Americans first encountered. As with many controversies that touch on issues of American Indians and the environment, other concerns often fuel the scholarly debate. Two popular beliefs about Native peoples' place in nature have denied the idea that the original inhabitants shaped the land in significant ways. Beginning with many of the first European observers of North America, many believed the indigenous landscape bore no or minimal marks of human influence. Europeans used this perception to justify their forcible taking of Indian lands. A corollary to this notion is that Native populations lacked sufficient technology, knowledge, or ability to manipulate the environment. Racial ideas thus deprived Natives a place as historic agents of change. The second belief that has often denied widespread Native use of fires is more recent. This image holds that American Indians possessed an innateness that tied them intimately to nature where they were somehow "at one with nature," living so harmoniously that there was virtually no division between nature and culture. Often called "the Noble Savage," this mythology of Native peoples suggests a population living during a Golden Age with no impact on the earth (Krech, 1999, 19–20). These erroneous and simplistic understandings of Native groups in the Pacific West, and throughout North America, severely limit and bias our understanding of indigenous resource management strategies.

Scholars no longer argue that American Indians did not burn their landscapes. The basic debate now surrounds simple questions: How much did Native groups burn? What were the ecological effects of anthropogenic fire? The answers reveal the extent to which the landscape at the time of contact was a humanized or domesticated one (Blackburn and Anderson 1993; Vale 2002). Understanding the human manipulation of flames and the ecological effects of fire regimes shows how culture and nature interact.

Fortunately, a significant amount of research concerning anthropogenic fire has centered on the Pacific West. The interpretations are predictable. Scholars in the humanities and social sciences, such as historians and anthropologists, have tended to see in the ecological, ethnographic, and historical evidence more instances of widespread anthropogenic burning with significant ecological impacts. On the other hand, physical scientists, such as geographers and ecologists, have tended to see greater evidence of natural fires (i.e., lighting ignited fires) with fewer cultural impacts on the land (Parker 2002, 259; Pyne 2003, 257). Most environmental historians believe too much historical and comparative evidence exists to discount or minimize Native peoples' widespread use of fire and its significant role in arranging large portions of the landscape. After all, outside the pages of academic journals and monographs, fires burned.

Fire is a complex element. The cultural and biological adaptations to patterns of burning create a fire regime (Pyne 2001, 16). Fire regimes exist even in uninhabited landscapes. Such natural fire regimes tend to create mosaics in forests of burned areas. Lightning-caused fires also typically burn hotter and larger than anthropogenic fire regimes. Generally, fire regimes depend on climate along with fuel and ignition sources. As humans have historically been unable to significantly alter climate, they have primarily shaped fuels and ignitions. Indians did not simply elaborate on natural fires; they changed the timing and added purpose. The general principle to anthropogenic burning, as historian Stephen J. Pyne has noted, is to "burn early, burn light, burn often," meaning fires before conditions became too dry, fires on a small scale, and fires habitually (Pyne 2001, 52). By taking relative control over ignitions, humans wrested away the power of wildfire and helped shape fuel loads and modify large ecosystems. Humans are not alone in shaping fuels, though. Animals, particularly grazers and browsers, can significantly change fuel loads. The disappearance of megafaunal grazers, for example, meant a significant consumer of fuel no longer consumed it. With that fuel gone, anthropogenic fire often took its place. Multiple factors plainly shaped the dynamics of fire's environmental history (Lewis 1993, 94; Lewis and Ferguson 1999, 164–65; Pyne 2001, 31).

Why did Native populations burn? Scholars have presented several interpretations. U.S. Forest Service historian Gerald Williams categorized eleven reasons for indigenous burning: hunting, crop management, insect collection, pest management, plant growth and yields, fireproof areas, warfare and signaling, economic extortion, clearing areas for travel, felling trees, and clearing riparian areas. More simply, they burned where they lived to improve their livelihoods (Williams 2002, 196, 209–10). The leading anthropologist who studied indigenous fire, Henry T. Lewis, has explained that Native groups typically maintained "fire yards" and "fire corridors" and burned forests that created "fire mosaics." Fire yards were clearings in forests maintained by fire, usually to promote particular floral and faunal species. Fire corridors represented basically the same purpose but were typically along watercourses, ridges, or trails and also facilitated travel. The fire mosaic grew out of natural and anthropogenic causes and resulted in patchy and fire-influenced environments (Lewis and Ferguson 1999, 164–169). To be sure, reasons for burning were not uniform everywhere. Native groups developed fire regimes particular to their environments. Four regional examples illustrate the diversity of Native fire regimes: Southern California's chaparral, inland ponderosa pine forests, lowland prairies, and coastal Douglas-fir forests. (One might include lodgepole pine forests here, as they frequently were the site of crown fires, a different fire type, but lodgepole pines do not constitute a major forest type in the Pacific West.)

Chaparral

Southern California's Mediterranean climate is prone to burning; indeed, environments, such as those in the Mediterranean, Australia, and California, have evolved with biota suited to periodic burning. Chaparral is one prominent example. Chaparral is a relatively imprecise term commonly used for the sclerophyllous brushlands that appear in some form in virtually all Mediterranean climates, although with different names. In California, chamise dominates chaparral with various oaks and manzanita as powerful components. It exists in several areas, including through some of the coniferous forests in northern California, the ecotone between the Sierra Nevada and the woodland-grass ecosystems, and the coastal mountains of Southern California, the focus here (Baumhoff 1978, 16, 18–19, 22–24; Davis 1998, 10–14, 95–147; Pyne 1997, 413–418).

The dry climate, low rainfall, and plant compositions combine to create a particularly pyrophytic environment. Fires help regenerate grasses and forbs, the seeds of which have lain dormant in the soil since the previous burn. The same proves true for some shrubs. The manzanita, for example, requires the heat generated by fire to germinate. The new growth results rapidly in highly nutritious plants for browsers. In addition, the postfire habitat improves populations of such small game species as rabbit or quail by a factor of as much as four-and-a-half (Bendix 2002, 269–272, 286–288; Lewis 1993, 69–74; Pyne 1997, 413).

Such ecological processes immeasurably benefited the hunting and gathering economy of local Native populations. The Chumash who lived along the Southern California coast used approximately 70 known plant species for food or medicinal purposes. About half of them occur after fire, and another 15 are most abundant the year after burning. Accordingly, the Chumash burned an area about every three to five years. They especially targeted the margins of the chaparral. Anthropogenic burning, then, transformed chaparral into a productive, useful environment. As an added benefit, the fires reduced the fuel in chaparral environments making lightning fires less damaging when they came. The removal of such fire regimes created significant fire hazards, to which modern Southern Californians can readily attest (Davis 1998, 95–147; Lewis 1993, 69–74; Timbrook, Johnson, and Earl 1993, 138).

Ponderosa-Pine Forests

In interior forests throughout the North American West, ponderosa pines often dominated or constituted a significant part of mixed-conifer forests. Representative examples include the eastern slopes of the Cascade Mountains and the Blue Mountains of northeastern Oregon. Fires were endemic here, flaring on average every seven to ten years. Ponderosa pines enjoyed protection from fire because of evolutionary adaptations. Its bark developed a protective layer by the time it was merely two inches in diameter that could withstand the forest's typical light burns. This defense measure allowed young pines to survive while other tree species, such as Douglas-fir, could not. Environmental historian Nancy Langston explained the environmental effects of frequent fires in the Blue Mountains: "These fires kept the stand open, killed off the youngest trees, reduced fuel loads, and released a surge of nitrogen into the soil that stimulated the growth of grasses and nitrogen-fixing shrubs such as ceanothus" (Langston 1995, 29). Ecologically, then, these fires accomplished much, from cycling nutrients through the ecosystem to maintaining a relatively open forest. Pockets that did not burn regularly faced stronger, stand-replacing blazes, but such fires proved the exception not the

Manzanita bush in flames. Fire triggers seed germination in Manzanita plants.
(iStockPhoto.com)

rule (Barrett and Arno 1999, 58–61; Langston 1995, 29–32; Whitlock and Knox 2002, 219–220).

Evidence strongly suggests that anthropogenic burning represented the major source for this fire regime. Routine, frequent fires, as evidenced in fire scars from the inland Northwest, indicate a regularity nature could not match. Several purposes exist for this anthropogenic fire regime. Fire uncluttered the undergrowth and unencumbered travel. Ungulates such as mule deer, moose, and especially elk favor various forest habitats shaped by fire. Thus, promoting these habitats improved hunting opportunities. Fires also promoted berry production, such as the black mountain huckleberry, that contributed an important food source to Plateau tribes. Fires set in late fall after rains had begun ideally ensured ample control of the fire and kept weedy trees and brush from overtaking mountain berry patches. In addition, Natives used fire to dry the huckleberries. Of course, lightning strikes sometimes ignited the pines, and Native-set fires occasionally burned out of control, but the managed open pine forests, especially near population centers and along travel routes, bore the unquestionable marks of a deliberate pattern of burning. As in chaparral, the removal of this anthropogenic fire regime thoroughly transformed the region's ponderosa pine forest (Barrett and Arno 1999, 58–61; Chatters 1998, 40; French 1999, 31–35; Hunn 1990, 132; Langston 1995, 46–49; Whitlock and Knox 2002, 221–222).

Lowland Prairies: Willamette Valley

Some of the most extensively burned areas lay in lowland valleys, such as western Oregon's Willamette Valley. This low-lying, broad valley nestled between the Coast and Cascade ranges in western Oregon includes several microenvironments in which a variety of species locally dominated. Some have called the Willamette the "Valley of the Long Grasses" after the variety of tall grasses that carpeted the area. Indeed, Native peoples called the collective grasses *kalapuya*, which Europeans used to label the valley residents (Boag 1992, 3). Perennial grasses, wildflowers, oaks, and bulbous plants constituted important biological communities in the valley. This open oak savanna grassland type appears to have evolved about 8000–4000 BP during a drier climate, while Native groups maintained it through anthropogenic fire (Boag 1992, 12–17; Boyd 1999, 95–98; Whitlock and Knox 2002, 203–06; Williams 2002, 190).

The Native residents, primarily the Kalapuyans but also the Umpqua and Rogue River tribes, routinely burned for a variety of subsistence purposes. Fire helped promote the ecology on which their broad dietary foundation of wild foods was based. Frequent burning aided the growth of some specific species, but it also eased gathering techniques and shaped habitats conducive to hunting. Evidence suggests that Natives burned alternate sides of the valley every fall when grasses were driest. Native women would then gather seeds from plants like the tarweed more easily because fire dried the pods and burned off the sticky resin on the plants' stalks. By torching dead grasses, fires made acorns easier to find and harvest and promoted regeneration of some grasses. By removing woody competition, too, fires increased the productivity of various plants, including acorns, wild onions, camas, and hazelnuts. Tobacco cultivation also depended on frequent flames. Natives planted tobacco on burned land, which furnished beneficial fertilizer. Besides these horticultural and agricultural purposes, fire might also be used to gather such insects as grasshoppers or crickets, which were used in making pemmican. As in other places, Natives used fire as an aid in promoting a favorable hunting habitat, as well as being used in the hunt itself. All told, the Kalapuya and other valley inhabitants impressively used fire to manage about 2 million acres in the Willamette Valley (Boag 1992, 13–17; Boyd 1999, 94–128; Bunting 1997, 13–15; Robbins 1997, 31; Williams 2002, 186–91).

Douglas-Fir Forests

Among the ecological types sampled here, Douglas-fir forests represent the least frequently fired. Although many inland forests include Douglas fir and include regimes similar to the Ponderosa pine forests, Douglas fir are most dominant on the wet west side of the Northwest. That climatic factor makes an enormous ecological difference, for it made natural ignitions less likely to strike and, if struck, less likely to take hold and erupt in a devastating crown fire. Normally, most people have not associated Douglas-fir forests with anthropogenic fire regimes because Native groups contributed to those regimes so differently from those of chaparral, Ponderosa pines, or valley grasslands. Nevertheless, these forests also bore the mark of fire.

As mature trees Douglas firs could withstand light burning; indeed, they are the most fire-resistant of all conifers on the west side of the Cascade Mountains. Young Douglas-fir trees, however, were not protected from flames as Ponderosa pines were. Moreover, as shade-intolerant species, they required relatively open areas in which to thrive, as well as a mineral seedbed for germination. Frequently, fires met those requirements, and burning reduced competitor trees, such as western hemlock or cedar. Thus, fires helped create the vast Douglas-fir bioregion west of the Cascades. Much of the Douglas-fir forests stood in even-aged stands, reflecting how the forest originated after some calamitous event, like fire. Indeed, after catastrophic fires Douglas-fir seedlings might return with 50,000 per acre, which would thin to about 500 after half a century. A mature Douglas-fir forest reconstitutes itself after about two centuries. The return of intense fires that recreated the process came sporadically, often after several centuries. The question becomes to what extent were these stand-replacing fires anthropogenic, and the answer remains unclear. Indian burning in the Douglas-fir bioregion is related to valley burning. When conditions were favorable, such as during an extended drought, a fire in the valley grasslands could escape into the forest and decimate thousands of acres. That is, Douglas-fir forests frequently abutted the lowland valleys in the western Northwest, which Indians routinely fired. Thus, much of the anthropogenic impact on Douglas-fir forests remained on the margins where forests met prairies (Bunting 1997, 15; Kruckeberg 1991, 185, 188–189; Langston 1995, 25, 30; Pyne 1997, 335; White 1992, 10).

As this survey demonstrates, while a ubiquitous presence, fire was not a generalized tool everywhere. Different environments responded to fire in unique ways. No less, different cultures manipulated flames for different purposes and to different effects. Nevertheless, the Far West's landscape, wherever inhabited, displayed evidence of anthropogenic burning. The removal of those fires, as will be shown in subsequent chapters and as is apparent in today's frequently out-of-control wildfires, caused enormous ecological changes to the Pacific West and testifies to the truism that fire removed is as powerful as fire applied.

Cultivation

The first Europeans to observe Far Western cultures arrived with preconceived ideas about agriculture. Because of their beliefs, Europeans did not recognize the extent to which western Natives cultivated the land. Indeed, "gathering" typically has been the term used to describe indigenous relationships to plants. The term, however, misrepresents as passive and haphazard the vastly complex and active role Native groups, primarily women, employed to manage, use, and collect a wide array of wild and domestic plants in the Pacific West.

In considering Native gathering, scholars first believed gathering merely augmented Native diets by relying on naturally occurring plants. However, the recovery of traditional ecological knowledge among contemporary Indian populations, along with a broader understanding and definition of agriculture among scholars, has significantly revised the notion that Far Western Natives merely gathered to augment their diets and were not true agriculturalists (Deur 1999, 139–140). Recently, anthropologists, geographers, historians, and Native peoples themselves have challenged such notions and presented abundant evidence of deliberate cultivation to improve productivity of an area, to transplant new species, or to expand cultivable land. Native people did not practice European-style agriculture with rectilinear fields, cereal crops, and many domesticated plants. Nonetheless, they applied sophisticated ecological and horticultural knowledge to numerous native wild plants and some domesticated species to create landscapes and ecological regimes dependent on careful cultivation. To varying degrees throughout the region, Natives consciously shaped the earth to produce various plants for subsistence and other purposes. In other words, they practiced a form of agriculture, resulting in what one anthropologist has termed "unusual gardens" (Marshall 1999).

Northwest Coast

Cultivation varied by geography. On the Northwest Coast, cultivation expanded perhaps 4,000 years ago as part of generalized resource intensification. At this time, coastal Natives began settling into more sedentary villages, a process facilitated by new technologies that allowed for the more efficient taking and storing of food. An increasing population also required families and groups to specialize their resource production. To support expanding populations and to complement dietary needs, Northwest Coast Natives cultivated at least two domestic plants: tobacco and potatoes. Both plants were likely imports from interior locales and arrived by intertribal trade before direct contact with Europeans; the potato arrived soon before Europeans and was rapidly assimilated into Native cultivation patterns. Native cultivators, usually women, gathered tobacco seeds and carefully replanted (Deur 1999, 140). On the southern Northwest Coast, Natives smoked tobacco, while in the northern reaches they mixed it with lime and chewed it (Suttles 1990, 24). To enhance productivity, they fertilized potatoes in "piles of marine detritus" (Deur 1999, 140). Natives grew both plants in clearings made by significant human labor and by fire. The management of tobacco and potatoes clearly demonstrates domestic agricultural practices.

In addition, coastal tribes tended native plants. Camas was one of the most important plants among all Northwestern tribes. These lilies with edible bulbs thrive in meadows. Not surprisingly, then, Natives used fire to promote these camas prairie habitats. The fires improved soil fertility and removed competitor species. Some prime camas habitat, such as the vast Willamette Valley, could be hundreds of square miles and include many other edible plants besides camas. Along the coast, however, Natives cultivated much smaller patches exclusively for camas production, usually along a moist forest's margins and on south-facing slopes. In these wet areas, the anthropogenic burning appears to be solely for camas production and not for one of the other common uses of fire. Coastal tribes not only promoted camas' growth by preparing habitat, but they are

Camassia quamash, *also known as small camas, photographed on Yellow Island, San Juan Islands, Washington. (Wolfgang Kaehler/Corbis)*

likely to be responsible for its presence on the coast to begin with. Camas is native to the interior, not the coast, so trade networks and conscious transplanting demonstrates the importance of this plant in coastal horticulture (Kruckeberg 1999, 66–67; Deur 1999, 140–142; White 1992, 21).

Another impressive example of the cultivation practices by coastal tribes was rhizome cultivation in estuaries. Natives cultivated two endemic plants—spring-bank clover and Pacific silverweed—in gardens along coastal estuaries for their underground stems, or rhizomes. Although archaeological evidence is scarce, it appears that coastal groups laboriously removed rocks, constructed dam-like fence structures and carefully weeded these plots to enlarge estuary habitat and to promote growth. Ecologically, the estuarine sites were located in a particularly rich environment, regularly gaining nutrients from marine detritus. The wall-like structures they built around the garden plots trapped nutrient-rich detritus but kept out excessive sediment to purposefully create rich soil; indeed, the Kwakiutl term for this earth translates as "manufactured soil." Moreover, enough salt came to keep some plants from growing but not too much to kill off these particular rhizomes. Although evidence is inconclusive, it is possible that Native peoples domesticated these rhizomes as different varieties developed in coastal locales. Such labor-intensive work indicates the importance of these plants for both subsistence and ceremonial purposes (Deur 1999, 142–146).

Intensified cultivation arose with generalized resource specialization along the coast between two and four millennia ago. An important result was increased social complexity. The relationship between environmental exploitation and hierarchical society seems clear. Communities became more sedentary and manipulated resources more intensively, which required increased social networks to redistribute resources. Moreover, the intricacy required to maintain territorial resources increased intervillage conflict and gave rise to elites that organized resource production and territorial defense. This hierarchical society that developed to manage access to resources, as well as for other cultural purposes, represented a marked alternative from interior tribes who approached resource management, specifically horticultural practices, differently.

Columbia Plateau

In the interior, Native populations initiated similar cultivation practices, although geographic and cultural factors made horticulture on the Columbia Plateau look somewhat different from that practiced along the Coast. Perhaps the most important cultural difference between Coast and Interior was mobility. Interior tribes moved around to a much greater degree than the coastal groups. According to anthropologist Eugene S. Hunn, that transhumance was "patterned after the temporal and spatial distribution of key carbohydrate resources" (Hunn 1999, 168). That is, Plateau tribes followed vital plant resources as they ripened. Although salmon runs dictated when many Natives would be fishing at a particular site, plant harvesting was also a key determinant for movements. Plateau tribes practiced a highly developed seasonal round to follow resources, both plant and animal. Thus, theirs was not a random, haphazard migration pattern.

Plant use on the Columbia Plateau constituted a complex ecological regime. Plateau Natives used at least 100 plants as food, representing approximately 60 percent of the Plateau diet. Recognizing where and when these resources would be ready required intimate knowledge of local landscapes. In midwinter, Natives began gathering a series of plants generically called Indian celeries (Hunn 1999, 159). By late spring, they harvested "foods that are dug," including bitterroot, varieties of lomatium, and camas, from low elevation sites to which they traveled from their winter villages or fishing camps. At times, they moved nearly 80 kilometers to reach gathering grounds. Native women took these roots in quantities far exceeding immediate need and preserved them in caches or by drying and baking them for later consumption (Hunn 1999, 160). Thus, the harvesting and consumption of plants were spaced out. While many fished for salmon at places like Celilo Falls or Kettle Falls, even more camped in the mountains to collect huckleberries and other "foods that are picked," including other berries and lichens. Not surprisingly, anthropogenic fire improved berry productivity (Hunn 1999, 162). Plateau tribes harvested across time and space so that they might obtain sufficient plant and animal resources to last through the less abundant time of year from late fall to spring.

Camas again provides a good illustration of the Natives' purposeful cultivation practices. Camas required more moisture than was typically available on the Plateau, so it could be found in favored meadows. Although it was not obtainable everywhere, according to anthropologist Alan G. Marshall, the Nez Perce considered camas, or *qem'es*, their "premier plant food" (Marshall 1999, 177). Using traditional digging tools (called *tuuk'es* by the Nez Perce or *pitse'* by the Coeur d'Alene), Native women would harvest camas for one to several weeks in summer. Such a short harvest provided sufficient camas for the entire year; indeed, one report from the mid-nineteenth century indicates that a four-day harvest provided enough camas for an entire year. Meanwhile, the digging disturbed the habitat and aerated the soil, helping camas regenerate as harvest occurred once seeds were ready (Hunn 1991, 109, 170–179; Hunn 1999, 160; Marshall 1999, 177–179).

Camas and other plant harvesting provided, in addition to subsistence and ecological purposes, important social functions. As was typical of any animal or plant harvest, before taking camas from the grounds, Natives provided an offering and a prayer, relating to the world in an intimate way. Because the camas prairies were localized, tribal groups traveled to them seasonally, often combining other cultural exchanges such as gambling, trading, politics, and marriages. A fur trader witnessed one such gathering in the Kittitas Valley of central Washington that included an estimated 10,000 Native peoples, a number perhaps exaggerated. More typical may have been the routine 1,500-person gatherings reported near Moscow, Idaho. Gathering thus clearly served intercultural relations, but individuals could also use camas production to enhance one's status within the community. As geographer Douglas Deur explained, "oral traditions suggest that they were motivated to intensify these resources through the use of fire and the tending of camas plots" (1999, 141). Apparently, some Native women increased their production to enhance their social status. Finally, today, Northwest Native peoples still use camas practically and ceremonially, although habitat change by urban development and agriculture has made it increasingly rare (Frey 2001, 154–161; Hunn 1991, 127; Hunn 1999, 160; Marshall 1999, 180).

Like many Indians, Plateau groups were long depicted by observers simply as nomadic hunters and gatherers but that misrepresents and simplifies their society. Among southern Plateau groups, for instance, plant exploitation intensified about four millennia ago giving rise to village settlements and a semisedentary life. This semisedentism was not predicated on access to salmon, a main food source, but on the location of plant resources. Larger villages of consumption existed during winter, while smaller communities of production gathered in the summer. This settlement pattern suggests a social complexity greater than many scholars allowed for so-called nomads, including communal housing, ranked leadership positions, inegalitarian social organization, and large multiethnic encampments. It seems likely that the rise of this social complexity grew, at least in part, from the adoption of horticultural methods and greater plant specialization for various ethnic groups. Marshall hypothesized that this complexity developed to account for the ecological simplification inherent in specialized horticulture, which leads to more unstable food supplies. Clearly, even with mobility as a subsistence strategy, inland Northwest tribes required careful social and ecological planning to meet cultural and subsistence needs (Ames and Marshall 1980–81, 25–47; Marshall 1999, 173–84).

California

To the south in California, Native cultures followed similar themes, though with some variations. Early European observers perceived a lack of cultivation, a fact that seemed unusual in such an abundant environment and that Europeans interpreted as Natives lacking sophistication. However, recent scholars have challenged European notions that agriculture represents the highest level of social evolution. Instead, they have questioned why California Natives should have adopted intensive agriculture. As two anthropologists have explained, "agriculture was an unnecessary alternative for the California Indian because of an efficient, interlocking series of energy extraction processes, some of which were semiagricultural" (Bean and Lawton 1993, 52). In other words, California Native peoples already possessed a cultivation regime fully capable of meeting subsistence needs.

Native horticultural practices favored certain plants. Overall, according to Kat Anderson, a leading authority on pre-Columbian Californian land use, the relationship between California Natives and plants constituted a "partnership" (Anderson 1993, 152). It is clear that wild plant populations did not simply appear; Natives prepared sites, enlarged them, collected and planted seeds, pruned branches, and aerated soil while harvesting. Favored plants reproduced best on disturbed ground, and Natives obliged by furnishing a variety of disturbances.

Together, these practices meant the landscape bore a strong mark of human manipulation (Anderson 1993, 152–56).

Throughout California, Natives practiced sophisticated and labor-intensive cultivation that complemented their hunting and fishing regimes and that also supplied necessary plants for their material culture. While harvesting at a given site, Natives did not take all the plants. Moreover, they returned reproductive materials (e.g., bulblets, cormlets, fragments of tubers) to ensure regeneration. Grasses, an important component to Native life, were sown carefully after women collected seeds using two baskets and a seedbeater. Before leaving the site, Natives would broadcast some of the harvested seeds. Burning often helped prepare the ground as well. For the plants generically lumped as Indian potatoes (e.g., wild onions, yampah, brodiaeas), Natives harvested the roots and bulbs only selectively and thinned them to improve their growth. These practices also aerated the soil, staved off weed incursions, and prepared the soil for germination. Even while favored plants were growing, Natives shaped them. They might burn in a focused area, even a single plant or tree (Anderson 1993, 156–160). Native women might also prune branches to make them grow straighter, which would be of greater use to them for making various baskets, an important cultural and practical activity. Anderson explained the intricate process of preparing plants:

> Hundreds of straight rhizomes and thousands of straight branches were needed to make the baskets produced by a single village, yet a search in the wilds for long, straight, slender switches with no lateral branching is largely in vain. In order to gather sufficient suitable branches for making the many kinds of baskets produced by adolescent and adult females in various villages, Native Americans had to manage and maintain abundant populations of certain plants at what was virtually an industrial level (1993, 162).

Clearly, cultivation was not haphazard.

Undoubtedly, California's most important crop was acorns. California included nineteen species of oaks, eight of them reaching tree size. Natives preferred the tan oak, but that species was available in only about 8 percent of the region, while the next preferred species, black oak, was common over nearly 20 percent of California. Fortunately, oak species overlapped in many areas, allowing Natives a choice. Evidence strongly indicates that Native Californians actively managed the oaks to make the larger environment more conducive to oak growth, to increase the number of productive trees in a specific area, and to maximize crop production (McCarthy 1993, 213–214).

Native management was broad. It seems unlikely that planting was a significant part of the management regime. The black oak, for instance, would not produce acorns for 30 years, yielded no bumper crops for 80 years, and reached maturity at 175 years. Moreover, individual trees possess unique characteristics, so it would be unclear whether one were planting an oak that would bear plentiful large acorns or few small acorns. Besides, according to a contemporary Native informant, that was the blue jays' job (McCarthy 1993, 217). Rather than active planting, knocking was a common method to aid acorn harvests. Its effects resembled pruning. Taking a long pole, Natives would knock acorns off the tree, breaking branches in the process that would increase canopy space and improve acorn production. Moreover, it allowed harvesters to obtain acorns before they fell naturally, which allowed them to then travel to more distant oaks effectively lengthening the harvest period. Finally, Natives burned. Fire improved acorn production in several ways. First, it improved the habitat for oaks. It reduced competition by killing sprouting conifers and allowing the shade-intolerant oaks to flourish. Eliminating competitors also meant greater access to often scarce groundwater for oak and acorn production. Second, it helped individual trees. Karok Natives reported that oaks were better if burned annually. The reason is that fires kill pests, including the filbert weevil and filbert worm, that harm acorns. Last, fires made harvesting easier by removing groundcover. Fire profoundly shaped the productivity and distribution of oaks. Clearly, the oak groves of California bore strong cultural imprints (McCarthy 1993, 213–228).

Through these various processes of cultivation, Natives developed a conservation ethos. Kat Anderson categorized seven principles of conservation practiced by California Natives:

1. The quantity taken does not exceed the biological capacity of the plant population to regenerate or recover.

2. Gathering techniques sometimes mimic a parallel natural disturbance with which the plant has coevolved, thus maintaining and sometimes enhancing plant production.

3. The tool used is appropriate to the resource. It does not deplete the plant population of interest.

4. Horticultural techniques are used to give plants a competitive edge and put resources back into the system.

5. Often plants are chosen that exhibit remarkable vegetative reproduction.

6. Management is frequently at a scale that maintains the integrity of the plant community.

7. Taboos, codes, or other social constraints are in place to discourage depletion or over-exploitation and avoid waste—thus reinforcing conservation-minded behaviors (Anderson 1993, 170).

These values represent the culmination of experience and traditional environmental knowledge. Furthermore, they illustrate the deep interconnection between human societies and the natural world. Moreover, although drawn from California, these principles are broadly applicable to the entire region.

CONCLUSION

Native cultivators modified the landscape around them, transplanting species and promoting various plant crops. Adapting a seasonal round coordinated with their hunting and fishing practices, Indian populations drew from nature even as they shaped it through fire and by manipulating species. After several millennia, Native peoples in the Pacific West established generally sustainable ecological regimes. Through trial and error, cultural prescriptions and practices, and close interaction with and understanding of the environments in which they lived, Native cultures engaged in subsistence economies that could take much from nature but kept it largely self-perpetuating with some important exceptions related to hunting.

The arrival of Europeans radically transformed those ecological regimes, disrupting the work of millennia. The landscapes Native groups created, as well as the people themselves, faced a series of enormous shocks beginning in the seventeenth century and accelerating at the end of the eighteenth and beginning of the nineteenth. The story of those ecological shocks constitutes the next chapter.

REFERENCES

Ames, Kenneth M., and Alan G. Marshall. "Villages, Demography and Subsistence Intensification on the Southern Plateau." *North American Archaeologist* 2 (1980–81): 25–52.

Anderson, Kat. "Native Californians as Ancient and Contemporary Cultivators." In *Before the Wilderness: Environmental Management by Native Californians*, edited by Thomas C. Blackburn and Kat Anderson, 151–174. Menlo Park, CA: Ballena Press, 1993.

Arnold, Jeanne E. "Transportation Innovation and Social Complexity among Maritime Hunter-Gatherer Societies." *American Anthropologist* 97 (December 1995): 733–747.

Barrett, Stephen W., and Stephen F. Arno. "Indian Fires in the Northern Rockies: Ethnohistory and Ecology." In *Indians, Fire, and the Land in the Pacific Northwest*, edited by Robert Boyd, 50–64. Corvallis: Oregon State University Press, 1999.

Baumhoff, Martin A. "Environmental Background." In *Handbook of North American Indians*. Vol. 8, *California*, edited by Robert F. Heizer, 16–24. Washington, DC: Smithsonian Institution, 1978.

Bean, Lowell John, and Harry W. Lawton. "Some Explanations for the Rise of Cultural Complexity in Native California with Comments on Proto-Agriculture and Agriculture." In *Before the Wilderness: Environmental Management by Native Californians*, edited by Thomas C. Blackburn and Kat Anderson, 27–54. Menlo Park, CA: Ballena Press, 1993.

Bendix, Jacob. "Pre-European Fire in California Chaparral." In *Fire, Native Peoples, and the Natural Landscape,* edited by Thomas R. Vale, 269–293. Washington, DC: Island Press, 2002.

Blackburn, Thomas C., and Kat Anderson, eds. *Before the Wilderness: Environmental Management by Native Californians.* Menlo Park, CA: Ballena Press, 1993.

Boag, Peter G. *Environment and Experience: Settlement Culture in Nineteenth-Century Oregon.* Berkeley: University of California Press, 1992.

Boxberger, Daniel L. *To Fish in Common: The Ethnohistory of Lummi Indian Salmon Fishing.* Lincoln: University of Nebraska, 1989. Reprint, Seattle: University of Washington Press, 2000.

Boyd, Robert. *People of The Dalles: The Indians of Wascopam Mission: A Historical Ethnography Based on the Papers of the Methodist Missionaries.* Lincoln: University of Nebraska Press, 1996.

Boyd, Robert. "Strategies of Indian Burning in the Willamette Valley." In *Indians, Fire, and the Land in the Pacific Northwest,* edited by Robert Boyd, 94–138. Corvallis: Oregon State University Press, 1999.

Broughton, Jack M. "Pre-Columbian Human Impact on California Vertebrates: Evidence from Old Bones and Implications for Wilderness Policy." In *Wilderness and Political Ecology: Aboriginal Influences and the Original State of Nature.* Edited by Charles E. Kay and Randy T. Simmons, 44–71. Salt Lake City: University of Utah Press, 2002.

Bunting, Robert. *The Pacific Raincoast: Environment and Culture in an American Eden, 1778–1900.* Lawrence: University Press of Kansas, 1997.

Chatters, James C. "Environment." In *Handbook of North American Indians.* Vol. 12, *Plateau,* edited by Deward E. Walker, Jr., 29–48. Washington, DC: Smithsonian Institution, 1998.

Colten, Roger H., and Jeanne E. Arnold. "Prehistoric Marine Mammal Hunting on California's Channel Islands." *American Antiquity* 63 (October 1998): 679–701.

Crosby, Alfred W. *Ecological Imperialism: The Biological Expansion of Europe, 900–1900.* New York: Cambridge University Press, 1986.

Davis, Mike. *Ecology of Fear: Los Angeles and the Imagination of Disaster.* New York: Metropolitan Books, 1998.

Deur, Douglas. "Salmon, Sedentism, and Cultivation: Toward an Environmental Prehistory of the Northwest Coast." In *Northwest Lands, Northwest Peoples: Readings in Environmental History,* edited by Dale D. Goble and Paul W. Hirt, 129–155. Seattle: University of Washington Press, 1999.

Fagan, Brian M. *Ancient North America: The Archaeology of a Continent.* 3rd ed. New York: Thames and Hudson, 2000.

Flannery, Tim. *The Eternal Frontier: An Ecological History of North America and Its Peoples.* New York: Atlantic Monthly Press, 2001.

French, David. "Aboriginal Control of Huckleberry Yield in the Northwest." In *Indians, Fire, and the Land in the Pacific Northwest,* edited by Robert Boyd, 31–35. Corvallis: Oregon State University Press, 1999.

Frey, Rodney, in collaboration with the Schitsu'umsh. *Landscape Traveled by Coyote and Crane: The Worlds of the Schitsu'umsh (Coeur d'Alene Indians).* Seattle: University of Washington Press, 2001.

Harden, Blaine. *A River Lost: The Life and Death of the Columbia.* New York: W. W. Norton, 1996.

Heizer, Robert F., and Albert B. Elsasser. *The Natural World of the California Indians.* Berkeley: University of California Press, 1980.

Hildebrandt, William R., and Terry L. Jones. "Depletion of Prehistoric Pinniped Populations along the California and Oregon Coasts: Were Humans the Cause?" In *Wilderness and Political Ecology: Aboriginal Influences and the Original State of Nature,* edited by Charles E. Kay and Randy T. Simmons, 72–110. Salt Lake City: University of Utah Press, 2002.

Hunn, Eugene S. "Mobility as a Factor Limiting Resource Use on the Columbia Plateau." In *Northwest Lands, Northwest Peoples: Readings in Environmental History,* edited by Dale D. Goble and Paul W. Hirt, 156–172. Seattle: University of Washington Press, 1999.

Hunn, Eugene S., with James Selam and Family. *Nch'i-Wána "The Big River": Mid-Columbia Indians and Their Land.* Seattle: University of Washington Press, 1991.

Kay, Charles E. "Afterword: False Gods, Ecological Myths, and Biological Reality." In *Wilderness and Political Ecology: Aboriginal Influences and the Original State of Nature,* edited by Charles E. Kay and Randy T. Simmons, 238–261. Salt Lake City: University of Utah Press, 2002.

Krech, Shepard, III. *The Ecological Indian: Myth and History.* New York: W. W. Norton, 1999.

Kruckeberg, Arthur R. *The Natural History of Puget Sound Country.* Seattle: University of Washington Press, 1991.

Kruckeberg, Arthur R. "A Natural History of the Puget Sound Basin." In *Northwest Lands, Northwest Peoples: Readings in Environmental History,* edited by Dale D. Goble and Paul W. Hirt, 51–78. Seattle: University of Washington Press, 1999.

Langston, Nancy. *Forest Dreams, Forest Nightmares: The Paradox of Old Growth in the Inland West.* Seattle: University of Washington, 1995.

Lewis, Henry T. "Patterns of Indian Burning in California: Ecology and Ethnohistory." In *Before the Wilderness: Environmental Management by Native Californians,* edited by Thomas C. Blackburn and Kat Anderson, 55–116. Menlo Park, CA: Ballena Press, 1993.

Lewis, Henry T., and Theresa A. Ferguson. "Yards, Corridors, and Mosaics: How to Burn a Boreal Forest." In *Indians, Fire, and the Land in the Pacific Northwest*, edited by Robert Boyd, 164–184. Corvallis: Oregon State University Press, 1999.

Marshall, Alan G. "Unusual Gardens: The Nez Perce and Wild Horticulture on the Eastern Columbia Plateau." In *Northwest Lands, Northwest Peoples: Readings in Environmental History*, edited by Dale D. Goble and Paul W. Hirt, 173–187. Seattle: University of Washington Press, 1999.

Martin, Paul S. "Prehistoric Overkill: The Global Model." In *Quaternary Extinctions: A Prehistoric Revolution*, edited by Paul S. Martin and Richard G. Klein, 354–404. Tucson: University of Arizona Press, 1984.

Martin, Paul S., and Richard G. Klein, eds. *Quaternary Extinctions: A Prehistoric Revolution*. Tucson: University of Arizona Press, 1984.

McCarthy, Helen. "Managing Oaks and the Acorn Crop." In *Before the Wilderness: Environmental Management by Native Californians*, edited by Thomas C. Blackburn and Kat Anderson, 213–228. Menlo Park, CA: Ballena Press, 1993.

McEvoy, Arthur F. *The Fisherman's Problem: Ecology and Law in the California Fisheries, 1850–1980*. New York: Cambridge University Press, 1986.

Parker, Albert J. "Fire in Sierra Nevada Forests: Evaluating the Ecological Impact of Burning by Native Americans." In *Fire, Native Peoples, and the Natural Landscape*, edited by Thomas R. Vale, 233–267. Washington, DC: Island Press, 2002.

Pielou, E. C. *After the Ice Age: The Return of Life to Glaciated North America*. Chicago: The University of Chicago Press, 1991.

Preston, William S. "Post-Columbian Wildlife Irruptions in California: Implications for Cultural and Environmental Understanding." In *Wilderness and Political Ecology: Aboriginal Influences and the Original State of Nature*, edited by Charles E. Kay and Randy T. Simmons, 111–140. Salt Lake City: University of Utah Press, 2002.

Pyne, Stephen J. *Fire in America: A Cultural History of Wildland and Rural Fire*. Princeton, NJ: Princeton University Press, 1982. Reprint, Seattle: University of Washington Press, 1997.

Pyne, Stephen J. *Fire: A Brief History*. Seattle: University of Washington Press, 2001.

Pyne, Stephen J. Review of *Fire, Native Peoples, and the Natural Landscape*, edited by Thomas R. Vale. *Restoration Ecology* 11 (June 2003): 257–259.

Robbins, William G. *Landscapes of Promise: The Oregon Story, 1800–1940*. Seattle: University of Washington Press, 1997.

Steinberg, Ted. *Down to Earth: Nature's Role in American History*. New York: Oxford University Press, 2002.

Suttles, Wayne. "Environment." In *Handbook of North American Indians*. Vol. 7, *Northwest Coast*, edited by Wayne Suttles, 16–29. Washington, DC: Smithsonian Institution, 1990.

Swezey, Sean L., and Robert F. Heizer. "Ritual Management of Salmonid Fish Resources in California." In *Before the Wilderness: Environmental Management by Native Californians*, edited by Thomas C. Blackburn and Kat Anderson, 299–327. Menlo Park, CA: Ballena Press, 1993.

Taylor, Joesph E., III. *Making Salmon: An Environmental History of the Northwest Fisheries Crisis.* Seattle: University of Washington Press, 1999.

Timbrook, Jan, John R. Johnson, and David D. Earle, "Vegetation Burning by the Chumash." In *Before the Wilderness: Environmental Management by Native Californians*, edited by Thomas C. Blackburn and Kat Anderson, 117–149. Menlo Park, CA: Ballena Press, 1993.

Vale, Thomas R. "The Pre-European Landscape of the United States: Pristine or Humanized?" In *Fire, Native Peoples, and the Natural Landscape*, edited by Thomas R. Vale, 1–39. Washington, DC: Island Press, 2002.

White, Richard. *Land Use, Environment, and Social Change: The Shaping of Island County, Washington.* 1980. Reprint with a new foreword, Seattle: University of Washington, 1992.

Whitlock, Cathy, and Margaret A. Knox. "Prehistoric Burning in the Pacific Northwest: Human versus Climatic Influences." In *Fire, Native Peoples, and the Natural Landscape*, edited by Thomas R. Vale, 195–231. Washington, DC: Island Press, 2002.

Williams, Gerald W. "Aboriginal Use of Fire: Are There Any 'Natural' Plant Communities?" In *Wilderness and Political Ecology: Aboriginal Influences and the Original State of Nature*, edited by Charles E. Kay and Randy T. Simmons, 179–214. Salt Lake City: University of Utah Press, 2002.

3

DISPLACEMENT AND REPLACEMENT: ESTABLISHING NEW ECOLOGICAL REGIMES

In the opening paragraphs of her autobiography, Mourning Dove, a Colvile from the interior Northwest, explained that she was born during a time of great changes. Members of her tribe faced the monumental task of readjusting their economies to meet the new reality of changing cultures. Ecological changes compounded those challenges. She wrote:

> The vast forests that our Indian forebears had jealously guarded from white invasion lay in ruins. The wildlife, staple food for natives, was slowly and surely vanishing each year; all the game was disappearing.
>
> The first white invaders were the fur traders selling firearms to the Indians of the Northwest, who enthusiastically adopted their use. Together with their white neighbors, they destroyed their ancestral heritage by killing the wild game of the forest, the livelihood of their past.
>
> The appeal of cheap beads and bright cloth was too much of a lure for the innocent Indians of that period. They were enticed to hunt and to overkill game to such an extent that they eventually had to get food elsewhere (Miller, ed., 1990, 3–4).

This poignant lament revealed the intersection of dramatic changes in economy, environment, and culture. Although the portrayal of Native peoples makes them seem more passive than they were, the description contains important pieces of environmental history. Markets provided goods and incentives to change ecological behavior. Beyond that, new populations decimated wildlife and other native species, rendering traditional seasonal rounds and subsistence activities insufficient. This led to even more dependence on the market—a vicious cycle between market exchanges and ecological changes. Such were the dynamics involved at the time of contact between Europeans and American Indians.

ECOLOGICAL IMPERIALISM AND REVOLUTIONS

Europe's expansion to North America initiated profound changes in world history. For the Far West's environmental history those transformations overwhelmed established human-ecological regimes. Two concepts developed by prominent environmental historians help convey the enormity of postcontact environmental alterations. "Ecological imperialism," a term used by Alfred W. Crosby, and "ecological revolutions," a model developed by Carolyn Merchant, suggest the powerful consequences experienced once the indigenous populations of North America discovered Europeans plying the waters off the coast (Crosby 1986; Merchant 1989).

Crosby began his important book on the subject thoughtfully: "European emigrants and their descendants are all over the place, which requires explanation" (Crosby 1986, 2). Indeed, it does, and for generations historians schooled in ideas of European racial superiority assumed that Europeans' worldwide success stemmed from inherent cultural superiority along with military and technological dominance. Even observers not embracing the racial arguments maintained that Europeans were somehow more advanced in their political and economic capacities, which explained their rapid proliferation around the globe. Instead, Crosby argued that European imperialism consisted of an important biological component that was, perhaps, greater than any supposed military or technological balances of power. Europeans brought with them flora and fauna, intentionally and not, that rapidly took hold and proliferated at alarming rates and frequently displaced indigenous plants and animals. Three important consequences followed. First, the displacement, and in some cases replacement, of indigenous ecologies meant that the people who depended on native plants and animals found themselves with a radically impoverished subsistence base. No longer could Native peoples maintain the ecological regimes created and sustained over generations. Second, and directly contributing to the problems of the first, microbes migrated with Europeans with devastating results. Diseases unknown to these continents and peoples accompanied cultural contact and European settlement. In combination with other ecological and cultural consequences of European expansion, diseases destroyed much of the preexisting cultural landscape. Finally, the biota that accompanied Europeans had coevolved with Europeans and thus offered them a significant advantage as animals, such as cattle and sheep, and food crops, such as wheat and rye, thrived in the soils of these new lands. Hence, European imperialism was a biological one, too.

A complementary approach to this era and its environmental consequences comes from Merchant's theory of ecological revolutions. Although

Coastal Indians stand by a sweat lodge (a sauna used for religious rites, preparing for a hunt, or healing). Neither indigenous healing methods nor European medicine could stem the onslaught of Old World pathogens in the Americas. (Underwood & Underwood/Corbis)

developed for a study of New England, the model is widely applicable. Merchant explained "that ecological revolutions are major transformations in human relations with nonhuman nature. They arise from changes, tensions, and contradictions that develop between a society's mode of production and its ecology, and between its modes of production and reproduction. These dynamics in turn support the acceptance of new forms of consciousness, ideas, images, and worldviews" (Merchant 1989, 2–3). The four categories—ecology, production, reproduction, and consciousness—undeniably responded to and shaped the extreme transformations in the several decades after Europeans initially arrived in the Pacific West. For example, the ecological imperialism Crosby described modified the region's ecology so much that traditional economic practices could no longer be pursued in the same ways. Moreover, Native depopulation because of disease and warfare reduced Indians' abilities to reproduce biologically and socially. Finally, in response to those changes and accompanying the new populations that peopled the Far West, new ideas about nature competed with and often supplanted indigenous ones. Thus, after the eighteenth-century arrival of Europeans, the region experienced an ecological revolution from which the area's subsequent residents could not turn. Ecological imperialism and the ecological revolution subsequent to European exploration and initial settlement produced a deeply shocked landscape. The new world created by the replacement and displacement of native plants, animals, and humans found Native sustainability practices much more difficult to achieve.

EXPLORATIONS

Although the Norse undoubtedly preceded Christopher Columbus's arrival in North America in 1492, and although some claim the Chinese also preceded the famous Genoese sailor, Columbus must be credited with initiating the first sustained contact between populations in the Western hemisphere and those in the Eastern hemisphere. Explorers, sponsored by emerging European nation-states, increasingly traveled the world's oceans and continents. Purposes of this exploration varied. Spreading Christianity, capturing resources to fuel economic expansion, and extending geopolitical power surely topped the list of motivations, although putative scientific discovery also came to matter much to these exploring states.

Spain showed the first sustained interest the Pacific West. Sporadic exploration in the Pacific Ocean in the sixteenth century reached the West Coast of North America. Juan Rodríguez Cabrillo became the first European to discover what is now known as San Diego, and his crew, after Cabrillo's death, sailed north to the approximate present boundary between California and Oregon. The Spanish did relatively little to continue exploration in the seventeenth century, but in response to other imperial presences in the Pacific Basin in the eighteenth century, they initiated more exploration and settlement after 1769. The principal Spanish maritime explorers were Juan Pérez, Alejandro Malaspina, and Juan Francisco de la Bodega y Quadra. In 1774, Pérez made landfall at 54° 40′, approximately the current boundary between southern Alaska and British Columbia. Thereafter Spanish explorers made it even further north (Engstrand 1998; Weber 1998).

Not as aggressive as early as the Spanish, English explorers became the most important geopolitical and commercial presence by the end of the eighteenth century, at least in the Northwest. Captain James Cook's third voyage around the world began in 1776 and arrived on the Northwest Coast in 1778, at a place he called Nootka Sound on the west coast of Vancouver Island. Of the many consequences of Cook's voyage, the one that stands out most for this

William Clark and members of the Corps of Discovery expedition shoot at bears, ca. 1805. This illustration is from the book published by Patrick Gass, a member of the expedition, in 1807. (Library of Congress)

study was his initiation of the trade in furs between the Far West and China. Following Cook, other British explorers came, including George Vancouver in 1792–1794 who conducted the most detailed reconnaissance yet of the coast, including the area he named Puget Sound. The British explorations documented the coast more carefully and completely than any previous population. Although Russians and Americans also visited the area by sea, their contributions to ecological knowledge were comparatively small.

Besides the maritime explorers, land-based Euro-American travelers also navigated their way to the Far West. The most famous of these were Meriwether Lewis and William Clark whose entourage, the Corps of Discovery, traveled throughout the Pacific Northwest in 1805–1806. Although an American trader, Robert Gray, sailed into the Columbia River in 1792 and claimed the river for the United States, it was the Lewis and Clark expedition that established a stronger American claim to the region—a claim that was not finally established by treaty with Great Britain until 1846. As historian William L. Lang has recently noted, one of President Thomas Jefferson's principal instructions to Lewis and Clark was to observe the life and landscape of the West in the scientific tradition of the Enlightenment (Lang 2004,

362–368). When the writings appeared, they brought to a literate public all sorts of new information about animals, plants, and peoples who had remained unseen by Europeans until the nineteenth century. Among their scientific accomplishments, Lewis and Clark chronicled for the first time a number of new species. Moreover, in the Northwest, Lang claimed, Lewis and Clark began to do more than simply observe nature. Instead, they located Native peoples within it, describing the importance, for example, of camas to the Nez Perce (Lang 2004, 384–385). This point is important, for it furnishes historians with some of the first documentary evidence of human-nature interaction in the region.

By publishing their accounts, albeit often years after the expeditions, explorers such as Vancouver or Lewis and Clark introduced to Europe and America the people and environment of the Pacific West. These portraits depicted prominently the potential of nature, and these representations did much to shape Euro-American environmental expectations and perceptions. For example, Vancouver wrote of the Puget Sound region:

> "The serenity of the climate, the innumerable pleasing landscapes, and the abundant fertility that unassisted nature puts forth, required only to be enriched by the industry of man with villages, mansions, cottages, and other buildings, to render it the most lovely country that can be imagined; whilst the labour of the inhabitants would be amply rewarded, in the bounties which nature seems ready to bestow on cultivation" (quoted in Robbins 1997, 52).

Such images reflected a particular worldview and imposed it on the environment. Vancouver's remarks, like so many others, perceived a largely uninhabited landscape wasting away without sufficient economic development. In this view, nature was simply waiting for the labor of an agricultural population that would find the region's landscape economically rewarding. Such descriptions could not help but inspire settlement. Moreover, portrayals like this began the process of dispossessing the land from Native peoples by excluding them and erroneously suggesting that the landscape bore no imprint of Native labor.

These explorations furthered geopolitical goals and commercial expansion. Indeed, the two were inseparable, despite dispassionate scientific observation being a stated goal of some expeditions. To be certain, explorers published important scientific data about the Far West, its geographic resources, and its inhabitants' cultures, but that information was invariably embedded within the political and economic agendas of the explorers and their state sponsors (Clayton 2000). Reading their accounts, then, is to read about the Pacific West's potential for economic development and settlement by Euro-Americans. Indeed, exploration led directly to economic and cultural exchange.

EXCHANGES

Market Exchanges

Exchanges—economic and biological—characterized European exploration and the years immediately following cultural contact. First, economic transactions. Captain James Cook's voyage to Vancouver Island initiated an active maritime fur trade after his sailors profited greatly from sea otter pelts obtained by the Nuu-chah-nulth (or Nootkans). British and American traders, along with some Russian and Spanish entrepreneurs, sailed to the Northwest Coast for the next several decades trading with coastal tribes for animal hides. The trade developed because an important intellectual and economic process had occurred: commodification. Turning nature into commodities, that is something to be bought and sold with its value determined by market prices, increased significantly after the arrival of Europeans. Native peoples certainly traded products of nature before Euro-Americans sailed to the Pacific West. Nonetheless, the market in which Euro-Americans circulated their goods was vaster in both its size and in the materials it encompassed. In other words, Euro-Americans commodified more parts of nature and distributed those commodities through more distant and elaborate networks. In the process of commodifying nature, they extracted valuable components of ecological systems, placed a monetary value on them unrelated to their place or function in nature, and moved them across continents and oceans by a logic governed by global capitalism. The introduction of this global market profoundly disrupted the cultures and economies that prefigured European exploration.

For the first time since the break up of Pangaea, the Pacific West became connected to other continents, this time through what historian David Igler fittingly called "diseased goods." According to Igler, global exchanges in the century after 1750 knit together the Pacific Basin and created an emerging coherent region with trade networks through which commodities and pathogens easily flowed: "Commercial and epidemiological exchanges constituted a deadly set of relations during this period due to the way new trade contacts entangled native communities with foreign biological agents" (Igler 2004, 694). Igler emphasized exchange of goods and microbes, because disease was both a consequence and the context for market exchanges in this Pacific world. Disease most tangibly linked this process of the market and populations across vast space (Igler 2004, 693–697).

The trade relationships that developed along the Pacific Coast depended on Native labor and trading interests, as well as various commodities being extracted from natural systems. Until the 1820s or so, the main exports from the

Far West were animal pelts, primarily from sea otters and beavers. Because obtaining these furs in this era of initial contact relied on Natives, commercial exchange led to close contact that ultimately resulted in the exchange of deadly pathogens, and these contacts only increased over time.

Pathogenic Exchanges

Europeans and North American indigenes evolved within distinct germ pools. Europeans enjoyed two historical advantages when it came to disease. First, they had a long history of interaction with domestic animals, a main source of disease. Second, Europeans had a long interaction with populations from Asia and Africa. The interaction of human and animal populations, as well as the interaction among cultures from diverse continents, made germs more competitive (Preston 1998, 262–263). Moreover, in Europe certain diseases such as smallpox had become endemic by the eighteenth century. Typically, one either died as a child or survived and gained immunity to the disease. Europe had a constant supply of hosts for *Variola,* the virus causing smallpox, and its trade networks united the continent by disease, survival, and immunity (Fenn 2001, 28). Thus, the European explorers or traders who met Far Western tribes could carry the virus, pass it to others, and be immune from its dangerous, often deadly, effects.

Consequently, disease spread early in the contact era. Venereal disease was particular common, as sexual relationships often accompanied and facilitated trade between European men and Native women. Venereal diseases such as syphilis negatively affected overall Native health and particularly increased sterility and infant mortality. This spelled demographic disaster. Furthermore, because of their relatively high population density and frequent interaction with other Native groups, Northwest Coast tribes furnished ideal circumstances for spreading disease once present in indigenous communities. In addition, as European trade in the Pacific basin increased and sped up, the faster traveling allowed diseases with shorter incubation periods to proliferate as new circumstances allowed contact with more who were involved in the trade (Igler 2004, 703).

But trade was not the only context for exchanging dangerous pathogens. Even in the comparatively isolated missions of California, disease ran rampant. Sherburne Cook, a pioneer in historical epidemiology, suggested an approximate 50 percent mortality rate caused by disease in the missions. The most important disease was syphilis. Contemporary reports indicated that it was virtually everywhere. Cook concluded that "venereal disease constituted one of the prime factors not only in the actual decline, but also in the moral and social disintegration of the population" (Cook 1998, 58). Syphilis appeared early in

Spanish occupancy and spread easily. In 1777 the first mention of an epidemic came from Santa Clara. Sudden severe epidemics arose with thousands dying at a time in the isolated missions. Other diseases like pneumonia, diphtheria, and measles arrived by 1806. Children were particularly vulnerable; indeed, in San Francisco, the population under the age of ten was virtually extinguished. By the 1820s, new contact with trans-Pacific commerce brought new diseases. By the next decade, smallpox, scarlet fever, and cholera decimated California populations. Besides having a "virgin soil" population, the mission system itself increased mortality (Crosby 1976). Missions located close to the coast had a somewhat damper climate than where many of the interior tribes had originated, and missionaries blamed this climate for the illnesses. Most importantly, though, the large rooms where Natives congregated in close quarters provided ideal conditions for respiratory illnesses to spread. Tragically, the mission systems were vectors of death (Cook 1998, 55–59).

Although the initial spread of diseases required intimate contact between Europeans and Native populations, diseases could race ahead of the newcomers through Indian-Indian contacts. Smallpox in the Northwest illustrates this. A smallpox epidemic between 1775 and 1783 moved through the Northwest along the Columbia River after the Shoshones contracted smallpox on the Great Plains. Coincidentally, it was this same route by which horses and manufactured items reached the region. In other words, Native trade networks were extensive by land and facilitated the movement of goods, animals, and germs. Indigenous groups recognized the connection, as oral traditions of Nez Perce and Flatheads tell of crossing the Rockies and returning with the virus or dying on the Plains. In California a different route of the same continental epidemic followed the *camino real*, or royal road, from central Mexico north to the Pacific Coast. The imperatives of trade and mission colonization, newly sped by horses, distributed *Variola* across the continent, killing thousands in the Pacific West (Fenn 2001, 144–46, 204–209, 250–257; Harris 1997).

Venereal disease and smallpox certainly overwhelmed Native populations, but other diseases may have been equally devastating. In the Pacific Northwest, malaria—called fever and ague by Americans and intermittent fever by the British—became endemic after arriving in epidemic strength in 1830 and several subsequent summers (Boyd 1999, xiv). Anthropologist Robert Boyd has indicated that these epidemics were the most significant epidemiological event in Oregon's history (Boyd 1999, 84). By the mid-nineteenth century, measles also reduced Northwest populations, arriving overland from California in 1846–1847. Perhaps the most famous consequence of this epidemic was that it precipitated a violent uprising against American missionaries, Marcus and Narcissa Whitman, at Waiilatpu in southeastern Washington (Boyd 1999, 145–146).

By the end of the first century after European contact, Native populations were only 10 percent of their immediate precontact population. This sharp decline resulted overwhelmingly from new pathogens against which they possessed little immunity, health care traditions that were unable to stave the resulting illnesses, and the social and environmental stress and poverty that accompanied contact. Thus, indigenous populations were generally weakened and Natives' abilities to respond to epidemics compromised too. High mortality was not inevitable with the arrival of new pathogens, but these other conditions made it more likely (Crosby 1976; Jones 2003).

In the aftermath of these disease and depopulation episodes, Native peoples and their cultural and ecological worlds were overturned. Historian Elizabeth A. Fenn summarized the cultural effects:

> significant cultural voids, the loss of generations of unrecoverable knowledge. . . . households combined, kinship alliances annihilated, religious convictions altered or abandoned. As smallpox squeezed the life from thousands of victims, it extinguished the accumulated wisdom of generations, leaving those who survived without the familiar markers by which they organized their worlds and leaving the generations that followed with a mere shell of their former heritage (2001, 258).

Consequently, the cultures the Spanish, English, and Americans encountered in the Far West in the late eighteenth and early nineteenth centuries were only remnants of larger cultural complexes.

Besides being a cultural nightmare for the indigenous populations, depopulation severely disrupted the human-ecological regimes Native peoples had established by the time Europeans arrived. The cultivation system developed by Native peoples, including their widespread manipulation of fire, meant their depopulation and displacement upset plant and fire regimes. Species that required fire to reproduce or that benefited from fire to reduce competitors faced sharply shrunken habitats. It also meant these species became more vulnerable to exotic grasses and weeds that arrived with Euro-American migrants. Just as Native Californians served as a keystone species for plants, so too did they help shape animal populations. Indeed, it is likely that the high number of wildlife Europeans reported was a response to the declining population of their key predator—humans. Thus, disease not only devastated human cultures, but it also meant ecological changes in plant and animal distribution. In addition, illness and death meant traditional seasonal rounds could not be fulfilled and so death from starvation became more commonplace (Preston 1998, 269). Depopulation spelled not only a cultural tragedy but also an ecological one, as Native cultures and ecologies shifted in novel ways.

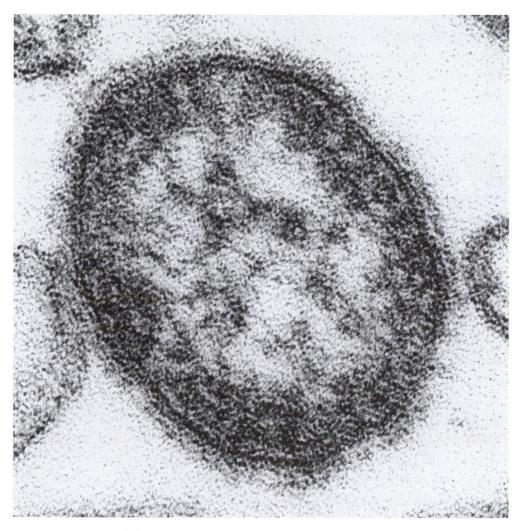

Microscopic view of the measles virus. Viruses causing diseases such as measles and smallpox decimated Native American populations following contact with Europeans. These viruses arrived as part of the Columbian exchange. (Cynthia Goldsmith/Centers for Disease Control)

Plant and Animal Exchanges

Besides bringing pathogens with them, Europeans traveled with domestic plants and animals exotic to the Pacific West. The animals and plants frequently went feral, just as diseases became epidemic, and thoroughly changed the biota of Western landscapes. These new species added enormous new shocks to the ecology of the indigenous Far West, and the consequences continue to ramify through the landscape today.

As with disease, plant and animal species on the American continents evolved independently of those native to the Eurasian and African landmasses. Eastern Hemisphere species often enjoyed significant advantages in an evolutionary sense, an advantage that wreaked havoc when these ecologies directly competed. The longer history of human occupation in the Eastern Hemisphere was the primary reason for this species advantage. Because of a longer history of biological exchanges in Asia, Africa, and Europe, various species of plants, animals, and microorganisms had evolved into stronger competitors than similar species in the Western hemisphere. Most importantly, the abundant domestic animals in the Eastern hemisphere meant that plant species like perennial grasses had coevolved with the pressures of heavy grazing. The grasses that arrived in the feed for Spanish cattle soon took a foothold in California soil and displaced native grasses, which were unaccustomed to grazing pressures. By the 1840s, only seven decades from the introduction of domestic cattle to California, invasive forage species dominated in many regions (Dasmann 1998, 196; Preston 1998, 262–263).

California was particularly susceptible to Mediterranean and Near Eastern species because of the similarity of climate. Geographer William Preston explained that because of this similarity, California was the most receptive colonial region to Mediterranean species, which accounts for their rapid proliferation (Preston 1998, 267). Consequently, the livestock that accompanied the Spanish to California, and the weeds that accompanied the livestock in their waste and feed, turned out to be singularly insidious ecological forces.

Domestic cattle were not the only animals to affect the Pacific West's landscape or culture. Horses arrived in the interior Northwest early in the eighteenth century through Shoshone traders, and cultural changes quickly followed. Some Native peoples acquired vast horse herds through trading and raiding. The most successful herders were the Nez Perce, Cayuse, Palus, and Yakama, all of whom lived in grassy interior valleys. Cultural adaptations followed. Several eastern Plateau groups readily adapted horses to bison hunting, crossing the Rocky Mountains to the Great Plains for seasonal hunts. This obviously expanded Natives' spatial impact on resources at the same time as it intensified the ecological demands placed on Columbia Plateau grasses. Moreover, it increased cultural conflict with northwestern Plains tribes such as the Blackfoot. At the same time, horses gave interior tribes such as the Klickitat new power when they ranged across the Cascades. The Klickitat adapted their policies of exchange, which were formerly about trade but became one of conquest (Bunting 1997, 31–32). The increased mobility of horse tribes further involved them in trade, and firearms increased along with horses changing regional power and trade dynamics. Finally, gender roles modified. Men owned horses, and the animals became a new avenue for accumulating wealth and then redistributing it. Furthermore, men used horses to hunt, but women were charged with processing hides. Thus, horses increased women's labor, but women also seemed to have gained some power over distributing those hides. For those tribes who adapted horses—and not all who could, did—culture changed to accommodate the new opportunities and responsibilities these domestic animals provided. And the environment responded too (Vibert 1997, 217–239).

Ecological changes accompanied cultural inventions. Some scholars have suggested that horses filled an ecological vacuum left by the Pleistocene extinctions of large grazers. Although not definitive, evidence seems to point to increased anthropogenic burning corresponding to the rise of equestrian adaptations. These fires renewed grasses and reduced encroaching brush, thus making the interior Northwest more amenable to grazing horses, who were readily incorporated into many groups' seasonal round. The increased burning following the arrival of horses may also have been a response to an environment showing its limits. That is, horses in the arid interior probably taxed existing forage resources. By making hunting easier, horses tipped subsistence strategies more toward hunting. On the Columbia Plateau, then, an exotic species introduced new ecological practices that began to alter nature nearly a century before Euro-Americans otherwise appeared and influenced the region's environmental history (Robbins 1997, 46–48; Vibert 1997, 217–239).

In the end, exchanges from the microscopic to the global scales wrought immense change for the region. Trade facilitated pathogen exchanges along with commerce, and the animals and plants that accompanied new populations found favorable habitat in which they thrived. Cumulatively, then, these exchanges interdependently reordered the Pacific West's cultural and ecological systems and prepared the way for new social and environmental regimes.

INITIAL SETTLEMENTS

More important culturally and environmentally than initial explorers were two groups of Euro-Americans who came to the Far West and stayed for longer periods. Although many of them lived out their remaining days in the region, they did not perceive themselves as residents. Instead, they were part of impermanent, but influential, cultural and economic developments. Despite the relatively small population in each group and their weak roots to the place, they initiated vast ecological changes. The first group was the Spanish missionaries who established a mission system in California beginning in 1769. Long celebrated and romanticized, the mission system has now been thoroughly implicated in some of European imperialism's worst practices, including forced labor, sexual exploitation, and appropriation of land, as well as the inadvertent spread of disease. The other group—Canadian and British fur traders from the North West Company and the Hudson's Bay Company—scattered through the Northwest beginning in the late eighteenth century and increased markedly after the two companies merged in 1821. Together, over the course of a half-century, these groups changed the trajectory of environmental history throughout the region.

Spanish Missions and Ranchos

Spanish missionaries arrived in 1769 and set about altering California's cultural and natural environments. By 1823, missionaries had established twenty-one missions between San Diego and Sonoma reaching 21,000 Indians at their greatest influence (Weber 1994, 66). Led by individuals such as Fray Junípero Serra and Juan Crespí, these missions sought a wholesale transformation in their Native subjects, or neophytes, and a central part of that change meant changing land-use patterns and traditions. Franciscans' utmost goal, of course, was religious conversion. However, they believed Native subjects required cultural conversion, too, namely that they needed to live in sedentary villages practicing agriculture by raising crops and animals. The missions, too, needed to obtain self-sufficiency, and accordingly, pursued agriculture with some success.

By its very nature, agriculture is ecologically disruptive. It seeks to make the environment yield species that it would not otherwise produce. Of course, Native Californians had long practiced horticulture and so were familiar with the basics of agriculture. However, mission farming was fundamentally different. Every plant and animal raised at the missions was exotic, and that meant indigenous plants and animals found their habitat shrinking and their use declining. Within six years, five missions combined to produce just less than 100,000 pounds of corn and 85,000 pounds of wheat. By 1805, total crops increased about thirty-fold amounted to approximately 94,500 bushels. (A bushel differs in weight from product to product, so a total weight cannot be given because of incomplete data.) Some missions became self-sufficient within just two years (Hackel 1998, 116, 138). This proliferation of foods was exceeded by the even more impressive expansion of domestic animals. In the initial move to California in 1769, Franciscans brought 104 head of cattle, 87 mules, and 53 horses (Preston 1998, 275). By 1805, the Spanish held more than "130,000 sheep, 95,000 cattle, 21,000 horses, 1,000 mules, 800 pigs, and 120 goats"

Chapel of Mission Santa Barbara, established in 1786 by Spanish Franciscans. From their bases at missions like these, the Spanish hoped to bring European-style agriculture and religion to Native peoples. (Julie Dunbar)

(Hackel 1998, 116). Not surprisingly, this boom in animals proved uncontainable and numerous mission animals went feral, joining herds of other ungulates and feeding the thriving grizzly bear population (Preston 1998, 279–280).

At times, the increase of animals interfered with the crops. Two Spanish priests, Joseph Antonio Murguía and Thomás de la Peña, noted in 1782 that livestock caused "unceasing damage." Moreover, they explained, "Indians will have to stop their field work, so as not to labor in vain; and they will have to rely for their food on the herbs and acorns they pick in the woods—just as they used to do before we came. This source of food supply, we might add, is now scarcer than it used to be, owing to the cattle" (quoted in Warren 2003, 100). The priests' complaints reveal a cyclic problem where cattle trampled domestic crops sending Indians to native plant sources only to find them similarly reduced by livestock. Clearly, the development of agriculture was not without problems.

If Spanish mission pastoralism inaugurated agricultural changes, the Mexican era from 1821 to 1848 modified ecosystems even further. Mexican independence led to secularization, a policy change that eliminated the missions and privatized their vast landholdings in grants. The Mexican government more readily than Spanish predecessors offered private land grants, or *mercedes*, to individuals who developed ranchos. Between 1833 and 1846, California governors granted 452 grants, encouraging private economic development (Isenberg 2005, 111).

Indeed, Mexico promoted trade much more than Spain had. California pastoralism offered virtually the only viable economic option before the gold rush. Besides the sea otter pelts Euro-Americans traded in China, cattle products were the first to circulate in the global market. California pastoralism developed a profitable hide and tallow trade, and livestock proliferated in total numbers and geographically. When mission pastoralism dominated, livestock populations predominated in Southern California; but ranchos from Mexican land grants eventually concentrated in the northern Central Valley. These larger landholdings and more open markets led to speculation in land and cattle. Consequently, as ranchos displaced missions, they induced ecological changes across larger areas (Jordan 1993, 163–166; Starrs 1998, 201–204).

The Californios were chronically short of capital and labor, so land became essentially their only avenue to economic growth. Rancheros, in the words of environmental historian Andrew C. Isenberg, were "ambitious and innovative participants in the international market economy and speculators in land and cattle" (2005, 110). In the lucrative hide and tallow trade with British and American merchants, rancheros exported 6 million hides between 1826 and 1848 (Isenberg 2005, 112). "What is incontestable," according to geographer Paul F. Starrs, "is the blossoming relationship between the hide and tallow industry, the rise of California ports and cities from which these commodities were shipped, and the role of ranching society in bringing newcomers to California in the 1830s and 1840s" (1998, 203). Cattle fit into the developing trade system of the Pacific Basin and encouraged greater market developments (Igler 2004; Starrs 1998, 203). The ranchero system was not as stable as such numbers suggest for unreliable financing and unpredictable nature could and periodically did upend the system.

To ameliorate the former problem, Californios consciously created kinship networks, forming an interrelated elite. Often Britons or Americans married into landed elite Mexican families. Abel Stearns exemplified this strategy. American born, Stearns made a small fortune in the hide and tallow trade and set about converting that capital to land. He became a Mexican citizen, married into an elite family, the Bandinis, and began buying out ranchos from individuals who were indebted to him. By the 1850s, his holdings reached nearly half a million acres in seventeen different ranchos. On just four of them, Stearns could count 20,000 cattle (Isenberg 2005, 115). However, vast holdings could not protect rancheros from all potential problems.

Unfortunately for Californios and their cattle, the environment remained unpredictable, and cultural changes weakened previous ecosystems. Frequent drought—about once a decade—devastated grasses and herds. One severe drought from 1828 to 1830 left 40,000 dead cattle (Isenberg 2005, 109). The extensive rancho system also overwhelmed native bunchgrasses that could not withstand the pressures of trampling and grazing. Their shallow roots easily gave way. Because the southern two-thirds of California roughly approximated the Mediterranean climate, the suite of seeds indigenous to the Mediterranean easily overtook native species. At the same time, native grasses' greatest biological allies were Native peoples and their various horticultural practices. Disease and the labor system that developed during and after the mission system made indigenous practices unreliable at best and impossible at worst (Preston 1998, 288).

Displacement of Native peoples, then, constituted more than a cultural crisis; it contributed to ecological changes that thoroughly transformed California's landscapes. Other habitats also transformed, largely because of the removal of Natives. Taking away anthropogenic fire resulted in greater brushlands, encroaching chaparral, and larger and entangled understories. The absence of regenerating fire and the presence of grazing meant some steep banks lost ground cover and eroded more easily. In sum, when Indians and their management left or were confined, the ecosystem declined in terms of species diversity and range (Preston 1998, 274). The greatest impact remained on the Coast, where missions had been concentrated and Native populations highest. However, many Natives who fled missions or ranchos inadvertently dispersed exotic plant and animal species through other parts of California. Also, Natives adapted to domestic animals, in part because traditional food sources declined and domestic cattle and sheep offered easy protein. Because of these declining food sources, raiding as a subsistence strategy markedly increased, much to the ire of Mexican officials and residents (Preston 1998, 275–276).

Thus, the ecological impacts of Spanish and Mexican settlement were far reaching and cascading. Some animals upon which Native Californians had preyed irrupted because of human population decline and displacement removed that hunting pressure. Right before the gold rush began, California's population was at a nadir; it had not been so low for possibly 2,000 years. Meanwhile, domestic cattle provided predators such as grizzly bears with easy meals, and grizzlies seem to have increased. Ungulates (e.g., deer, antelope, elk) were not bothered by the decimation of perennial native grasses, and easily converted to consuming the imported annuals. Rodents irrupted, too, because of the new domestic grains being grown. Some wild animals hunted the new livestock, increasing their population. New arrivals in California marveled at the region's wildlife abundance and diversity, not recognizing that it was a relatively new creation (Preston 1998, 278). California's shifting ecology in response to invasions represented but one area in the Far West undergoing an ecological revolution. Further north, other Europeans reacted to different cultural and economic imperatives and created other environmental transformations.

Fur Trade Posts and Nascent Agriculture

In the 1780s, the fur trade had begun affecting Northwestern environments with maritime activities. When the trade moved to land and especially after the Hudson's Bay Company (HBC) decided to invest in agriculture at Fort Vancouver in the 1820s, ecological change related to fur traders changed dramatically. The first case study in this volume explains the way fur trading affected animal populations and their ecological niches. Other environmental factors were associated with the fur trade, though. For instance, the HBC was the source for many of the plants and animals Euro-American farmers used to start their own farms. Indeed, the HBC's role in promoting agriculture instigated the first efforts at European-style agriculture, an economic and ecological change that would ultimately transform much of the Northwest.

The HBC recognized that fur trading could not sustain profits forever. Thus, leaders such as George Simpson and John McLoughlin developed new markets by commodifying and trading fish, forest, and farm products locally and throughout the Pacific Basin. This strategy was part of the HBC effort to make their posts self-sufficient and eliminate the wasteful practices of importing food. To get started, in 1824, the company introduced seventeen cattle from California, which rapidly grew into a herd. By 1839, the HBC formed a subsidiary, the Puget's Sound Agricultural Company, that operated a farm in the South Puget Sound region on the Nisqually Plain. At its zenith, this farm ran 17,000 sheep and 8,000 cattle, demonstrating the potential agricultural development on the west side of the Cascades. Agricultural efforts at Fort Vancouver were even more impressive. Although only 120 acres were cultivated in 1829, that increased to 1,000 acres in less than a decade and to about 1,500 in 1842. These efforts were not haphazard. Farmers carefully regularized the practices with systematic manuring and crop rotation. Yields by the mid-1830s could reach twenty bushels per acre of wheat and fifty bushels of oats. The company's

The Hudson's Bay Company's post at Fort Vancouver (on the Columbia River) in 1860. (Library of Congress)

efforts produced surpluses, keeping not only other HBC posts fed but allowing exports to California, Hawaii, and Russian Alaska. The western valleys served as nurseries for later expansion of livestock herds throughout the region. The drier interior Northwest and California demanded much larger acreages for productive herding, but intensive agriculture was perfectly suited for the Pacific slope (Bunting 1997, 37–38; Gibson 1985, 36–38; Jordan 1993, 245; Mackie 1997, 151–154; Oliphant 1968, 78; Robbins 1997, 55–56).

Hudson's Bay Company agriculture in the Northwest demonstrated several things. Perhaps most importantly, it demonstrated the land's potential for productive agriculture on Euro-American terms. This success encouraged further Euro-American settlement. Many HBC employees retired in the Willamette Valley, just south of Fort Vancouver, and succeeded agriculturally, ultimately making the Oregon Country synonymous in the mid-nineteenth century with an American Eden (Bunting 1997). Agriculture also displaced native species and introduced exotics that would come to dominate the region, just as they had in California. This displacement of indigenous environments came with the displacement of indigenous people. And although some American Indians in the region did adopt agriculture, most initially did not. Finally, HBC agriculture illustrated how from its beginning, agriculture and environmental change were linked to demands for commodities in markets that lay both within and outside the region. As with the Spanish missions, HBC posts initiated new ecological and economic practices that represented European ideas of land use. The HBC enacted an ecological imperialism to accompany the political claims of Spain, Mexico, Great Britain, and the United States. Both missions and fur trade posts induced further changes that extended the ecological revolution under way.

Hudson's Bay Company agriculture achieved much and presaged more as part of a larger economic strategy to support an intercontinental commercial empire. By contrast, the American farmers who began arriving and transforming the Far West in the 1830s and increasing in the 1840s operated with some different ideals, though with similar ecological effects. Two groups of American citizens began populating the Northwest in the 1830s, rapidly accelerating in the next decade. The first of these was the Protestant missionaries beginning with Jason Lee in the Willamette Valley in 1834 and the Whitman and Spaulding parties who arrived in the interior Northwest two years later. The second group, in part inspired by the missionaries' writings and examples, were middle-class farm migrants who traveled on the overland trail starting in 1842. Both of these groups were inspired by and perpetuated the emerging American myth of Oregon as a promised land. Boosters like Hall Jackson Kelley and Missouri Senator Thomas Hart Benton had long been calling on Americans to go to Oregon and secure it for the American republic and for Protestantism (Robbins 1997, 65). And Americans came and wrought change.

Environmental change built alongside both cultural conflict and adaptation. Missionaries introduced European-style agriculture to Native peoples but were generally unsuccessful at converting tribes wholesale to agriculture or to Christianity. Native peoples did assimilate some parts of agriculture quickly. For instance, the Salish of Puget Sound readily adopted potatoes, because cultivating them was so similar to the horticultural practices for camas (White 1992, 32–33). The missions on both sides of the Cascades quickly built flour mills and some even exported flour. In addition, the interior missionaries introduced irrigation apparently for the first time in the region (Gibson 1985, 151–177). However, not all worked well, as Marcus Whitman explained in 1844:

> Although the Indians have made and are making rapid advance in religious knowledge and civilization, yet it cannot be hoped that time will be allowed to mature either the work of Christianization or civilization before the white settlers will demand the soil and seek the removal of both the Indians and the Mission. What Americans desire of this kind they always effect, and it is equally useless to oppose or desire it otherwise. To guide, as far as can be done,

and direct these tendencies for the best, is evidently the part of wisdom. Indeed, I am fully convinced that when a people refuse or neglect to fill the designs of Providence, they ought not to complain at the results; and so it is equally useless for Christians to be anxious on their account. The Indians have in no case obeyed the command to multiply and replenish the earth, and they cannot stand in the way of others doing so (quoted in Gibson 1985, 174).

Whitman wrote prophetically in this justification. Americans were coming, and pressure on Native lands increased. The missionary argued that since Native peoples had not followed biblical injunctions to subdue the earth, then Americans, however regrettably, were justified in appropriating the land for those purposes. That was just what happened.

When Whitman led a group of overland emigrants to Oregon in 1843, the annual migrations on the fabled Oregon Trail began. These "instruments of transformation," as historian William G. Robbins aptly characterized them, displaced Native Americans, appropriated their land, carried deadly pathogens, brought domestic livestock and crops, and established an ideology of conquest of nature and people (1997, 69). Such an arsenal of material and intellectual baggage accelerated the transformations begun by fur traders and missionaries. An early farmer in Puget Sound declared in what was certainly a common perspective of the time that he wanted "to get the land subdued and the wilde [sic] nature out of it. When that is accomplished we can increase our crops to a very large amount and the high prices of every thing that is raised heare [sic] will make the cultivation of the soil a very profitable business" (quoted in White 1992, 35). These early farmers wanted to domesticate the land by replacing native species with domestic crops. Although Native peoples certainly shifted the distribution of species, Euro-Americans simply replaced them. Moreover, and this was central to the endeavor, they wished to profit. Just as fur traders commodified the pelts of certain fur-bearing animals, farmers commodified the products of the land. Subsistence farming may have been a goal or a result for some Americans in Oregon, but most sought broader and more profitable market exchanges (Bunting 1997, 36–50; Robbins 1997, 50–78; White 1992, 35–53).

These agricultural developments—from California to Washington, from cattle to wheat—represented change of immense environmental, economic, and cultural importance. The choice to farm and to trade was easy to make. The results, however, were difficult to predict. Time and again, markets declined and failed, and ecosystems occasionally crashed. Moreover, cross-cultural conflict caused by these deliberate choices increased, because the environment was at the center of Euro-Americans' colonial and economic endeavors. Repeatedly, differing perspectives about proper resource use and diverse traditions governing that use converged on specific resources. Water offers an excellent example.

TRANSFORMING WATER, CHANGING CULTURE

A significant portion of the Far West, primarily the interior Northwest and southern and central California, is semiarid. The lack of abundant, reliable fresh water makes living there precarious. Nevertheless, various societies have forged homes in these places and have adapted multiple economic, political, and legal strategies to accommodate this basic geographic fact. Consider Southern California. The arrival of Spanish missionaries, soldiers, and colonizers transformed the California waterscape and legal and cultural milieu concerning water allocation in yet another example of the ecological revolution spawned by the arrival of these new populations. Water use and traditions deserve this lengthy discussion because subsequent developments and conflicts in the dry Pacific West often centered on this scarce resource.

Compared with the Spanish and later the American populations in California, Native Californians did relatively little to change the region's waterscape. Only in the eastern deserts did indigenous cultures irrigate on a wide scale. One of their primary strategies to ensure survival in this dry land was to limit their communities' size so that they would not overuse the precious resource. Unlike subsequent populations, Native peoples generally traveled to water, rather than making water move to them. Thus, a group's territory tended to encompass entire watersheds as part of their own space, encompassing several elevation zones and diverse plant and animal life communities within them. Indigenous groups recognized that their survival depended on careful use of water resources, and they guarded such places as important community resources. They did not, however, imbue water with property rights (Hundley 2001, 1–26).

The Spanish who arrived in California in the eighteenth century also placed a primacy on water as a community resource. In their instructions to missionaries and other colonists, Spanish authorities made water a central feature to their plans. Indeed, securing water was central to successful imperial control in California. When searching for potential locations for missions, forts, and towns, colonists were instructed to locate where reliable water sources, along with sufficient timber, were available. In expedition journals, searching for and finding water, as well as descriptions of the water source's characteristics (e.g., quantity, length of stream), dominated the accounts. Yet, despite the importance of water planning to colonists' strategy, managing water remained tricky. For instance, at the first permanent European town in California at San Diego, the Spanish planted crops in a fertile and well-watered valley. Heavy rains, though, flooded the fields. When replanting the following year, determined not to make the same mistake, the colonists set their fields on higher ground. This time, however, drought made it so that irrigation systems could not deliver water to the parched soil. Although Spanish colonists recognized the principal importance of water, a fluctuating climate and difficult geography made establishing agriculture a challenge in California (Hundley 2001, 31–38).

Nevertheless, because of a similar climate in Europe, Spanish law had evolved in part to deal with such difficulties and to ensure equity among community water users. Although Spanish traditions focused on community survival much as did Native communities, the details were different and represented an important new presence. While satisfying local communities' needs remained the main concern and the local community constituted the basic unit for water management, power ultimately rested with a centralized authority, specifically the Spanish Crown and its agents in the colonies. This legal position put final control far from the local ecosystem, a change of real significance (Hundley 2001, 38–43).

Such a large political structure required responsibility to be distributed among various levels. While the Crown claimed ownership of the land and water in its California colonies, it occasionally could and did grant individual ownership. More often it granted temporary use rights only and defined water along with pastures and woods as common resources. Indeed, the water right was vested in the municipality, not with individuals. The town had authority of all water uses, not just irrigation. In this system, no one held superior rights and allocation needed to be conducted fairly. Historian Norris Hundley, Jr. explained how the Hispanic system allocated water, following a principle of "'proportionality'—people obtaining an amount in proportion to their legitimate needs and in proportion to the volume of water available" (2001, 30). This ideal of proportionality extended to maintenance of the irrigation apparatus. The amount of time one had to work on the hydraulic system was proportional to the amount of water one used (Hundley 2001, 38–50).

To ensure that this principle of proportionality and equity worked, Spanish colonial leaders drew on complex Iberian legal traditions, honed in their American colonies, and instituted clear political mechanisms to govern water use. At the local level, an elected council, the *ayuntamiento*, was charged with ensuring a fair distribution of water, while an administrator, a *zanjero*, oversaw the maintenance irrigation system. If these agents of power failed to discharge their office, they would be replaced. Similarly, water users would be punished if they wasted water or jeopardized the quality of water in the ditches designed to carry water for domestic and agricultural purposes. The legal system clearly established water and its care as a communal task. Failure to labor on behalf of the irrigation system resulted in severe punishment. Together, communities built and maintained the main ditch, *zanja madre*, and associated canals. So central were these waterworks to colonial plans and survival, colonists built them before churches and homes. This hydraulic structure demanded constant attention so that for years, many communities' additional built environment remained meager by comparison. Such an effort, though, demonstrated how the communities and their institutions functioned cooperatively to ensure full and equitable use of the water (Hundley 2001, 43–45).

These principles of fairness even extended to those in adjacent communities where conflict over this scarce resource sometimes occurred. Discord emerged most often when mission communities and other settlements were too close to one another. Nevertheless, Hispanic water laws were flexible and emphasized sharing water, so competing communities generally managed to find mutual accommodation. For example, three times in the Spanish and Mexican period, the pueblo of Los Angeles raised concerns about activities at the mission at San Fernando that threatened the pueblo's water supply. In 1810, neophytes (i.e., baptized Natives) at the mission built a diversion dam for irrigation to support a fairly thriving agricultural operation, comprised of 11,500 cattle and producing 9,700 bushels of grains and other crops in a year. Worried about their ability to continue their own domestic and agricultural practices, officials in Los Angeles requested the removal of the dam. After an investigation, mission officials agreed to remove the dam as they only wanted to irrigate a small area and did not wish to harm the pueblo. This example reflected the general desire and ability to work out solutions. Similar conflicts in northern California between the pueblo of San José and the mission at Santa Clara developed with a similar judgment that accommodated both communities. The practice of trying to reach a common good—*bien procumunal*—was typical throughout the Spanish Empire in North America and was decidedly different from water struggles that emerged in the subsequent American period (Hundley 2001, 50–60).

Unfortunately, the ideal of common good extended only to the Spanish colonists. Indians provided the labor that built the water system. Indeed, from the perspective of a large portion of the California indigenous population, the manipulation of water meant a concomitant manipulation of labor. In this system, then, a new ecological regime was used both to control a population and to serve the interests of the newly dominant. As Hundley remarked, the "hydraulic system [was] . . . an instrument of coercion" (2001, 63). In Spanish missions, Europeans forced Natives to work at the agricultural enterprises the Fransciscans planned. After the Mexican government secularized the missions in the 1830s, many neophytes who were now without their homes found work on the ranchos as laborers in an agricultural economy increasingly devoted to serving a global market in the hide and tallow trade. Besides turning to rancho labor, Natives frequently turned to alcohol to salve the profound disruption of their lives. Meanwhile, town councils adopted various legal ways to imprison

California Indians and allowed prisoners to work off part of their sentence, for spurious crimes such as vagrancy. In addition, a system of debt peonage developed on the ranchos where Indian workers without access to subsistence practices incurred debts to rancheros. The legal system thus used unfree Native labor to build and maintain the infrastructure needed for the emerging market economy in California, a process that changed the California environment in local areas to a notable degree in new directions and for different purposes (Hundley 2001, 60–64).

By the end of the Mexican era in 1848, about 10,000 acres had been cultivated in California missions alone, not to mention crops and animals from ranchos and pueblos (Hundley 2001, 61). Of course, virtually all the domesticated plants raised in this emerging agricultural system came from Eurasian species. Thus, irrigated agriculture functioned not only to manipulate water systems but also to further the expansion of various biological and market revolutions. As the agricultural system developed and produced surpluses, agriculturalists capitalized on market exchanges that crossed continents and oceans. Central to the cultural change occurring in California were the ecological alterations in the region's ecosystems. Arguably, those changes in California ecosystems represented the most important factor in allowing a European population to gain a foothold there.

Compared with the devastating effects of new microbes, these changes in water manipulation and legal customs do not seem to have created the same shock to the Far Western landscape. However, they are significant in several ways. First, although Hispanic law required new uses to not interfere with established practices, Natives' claims to water seldom were respected. Thus, Spanish community practices effectively displaced time-honored indigenous practices and replaced management customs. Second, the imposition of centralized, state authority over these streams and rivers announced a new period in which the state established much stricter rules governing human manipulation of the environment. Compared with subsequent influence of the state, these initial laws and practices seem rather subdued. Nonetheless, its arrival marked a new era. The onset of state power, after all, constituted traditional political imperialism, and when combined with ecological imperialism, fundamentally changed the region's environmental, political, and cultural histories. Third, these systemic transformations by which water was used and managed were simply the first of many revolutions in water politics and ecology. Once Americans arrived in the Pacific West, new laws and practices transformed multiple times. Indeed, water and its associated ecology take on central importance in the Far West's ongoing environmental history. Taken together, then, this shock signified an ecological revolution. Ecology changed when irrigation was implemented in new locales. The production of new crops to be consumed locally and eventually to be traded in the global market linked these ecosystems to the market in new ways, while European communities reproduced social and legal traditions in new forms representing further evidence of a change in environmental understanding and Native cultures found replicating their traditional use of water and other resources almost impossible in areas where Europeans gathered. A first revolution in water was complete.

ENVIRONMENTAL CONFLICT

The 1840s found the Pacific West in a state of flux. Profound ecological change produced a shocked landscape throughout much of the region. Native floral species receded everywhere Europeans had arrived. Grasses associated with grazing animals took root in pastures and open ranges. Wheat and weeds displaced camas and bunchgrasses. Livestock best illustrate the new state of ecological instability. They reproduced well in the region and became central to the early economy, especially of California. They were vectors of environmental change. Not only did they consume native grasses and help give a foothold to exotics, but they also went feral and helped feed a growing grizzly population. Thus, on the land, species were changing and their populations were redistributing, displacing and replacing patterns developed over many generations.

Most importantly, the Native peoples confronted the ecological imperialism of Euro-Americans and bore the brunt of it. Many of the ecological regimes indigenous populations had established became completely disturbed with the arrival of new species. Disease, however, made the biggest difference. With depopulation rates at 90 percent in many areas, and a changing ecological repertoire with which they had minimal experience, Native peoples faced cultural and environmental disorder. The largely sustainable practices they had adapted over millennia became nearly impossible to maintain with reduced numbers and changing ecological dynamics. Indeed, their displacement on missions and away from new Euro-American settlements did as much as anything to change the ecology of the Pacific West, for Native practices had been central to the landscape before the eighteenth century.

Native peoples recognized the importance of ecology in extending Euro-American preeminence in the region. For instance, although indigenous populations did not understand the nature of disease (and neither did Euro-Americans), they associated it with Euro-American traders. For instance, the term some Pacific Islanders used for these new illnesses translated as "I am shippy," and one tribe on the Northwest Coast called smallpox "Tom Dyer," commemorating a

European sailor they believed infected them (Igler 2004, 704). Moreover, conflict between Euro-Americans and Native peoples often erupted along environmental fault lines. Not to minimize cultural conflict associated with religion or politics, but quarrels involving resources were equally important and common.

Several examples illuminate this phenomenon. First, ecological changes made traditional subsistence practices difficult from the beginning of the mission era. For instance, Spanish authorities bent on colonizing Southern California drove cattle north from Mexico. Where they crossed the Colorado River near present-day Yuma, Arizona, the cattle trampled Indian crops. In response, in July 1781 the Yumans killed Fray Francisco Tomás Hermenegildo Garcés and his soldiers, who had established a mission that was the stop on the travelers' journey to Los Angeles (Fenn 2001, 153). Four years later, an uprising at Mission San Gabriel demonstrated that Indians did not simply passively accept the subordinate place in Spanish missions. Among the many sources of conflict, historian Steven W. Hackel has concluded that a significant one rested with intertribal rivalries over increasingly scarce resources. As Mission San Gabriel's agricultural output and spatial needs increased along with its Native populations, competition for food among nonmission Indians increased. The Gabrielinos who rose up against mission authorities were doing so, in part, because neophytes were displacing traditional subsistence practices (Hackel 2003, 656–660). A final example of this type occurred much later, in the 1840s, at the Whitman mission on the Columbia Plateau. Marcus Whitman achieved minimal success at converting Natives to either Christianity or the sedentary life of farmers. Moreover, personality problems beset the mission almost from the beginning. In 1845, in a symbolic act of violence, the Cayuse destroyed the irrigation works created at the mission. Although the Cayuse would later kill thirteen Americans at the mission, this earlier protest has as much symbolic power, for it was destroying the very apparatus necessary to manipulate the arid environment into a productive agricultural land (Addis 2005, 241). These violent examples demonstrate that Indians recognized that ecological changes were making their lives more difficult and they were willing to fight back to regain ecological regimes and disrupt the incipient ones.

If Natives recognized that they must sometimes fight to keep or regain control of their environment, Euro-Americans understood that their own control of the region's ecology could be used to manipulate the Native populations. In California, the Spanish and Americans instituted a practice of raiding Indians during times of food preparation, destroying their food and thus their ability to maintain their autonomy. This, then, forced Natives to work for ranchos to earn money or food necessary to survive (McEvoy 1986, 44). In the Northwest, conflicts between the U.S. Army and Natives similarly centered on

A Cayuse Indian attacks missionary Marcus Whitman on November 29, 1847, near present-day Walla Walla, Washington. The Cayuse killed thirteen other people. Earlier, the Cayuse destroyed the irrigation works built by the missionaries as a protest. (Getty Archives)

subsistence. Soldiers might destroy fishing sites or prevent access to them (Taylor 1999, 134). In one of the most punitive retaliatory actions by the army, Colonel George Wright destroyed horse herds numbering more than 700 and eliminated Native modes of transportation, means of trade, and sources of wealth and cultural pride (Schwantes 1996, 148). The nature of these conflicts reveals the centrality of nature to the lives of far westerners and the clear recognition of that fact on all sides.

CONCLUSION

The environmental changes induced by contact between indigenous and Euro-American cultures represented a local example of a global process whereby markets were reshaping the world's ecosystems. Global market capitalism expanded and drove fundamental changes in the Far West's human and natural communities (Robbins 1997, 57). The capitalist logic that nature was a collection of commodities challenged and ultimately dominated other ideas about nature. Capitalism redefined the region, as well as the relationship between people and nature. It was one of the informing, though often unspoken, principles for early exploration. It certainly framed the fur trade and the market in hides and tallow. Finally, it propelled the ranchos and early Euro-American farmers coming to the region. Although capitalism may not have framed the missionary efforts as much as evangelical religious beliefs, missionaries certainly brought new understandings of nature and the relationships they hoped to forge between Indians and the land resembled capitalism more than anything else. Capitalism, then, mediated these relationships between natives and newcomers—human and nonhuman alike. This economic transition, of course, was built on a dynamic ecological foundation. This underpinning proved unstable and unpredictable. Euro-Americans set out to control that unpredictability.

Ecological imperialism enabled an ecological revolution that assisted the invading Euro-Americans. From this perspective the era of displacement and replacement began when European species and germs found their way into the pathogenically isolated Far West. Introduced species displaced and replaced the ecosystems that had evolved in the region with careful management by various Native groups. These environmental changes deeply challenged the prevailing economies and cultures of the region. Unable to maintain population levels, Natives lost cultural traditions and often combined with other groups to survive. Meanwhile, refused access to traditional places to fish or burn, they faced subsistence crises that exacerbated social problems. Finally, new ideas, capitalism and Christianity being two of the most important, ushered in new consciousness in the relationship to the Far West's ecology. New institutions developed; the most elaborate being the irrigation structure adapted by the Spanish. The tensions between ecology, production, reproduction, and consciousness necessitated and produced an ecological revolution—the first since humans arrived and the Pleistocene extinctions occurred. By the 1840s, one revolution was clearly accomplished. In the coming decades, the ecological changes initiated in this colonial period paled compared with the ecological reordering evident with industrialism. Such forces would render the Far West an entirely new place.

REFERENCES

Addis, Cameron. "The Whitman Massacre: Religion and Manifest Destiny on the Columbia Plateau, 1809–1858." *Journal of the Early Republic* 25 (Summer 2005): 221–258.

Boyd, Robert. *The Coming of the Spirit of Pestilence: Introduced Infectious Diseases and Population Decline among Northwest Coast Indians, 1774–1874.* Seattle: University of Washington Press, 1999.

Bunting, Robert. *The Pacific Raincoast: Environment and Culture in an American Eden, 1778–1900.* Lawrence: University Press of Kansas, 1997.

Clayton, Daniel W. *Islands of Truth: The Imperial Fashioning of Vancouver Island.* Vancouver: University of British Columbia Press, 2000.

Cook, Sherburne F. "The Impact of Disease." In *Green Versus Gold: Sources in California's Environmental History,* edited by Carolyn Merchant, 55–59. Washington, DC: Island Press, 1998.

Crosby, Alfred W. "Virgin Soil Epidemics as a Factor in the Aboriginal Depopulation in America." *William and Mary Quarterly* 33 (April 1976): 289–299.

Crosby, Alfred W. *Ecological Imperialism: The Biological Expansion of Europe, 900–1900.* New York: Cambridge University Press, 1986.

Dasmann, Raymond F. "The Rangelands." In *Green Versus Gold: Sources in California's Environmental History,* edited by Carolyn Merchant, 194–199. Washington, DC: Island Press, 1998.

Engstrand, Iris H. W. "Seekers of the 'Northern Mystery': European Exploration of California and the Pacific." In *Contested Eden: California before the Gold Rush,* edited by Ramón A. Gutiérrez and Richard J. Orsi, 78–110. Berkeley: University of California Press, 1998.

Fenn, Elizabeth A. *Pox Americana: The Great Smallpox Epidemic of 1775–82.* New York: Hill and Wang, 2001.

Gibson, James R. *Farming the Frontier: The Agricultural Opening of the Oregon Country, 1786–1846.* Seattle: University of Washington Press, 1985.

Hackel, Steven W. "Land, Labor, and Production: The Colonial Economy of Spanish and Mexican California." In *Contested Eden: California before the Gold Rush,* edited by Ramón A. Gutiérrez and Richard J. Orsi, 111–146. Berkeley: University of California Press, 1998.

Hackel, Steven W. "Sources of Rebellion: Indian Testimony and the Mission San Gabriel Uprising of 1785." *Ethnohistory* 50 (Fall 2003): 643–669.

Harris, Cole. *The Resettlement of British Columbia: Essays on Colonialism and Geographical Change.* Vancouver: UBC Press, 1997.

Hundley, Norris, Jr. *The Great Thirst: Californians and Water: A History*. Revised ed. Berkeley: University of California Press, 2001.

Igler, David. "Diseased Goods: Global Exchanges in the Eastern Pacific Basin, 1770–1850." *American Historical Review* 109 (June 2004): 692–719.

Isenberg, Andrew C. *Mining California: An Ecological History*. New York: Hill and Wang, 2005.

Jones, David S. "Virgin Soils Revisited." *William and Mary Quarterly* 60 (October 2003): 703–742.

Jordan, Terry G. *North American Cattle-Ranching Frontiers: Origins, Diffusion, and Differentiation*. Albuquerque: University of New Mexico, 1993.

Lang, William L. "Describing a New Environment: Lewis and Clark and Enlightenment Science in the Columbia River Basin." *Oregon Historical Quarterly* 105 (Fall 2004): 360–389.

Mackie, Richard Somerset. *Trading Beyond the Mountains: The British Fur Trade on the Pacific, 1793–1843*. Vancouver: UBC Press, 1997.

McEvoy, Arthur F. *The Fisherman's Problem: Ecology and Law in the California Fisheries, 1850–1980*. New York: Cambridge University Press, 1986.

Merchant, Carolyn. *Ecological Revolutions: Nature, Gender, and Science in New England*. Chapel Hill: University of North Carolina Press, 1989.

Miller, Jay, ed. *Mourning Dove: A Salishan Autobiography*. Lincoln: University of Nebraska Press, 1990.

Oliphant, J. Orin. *On the Cattle Ranges of Oregon Country*. Seattle: University of Washington Press, 1968.

Preston, William. "Serpent in the Garden: Environmental Change in Colonial California." In *Contested Eden: California before the Gold Rush*, edited by Ramón A. Gutiérrez and Richard J. Orsi, 260–298. Berkeley: University of California Press, 1998.

Robbins, William G. *Landscapes of Promise: The Oregon Story, 1800–1940*. Seattle: University of Washington Press, 1997.

Schwantes, Carlos Arnaldo. *The Pacific Northwest: An Interpretive History*. Revised and enlarged ed. Lincoln: University of Nebraska Press, 1996.

Starrs, Paul F. "California's Grazed Ecosystems." In *Green Versus Gold: Sources in California's Environmental History*, edited by Carolyn Merchant, 199–205. Washington, DC: Island Press, 1998.

Taylor, Joseph E., III. *Making Salmon: An Environmental History of the Northwest Fisheries Crisis*. Seattle: University of Washington Press, 1999.

Vibert, Elizabeth. *Traders' Tales: Narratives of Cultural Encounters in the Columbia Plateau, 1807–1846*. Norman: University of Oklahoma Press, 1997.

Warren, Louis S., ed. *American Environmental History*. Malden, MA: Blackwell Publishing, 2003.

Weber, David J. "The Spanish-Mexican Rim." In *The Oxford History of the American West*, edited by Clyde A. Milner II, Carol A. O'Connor, and Martha A. Sandweiss, 45–77. New York: Oxford University Press, 1994.

Weber, David J. "The Spanish Moment in the Pacific Northwest." In *Terra Pacifica: People and Place in the Northwest States and Western Canada*, edited by Paul W. Hirt, 3–24. Pullman: Washington State University, 1998.

White, Richard. *Land Use, Environment, and Social Change: The Shaping of Island County, Washington*. 1980. Reprint with a new foreword, Seattle: University of Washington, 1992.

4

INDUSTRIALIZING NATURE

Westerners enjoy their myths. In them, lone miners pan for gold in beautiful mountain streams; farm families battle heroically and successfully to control nature and eke out a subsistence that provides for all their needs and leads to self-suffiency; lumberjacks chop at trees with oxen and deer as companions. These characters assume great importance in regional folklore. After all, they are undeniably far more interesting and colorful than wage-working miners gradually poisoned by exposure to quicksilver, immigrant capitalists acquiring millions of acres of land that they promptly engineered and integrated to achieve near-monopoly status, or engineers who devised technologies that cut trees or canned salmon with terrifying efficiency that left little behind. These, however, resemble the Far West's environmental history far more accurately than the miners, farmers, and loggers of storybooks.

The Pacific West industrialized much faster than the rest of North America. Almost immediately after the ecological shocks of contact, the newly dominant populations, relying on capitalist imperatives and the support of political, economic, and legal structures, transformed Western ecosystems with impressive rapidity and devastating environmental consequences. Harnessing industrial technologies, these commercial farmers, loggers, fishers, and miners fundamentally transformed the region's economy and ecology, thoroughly disrupting previous ecosystems.

INSTITUTIONAL AND INTELLECTUAL FRAMEWORKS

Americans industrialized nature in the last half of the nineteenth century through various supportive institutional and intellectual frameworks. The American political system created laws amenable to economic development. Throughout the nineteenth century, Americans in positions of political and economic power ensured that their institutions worked to secure what they believed to be a proper social order. Thus, the law and the market developed in ways to promote a specific historical and cultural worldview concerning the proper use of nature. Needless to say, these perspectives differed enormously from indigenous ways, and even vastly altered the Hispanic structure in place in early nineteenth-century California and the British and American foundations of the Northwest. Consequently, large numbers of Americans arrived in the Far West equipped with a government on their side that was willing to use its power to foster economic growth.

Perhaps better than any other scholar, legal theorist James Willard Hurst has explained the nineteenth-century use of law in his classic, *Law and the Conditions of Freedom in the Nineteenth-Century United States.* According to Hurst, crafters of law in the nineteenth century believed that humans were inherently creative and required freedom to use that ingenuity (1956, 5–6). These attributes were particularly important to the United States, because "here unclaimed natural abundance together with the promise of new technical command of nature dictate that men should realize their creative energy and exercise their liberty peculiarly in the realm of the economy to the enhancement of other human values" (6). In addition, Hurst described how law embodied these assumptions, ensuring that individuals would not be hampered by undue restrictions on their opportunity while unleashing their economic and technological resourcefulness. Such a perspective belies the common claim that the nineteenth-century governing philosophy was simply laissez faire, or a hands-off approach. Politicians and judges readily "interfered" if such "interference" seemed likely to prompt a greater release of creative economic energy (Hurst 1956, 5–8). As the nineteenth century progressed, law further evolved to protect private property that had been released for use by American economic energy. For the United States at the time, private property assumed great importance in providing economic security, and U.S. courts sought to impose order on and protection for these property rights (Hurst 1956, 8–9). Concern for property—obtaining it and protecting it—lay at the heart of much of the era's environmental history.

Getting land into private hands became an increasingly important concern for nineteenth-century governments. Moreover, as Hurst pointed out, as law developed to allow the release of creative energy, it allowed a flourishing of individual economic liberty and opportunity, at least for Euro-American men and eventually for other social and cultural groups. Politicians maintained that policies should promote broad economic development. In this light, land disposal laws, railroad land grants, and natural resource policies exemplified the positive role nineteenth-century law played in promoting development (Hurst 1956, 33–70). Hurst's important claims demonstrated that economic change influenced legal development and vice versa. All of this is important to environmental history, for these crucial institutional frameworks supported the transformation of nature into private property from which individuals and groups would appropriately benefit.

If this was the policy foundation for the late nineteenth century, Americans' attitudes about government constituted another important context. Historians Norris Hundley Jr. and Donald J. Pisani in their respective studies of Western water have shown that Americans distrusted government, especially centralized power. Distrusting government, however, did not mean that westerners and others did not turn to it for assistance in developing natural resources. Indeed, Americans expected government to create an environment in which individuals could pursue their economic activities with minimal restrictions. Because national and state legislatures offered few restrictions, regulatory power largely devolved to the courts. In this context, long-term planning, community well-being, and thoughts to environmental costs were luxuries not needed, at best, or anathema to economic liberty, at worst (Hundley 2001, 65–69; Hurst 1956; Pisani 1992, 1–10). In the end, the political culture confirmed the economic culture.

Finally, Americans' broader cultural ideas of nature also contributed to the context in which nature's industrialization proceeded. The majority of Americans shared certain ideas about nature. Simply put, it was something to be used to fuel economic expansion and to promote individuals' pursuit of wealth. Because of North America's vast land and resources, most Americans posited that the resources were inexhaustible. Fundamentally, Americans saw nature instrumentally; that is, nature constituted a source from which humans were to extract value for human benefit. Few saw nature as anything like a holistic connection of interrelated components on which other elements depended. To be sure, some Americans, intellectuals such as Henry David Thoreau and George Perkins Marsh in New England, articulated a deeper spiritual meaning of nature and nascent ecological ideas that might be found here and there. However, such ideas remained a distinct minority view. Instead, the idea of a nature subordinate to human designs, existing to improve human economies, dominated.

All of these American contexts mattered, because by the mid-nineteenth century the Far West had finally been incorporated into the United States. The process, which began slowly with explorers such as Robert Gray claiming the Columbia River and the Lewis and Clark expedition and merchants trading for hides and tallow in California, moved rapidly toward a national resolution in the 1840s (see Table 1). Besides establishing territorial jurisdiction, the federal government enacted a series of laws and policies that vastly accelerated the movement of American citizens into the region (see Table 2). These laws by themselves demonstrate Hurst's argument that laws facilitated market growth and illustrate the high priority U.S. policy placed on getting the public domain into private hands that might generate a profit. Of course, the national government first had to obtain title to those lands, which it ostensibly did in the 1850s

TABLE 1
Establishing U.S. Jurisdiction

State	Date of Initial U.S. Claim and Source of Claim	Date of Statehood
California	1848: Treaty of Guadalupe Hidalgo with Mexico	1850
Oregon	1846: Oregon Treaty with Britain*	1859
Washington	1846: Oregon Treaty with Britain	1889
Idaho	1846: Oregon Treaty with Britain	1890

*In 1818, the United States and Great Britain agreed to jointly occupy the Oregon Country, a region north of the 42nd parallel, west of the Rocky Mountains, and north to 54°40'. The agreement was reaffirmed in 1827. Under these agreements, neither nation would assert sovereignty. The 1846 treaty divided the larger Oregon Country into separate jurisdictions at the 49th parallel.

with a series of treaties with various Native American tribes in California and throughout the Pacific Northwest (White 1991, 101). These treaties were fraught with problems, including the likelihood that the U.S. Senate would not ratify them, the lack of representation from countless Native groups at treaty proceedings, and frequent neglect in enforcing the rights contained within these documents (Milner 1994, 175). Nonetheless in the eyes of the U.S. government and citizens, treaties extinguished Native title to vast land holdings, clearing the way for an American populace to acquire land, although American Indians maintained their traditional rights to hunt and fish. Collectively, these policies not only illustrated the American preference for private property but also the fact that land and other resources were to be treated as commodities for exchange in the marketplace (Robbins 1997, 83). Such values toward the natural world represented a sharp disjunction with indigenous values but confirmed and extended the values introduced by earlier Euro-Americans.

INDUSTRIALIZING MINING

Americans rapidly put these ideals to work in transforming nature beginning with the California Gold Rush. In 1848, working at John Sutter's mill in the Sierra Nevada foothills, James Marshall found gold. In the next decade alone, California produced more than half a billion dollars worth of gold (Isenberg 2005, 23). This economic boom initiated nothing less than a wholesale ecological transformation with effects far beyond the Sierras and the local mining industry.

TABLE 2
Select U.S. Land Laws

Name of Law (Date)	Size of Land Grant (In Acres)	Cost of Land (Per Acre)	Conditions
Oregon Land Donation Act (1850)	320; if married, each partner had title to 320	Free	Live on land for four years and make improvements
Homestead Act (1862)	160	Free	Pay entry fee and live on land for five years; could pay for land after six months
Railroad Land Grants (various years between 1862 and 1872)	Variable, total reached approximately 223 million acres	Free	Companies could sell their land to finance the construction of rail lines; theoretically, railroads received odd numbered sections of land stretching 20 miles on either side of the line's route without disrupting preexisting land claims
Timber Culture Act (1873)	160	Free	Planting trees on one quarter of the land gave title to 160 acres; after 1878 amendments required only one-sixth of the area to be cultivated in trees
Desert Land Act (1877)	640 (reduced to 320 in 1890)	$1.25	Settler must irrigate within three years
Timber and Stone Act (1878)	160	$2.50	Sale of land valuable in timber and stones to authentic settler or miner

Source: Opie 1998, 104–105; Schwantes 1996, 121; White 1991, 145–146.

It was, simply, the seminal environmental event of the Far West of the time, and perhaps ever. The Gold Rush established a pattern repeated in other regions where Americans exploited minerals or other resources. The California Gold Rush refined technology that created other environmental changes in the West's other mines; it encouraged and helped, complicated and harmed agricultural development; it devastated vegetation and aquatic life; it accelerated the cutting of trees and damaged forests; and it enticed so many newcomers to the region that American Indians were almost entirely displaced, Hispanic cultures were sharply circumscribed, and earlier environmental practices would never again be dominant.

Although the U.S. federal government had known since 1843 that California had gold deposits, Marshall's discovery initiated an unprecedented economic boom and ecological bust (Steinberg 2002, 118). From the beginning of the Gold Rush, change of all kinds was pronounced. In the first few years of the boom, miners swarmed through the Sierra Nevada foothills and scoured streams for gold, the product of millions of years of geologic activity. Magma forced gold in a silica solution to the surface where it ultimately stuck in quartz veins within granite. Then, millions of years of erosion by Sierra water made it available in gravel streambeds flowing out of the mountains (Isenberg 2005, 26).

Mining the gold occurred in various ways and with diverse environmental effects. Surface placer deposits were so easily taken from streams that individuals with simple technology readily acquired gold. Miners swarmed through watersheds along the Sierra Nevada foothills searching for easy gold, some obtaining $8,000 of gold a day (Bryant 1994, 200). Their tools were rudimentary, including spoons and pans and simple sluices, but their energies rapidly depleted the surface deposits, so next these individuals combined to work more difficult claims, such as those riverbeds and deposits in alluvial gravel. These cooperative endeavors employed more sophisticated technology. For instance, they built small dams and larger sluices to divert river water temporarily and to move greater volumes of gravel (Beesley 2004, 47).

Rivers proved great hindrances and terrific aids to miners. With so much gold in riverbeds, water covered valuable deposits. However, by building dams and sluices, miners could marshal water for their own use washing out gravel and leaving behind valuable minerals. Unfortunately, dams and sluices broke easily, and floods were often too powerful for the flimsy structures. Thus, diverted water frequently broke back into its natural channel and wiped out much of the miners' work. As environmental historian Andrew C. Isenberg has pointed out, nature's unpredictability made mining difficult labor and slowed potential investment in the mines (Isenberg 2005, 27–28). Meanwhile, as miners constructed dams and sluices, as well as burned wood as fuel, they rapidly

Miners use hydraulic equipment to wash gold from the mountains near Dutch Flat, California, ca. 1868. The erosion caused by these activities led to enormous ecological changes. (Andrew J. Russell/Library of Congress)

depleted local timber supplies (Rohe 1998, 126–127). This depleted vegetation, accelerated erosion, and worsened floods. By 1852, a mere four years after Marshall's discovery, cumulative gold output reached $80 million and had significantly reordered the mountains' ecology (Bryant 1994, 200). But those ecological and economic results would be dwarfed by the next stage of extraction: hydraulic mining.

Hydraulic mining required technological sophistication that depended on large capital sums. The practice created long-term and long-range environmental changes. Moreover, as hydraulic technology developed, corporations displaced individual and small groups of miners who could no longer compete economically. It caused a revolution in the goldfields in every way.

Raising the requisite capital to finance hydraulic operations required miners to search for outside sources. When groups of miners succeeded in damming streams and building reservoirs and ditches, they demonstrated to outside investors that some control might be achieved over an unpredictable environment. This attracted investment to the California goldfields and helped develop and implement extensive hydraulic systems (Isenberg 2005, 21, 28, 35). Besides displacing lone miners with corporations, outside investment in corporate projects helped nurse a feeling, increasingly common throughout the next century, that far westerners were dominated by outsiders, including Eastern and European investors and a distant and sometimes hostile federal government. Despite such resentment, westerners could not succeed without this outside capital that developed the technology necessary for hydraulicking. Eventually by 1862, another strategy for raising money included forming the San Francisco Stock and Exchange Board, or colloquially the 'Change. The colorful but accurate words of a recent study illuminate its function: "By rendering nature in the abstract and interchangeable units of the marketplace, the 'Change succeeded in dividing and distancing it, as a slaughterhouse rendered the carcasses of animals into precise units of tinned meat" (Brechin 1999, 38). The statement suggested how commodified and abstract nature had become and the distance between economic decisions and ecological consequences. Fundamental to that problem of distance was the rise of cities, in this case San Francisco, in propelling nature's industrialization.

Human labor applied to new technology reorganized nature. These hydraulic systems required the control of water, so laborers constructed elaborate dams, flumes, and ditches designed to harness waterpower. These systems connected to highly pressured hoses and nozzles, called water cannons or monitors, that wageworkers, often Chinese immigrants, aimed at hillsides. The water washed soil away to find and extract hidden pockets of gold. One of the first ecological impacts was diversion. By 1855, 1,159 miles of ditches reworked Sierra waterways; by 1882, more than 6,000 miles existed. Such work was costly. Constructing ditches cost between $4,000 and $17,000 per mile (Isenberg 2005, 30). Reservoirs also caused ecological change. Some storage reservoirs on a few of the main rivers held up to 6 billion cubic feet (Rohe 1998, 131–32). Several individual companies impounded more than a billion cubic feet of water, suggesting the immense ecological power these companies could bring to bear on California (Isenberg 2005, 30). One large company, the North Bloomfield Company, alone managed more than 1,500 acres, constructed two large reservoirs and more than 100 miles of ditches and canals, and employed more than 500 workers. It spent more than $2 million to construct its water system. Its largest dam, completed in 1872, created a reservoir that covered more than 500 acres and was 2.5 miles long (Pisani 1992, 19). Furthermore, technological improvements in hoses and nozzles allowed water to be shot at 150 feet per second, or more than 100 miles per hour. Eventually, too, corporations kept their employees working in shifts twenty-four hours a day washing away the soil (Isenberg 2005, 35). Not surprisingly, this labor exacted significant environmental changes.

Once the power of the dams, flumes, ditches, and monitors was marshaled and unleashed on California's environment, the ecological changes vastly accelerated. Most dramatic was erosion. With water cannons aimed at Sierra soil, debris washed away in amazing quantities and with disastrous effects as far away as San Francisco Bay. Boulders, sand, soil, and gravel washed through sluices into mountain streams and eventually the rivers. This mining debris wreaked enormous havoc. In 1917, Grove Karl Gilbert, a geologist working for the federal government, concluded that approximately 1.5 billion cubic yards of debris washed out of the Sierra foothills during the three decades of intense hydraulic mining, from the mid-1850s to the mid-1880s. For some perspective, that volume amounted to about eight times the earth moved for the construction of the Panama Canal (Beesley 2004, 53).

This debris fundamentally changed riverine environments and affected broad ecological and economic relationships of California. To begin with, the debris settled on riverbeds and thus raised them. Some continued to rise so that they were actually higher than the surrounding landscape. For instance, the Bear River bed raised 97 feet in just three years after 1870 (Rohe 1998, 132). With these rivers filled with debris, they could not contain normal rainfall, much less any season with unusually high precipitation. Consequently, floods increased in frequency and devastation. In one flood, the debris spread throughout a basin from 1 to 3 miles, covering 25 square miles with half a million cubic yards (Rohe 1998, 133). The debris also created a combination of sand and gravel known as slickens that spread out over valuable agricultural bottomlands. One farmer on the Yuba River reported that 8 feet of slickens poisoned his land. Farm values throughout the Sacramento Valley declined by as much as 80 percent (Isenberg 2005, 45). The debris that flooded and polluted the farmland created, in the words of one scholar, a "barren wasteland" (Kelley 1998, 120).

Urbanization complicated these ecological changes. Surrounding communities and farms required protection from these erratic rivers. Marysville, a community at the conjunction of the Yuba and Feather rivers, had been built 25 feet above the river, but by 1879 its streets were below the river level (Rohe 1998, 133). Sacramento faced even worse problems. This capital city, also built at the conjunction of two rivers (the Sacramento and the American), always faced uncertainty because of floods. With the increased flooding, Sacramento reeled from devastating floods almost annually. In 1852, a flood was so severe that the American River rose by a foot per hour. The floods in the winter of 1861–1862 were even worse. These floods caused additional property damage because the Gold Rush brought thousands of new residents to the city. Moreover, disease outbreaks contributed tragic side effects to nineteenth-century urbanization. With the concentrated population in a city unprepared to deal with sanitation, Sacramento confronted a cholera epidemic in 1850, the result of poor public works and unsafe water supplies. Civic leaders approached flooding and public health crises in similar ways: They raised taxes to improve public works. To contain the rising rivers, local farmers and civic leaders built levees. And when those still did not stop rising waters, they built them higher. In 1868, the city funded a canal that would divert floodwaters from the city. A decade later, however, floods returned. Ironically, despite the inability to entirely hold floodwaters at bay, Sacramento civic leaders gained political legitimacy for their practical efforts at flood control and other public works projects (Isenberg 2005, 53–74). Like downstream farmers, valley cities bore the ecological costs of mining in the mountains.

Besides flooding, mining changed water quality so much that it hampered other activities normally pursued on the rivers and in coastal waters. First, salmon could no longer spawn. Because these fish require cool, clear water and gravel riverbeds, the rivers after mining could no longer function as spawning grounds. For example, the Sacramento River experienced its last healthy spawning run in 1852 (McEvoy 1986, 83). Second, river navigability declined. The debris from mining and floods, as well as the propensity of channels to change positions because of the violent changes in riverine environments, made navigating the inland rivers almost impossible by 1880 (Rohe 1998, 133–34). Finally, San Francisco Bay began to silt in. In the 1870s, observers could see the sediment from the mines working its way into the bay. In the same decade, naval officials complained of shoaling. Some feared that San Francisco Bay might eventually be destroyed as a harbor. Besides the commercial implications of such aquatic changes, oyster beds smothered to death in the silt, ruining that important fishery (Isenberg 2005, 47; McEvoy 1986, 85–87; Rohe 1998, 134).

Other changes affected California's environment. Deforestation accelerated because of the multiple demands of wood for fuel and building material. A flume for one mine was 47,520 feet long, 6 feet wide, and 3 feet high; such a flume, multiplied many times over, meant significant wood consumption. An 1870 estimate suggested that one-third of California's commercial timber was already exhausted (Isenberg 2005, 41–42). Because of mining's timber needs, California turned to more distant forests along the northern California coast and in the Northwest to meet their demands. Thus, the ecological impacts of California mining were more than local.

Perhaps the most deadly effect associated with mining involved mercury. Mining companies added quicksilver to sluices where it would amalgamate with gold, allowing easier gold collection. Its use was prodigious: one company used more than half a ton of mercury every twelve days. Fortunately for mine companies, a mercury source was nearby in deposits in the Coast Range. From 1850 to 1885, the New Almaden mine south of San Francisco produced an average of 1.7 million pounds annually. By the mid-1860s, the mine was just one component of an enterprise that employed more than 2,000 workers, largely Mexican American, and used six furnaces and 400 buildings. Quicksilver mining was dangerous. Explosions below ground killed or crippled workers, and vapors from the furnace above ground were toxic. An observer in 1863 reported nothing green left near the furnace: "Every tree on the mountain-side is dead." The observer continued, describing the workers who worked only one week a month and the air: "Pale cadaverous faces and leaden eyes are the consequences of even these short spells. To such a degree is the air filled with the volatile poison that gold coins and watches on the persons of those engaged about the furnaces become galvanized and turn white. In such an atmosphere, one would seem to inhale death with every respiration" (quoted in Isenberg 2005, 48–49). This situation resulted from the human body absorbing 75 to 85 percent of the mercury, which also washed into streams where other organisms absorb it. In the process of moving through complex food webs, the concentrations of methyl mercury increases. This process, called bioaccumulation, meant higher order animals can have mercury concentrated 100,000 times the level in its surrounding environment. And it stayed there. Much of the mercury was never recaptured. One company, for example, believed it lost about eleven tons in only six years in the Yuba River. Thus, mercury is still embedded in the environment. Even more troubling, people recognized the deadly effects of mercury at the time. The case of mercury demonstrates both the widespread ripple effects of California's Gold Rush, as well as the long-lasting nature of some environmental change (Beesley 2004, 56–57; Brechin 1999, 61–63; Isenberg 2005, 47–50).

Indeed, the Gold Rush's ecological effects were not even limited to California or the West Coast. About 80,000 migrants crossed the Plains in 1849—the storied forty-niners. The ecological demands of their animals damaged the grasslands through which they traveled. Their wagons similarly damaged the Great Plains, and they cut down the already scarce timber along watercourses. Migrants heavily used river valleys for travel, and this overuse spelled problems, especially for Natives who relied on them after the migrants moved on (Steinberg 2002, 122–124).

Legal, political, and economic strategies allowed, and even encouraged, these developments. Policies that governed mining did not so much establish rules as they confirmed the practices miners instituted themselves. The practices came from deeply rooted American ideals that accompanied migrants to the goldfields. Historian Donald J. Pisani enumerated miners' beliefs:

> first, that in the absence of a formal legal structure, or when that structure broke down, "popular sovereignty"—in this case "squatter sovereignty"—should prevail; second, that "liberty" meant freedom from government interference in the individual search for wealth; third, that the right to pursue wealth was not a gift of the state, but, rather, derived from an individual's inalienable right to the product of his labor; fourth, that no person should be able to secure more land or water than he or she could make use of; and fifth, that the initiative and energy needed to develop a natural resource did more to create property than a formal title (Pisani 1992, 22).

In practice, these beliefs promoted an institutional structure that encouraged individual economic development, protected individual self-interest, and did not promote the general welfare. This approach clearly differed from that which preceded it in California. Although Americans shared with their Spanish-Mexican predecessors a belief in using resources, American customs focused much more on individuals (Hundley 2001, 65–69; Pisani 1992, 11–32; Wilkinson 1992, 38–40).

The most important practice institutionalized in law was prior appropriation. This doctrine applied to mining claims, but more importantly, it would guide water rights in the Far West. Pisani called the doctrine of prior appropriation the West's "greatest legal innovation" (1992, 11). Known by many in the shorthand descriptive form, "first in time, first in right," prior appropriation provided water rights to the first person to put water to productive use. Miners, or irrigators, only had to continue using the water to keep the right to it. Riparian rights, the dominant legal model in the East, granted use rights to water that ran adjacent to or through one's property, but one could not significantly diminish the quantity or quality of that water. Prior appropriation followed different rules and essentially made water private property. So not only was gold commodified, but mining corporations succeeded in modifying water law so that water itself became something to be bought and sold. Prior appropriation did not survive legal challenges unqualified. Numerous decisions by the California Supreme Court modified and restricted prior appropriation, but generally the doctrine remained. And it was not without costs. Prior appropriation gave nothing back to the state and often led to concentrated corporate control of water rights, dangerous monopolies, and endless litigation. Like so much else with the California Gold Rush, prior appropriation profoundly influenced other industries and other places (Isenberg 2005, 32–35; Pisani 1992, 11–32).

Although the legal system facilitated this environmental and economic exploitation, it could also stop it. Farmers in the Central Valley, involved in their own massive reconfiguration of the region's ecology, clashed with miners over mining's downstream impact. Communities joined farmers in their campaign against hydraulic mining. Initial attempts to get the California legislature to do anything about it met with resistance or indifference. Consequently, they turned to the courts, hoping for an injunction against dumping debris into rivers, which would have effectively halted hydraulic mining. A district court ruled in favor of the farmers, but the state Supreme Court disagreed on the basis that they could not lump all hydraulic miners together. Entering into several individual lawsuits directed at every hydraulic mining enterprise would not have been manageable, however, so the antidebris forces seemed stymied (Kelley 1998, 120–125).

Meanwhile, some farmers and supporting communities believed they could engineer their way out of the problem by creating a vast flood control system. For instance, debris dams would catch the large debris, while levees would stave the floods and by narrowing the stream would allow the river to scour its own bed. The Drainage Act of 1880 put this plan in action, but floods the following winter were disastrous, showing the futility of engineering solutions. In June 1882, Judge Jackson Temple ruled that miners could not dump "coarse" debris into the stream. Fortunately for miners, "coarse" was a vague term that did little to slow their work. Finally, Edward Woodruff, a New Yorker who owned property in Marysville, initiated a suit in the Ninth U.S. Circuit Court seeking a perpetual injunction. Going to a federal court perhaps reflected a desire to move away from local prejudices that favored the mining industry. After a year and a half of intense legal investigation and testimony, Judge Lorenzo Sawyer ruled against the mining companies and effectively halted legal hydraulic mining. Farmers continued fighting for nearly a decade to ensure that miners were not still dumping debris illegally into the rivers. This decision marked a significant economic change, and contemporary newspapers were resigned to it and noted that at least agriculture was not as transitory an industry as mining. Observers have suggested that the decision reflected the declining fortunes of the mining industry and the rising position of agriculture in the state's economy. By 1900, farmers became the dominant force in the Sacramento Valley, and they were already well under way in their own endeavors to rework nature for massive commercial success (Isenberg 2005, 171; Kelley 1998, 120–125).

Despite the end of hydraulic mining in California, it was not the end of hydraulic mining in the Pacific West. Indeed, a diaspora of mining engineers and technology forged in the California Gold Rush spread throughout the West and abroad to Africa and Latin America (Brechin 1999, 54–56). In the Northwest, mining suffered through a series of booms and busts. In southern Oregon's Rogue Valley, mining began in the 1850s as California miners drifted north. Hydraulic mining arrived by the late 1860s and quickly repeated the ecological havoc similar to California's mined landscapes: canyons filled with debris, streams clouded with soil, erosion increased, trees and other vegetation were removed, and the salmon habitat was severely compromised. Some mines moved 800 cubic yards of soil daily. Gold strikes in the interior Northwest initiated rushes to mountains and valleys that were previously home only to Plateau tribes. Besides the significant influx of new populations and the growth of towns to serve the miners, these mining booms instigated deforestation in inland forests to supply miners with construction material. Hydraulicking wreaked its havoc and reorganized stream systems, causing disproportionate problems in this poorly watered region. Mining booms grew during the 1860s and later introduced important agents of environmental change to the interior Northwest: new populations and external market demands for Northwestern commodities (Robbins 1997, 128–130, 142–145).

For example, mines in Idaho brought Euro-Americans in significant numbers for the first time, disturbing the ecology and Native populations. In the 1860s, more than 20,000 miners took out $57 million of gold and silver. In a mere six years in the Boise Basin, miners extracted 1.4 million ounces of gold. Hydraulic mining arrived by the mid-1860s (Arrington 1994, 231, 234). Clearly, there was gold to mine, but political problems required solutions. In one rush, minerals found up the Clearwater River from Lewiston were on the Nez Perce Reservation where Euro-American entry was illegal. Nevertheless, non-Natives invaded the reservation. The U.S. government accommodated the miners by severely reducing the size of the reservation by approximately 90 percent, although the renegotiated treaty did not meet nearly all the Nez Perce's approval and remains controversial (Schwantes 1996, 72). The episode exemplified two important themes. First, it showed the willingness of the federal government to use its power to displace American Indians if minerals or other profitable commodities might be obtained. Second, it demonstrated how old and new land-use patterns often brought new populations into contact and conflict with each other. Although the Nez Perce land reduction was most dramatic, other conflicts over mineral and other resources routinely plagued the Northwest and parts of California for most of the nineteenth century.

The 1880s found another round of mining rushes in Idaho. This time lead and silver were the principal minerals, and industrial methods dominated.

A view of the Nez Perce Agency in Idaho in 1879. The Nez Perce lived in what is now Idaho, northeastern Oregon, and southeastern Washington. The Nez Perce repeatedly found themselves forcibly displaced by ranchers and miners in the nineteenth century. (National Archives and Records Administration)

Discoveries in 1883–1884 along the Coeur d'Alene River in northern Idaho attracted 10,000 miners. Placer mining proved relatively unprofitable because the ores were too difficult to obtain, so lode mining became the method of choice. Hard-rock mining, though, required specialized technology, which demanded significant investment capital that necessarily came from outside the region and in some cases outside the country. It also depended on a large labor force of wage-workers. All of these factors inaugurated a highly industrialized enterprise complete with infamous labor violence. However, the significant ecological effects in hard-rock mining, including air and water pollution, belong in the twentieth century as industrialism consolidated and elaborated its hold among Far Western resources (Aiken 2005, 3–39; Arrington 1994, 351–360; Schwantes 1996, 94–95).

The great mining rushes of the Far West at the end of the nineteenth century typified this era's and region's pattern of resource exploitation. Individuals and small groups identified a resource and began to exploit or develop it. Unable to maintain high production or to increase profitability, capital from outside the region industrialized the exploitation of the resource with new technologies, vastly accelerating the ecological consequences. Those environmental changes severely disrupted existing ecosystems. Moreover, these disturbances rarely stayed confined to the immediate environment; instead, ecological consequences rippled outward throughout the landscape affecting distant ecological, economic, and cultural systems. The pattern repeated itself in agriculture, fishing, and logging. Among the most important impetuses for these industries was the demand created by the California Gold Rush. Thus, an interdependent dynamic wreaked regional havoc.

INDUSTRIALIZING AGRICULTURE

Before the Gold Rush, Euro-American agriculture had developed sporadically through the Pacific West at missions, fur trade posts, and some isolated Euro-American communities. However, mineral rushes throughout the region facilitated agricultural expansion, even as they sometimes harmed agricultural development with debris, pollution, and habitat destruction. Farming developed at an accelerating pace in the last half of the nineteenth century and in the process changed the Far West's ecology and the nature of agriculture.

Important agricultural divisions were at work. Environmental historian Donald Worster characterized two main ecological modes in Western agriculture after Euro-American settlement. In the first, the pastoral mode, ranching prevailed and individual herders dominated the mythology. In the second, the hydraulic mode, irrigation governed and water engineers shaped the land and waterscapes (Worster 1992, 28–29). Although these two modes describe much of the region, some commercial agriculture existed independent of irrigation. Smaller productive farms in the temperate coastal regions and massive dryland wheat farms in some interior locations contributed a third mode that did not rely on extensive pastoralism or the hydraulic infrastructures necessary in the other modes. Large-scale commercial agriculture preceded irrigation in some locations. Despite their differences, all three of these types became increasingly industrialized as the nineteenth century closed. Besides this division of agricultural types, two processes dominated in this era: enclosure and the move from extensive to intensive agriculture. The convergence of all these factors spelled monumental change for the region.

Commercial Farming in the Far West

Of course, not all agriculture in the Pacific West was industrial, especially in the mid-nineteenth century. But all depended on markets and promoted a commercialization that led easily to industrial influences, influences so great that they dominated by the turn of the twentieth century. An environmental history of market agriculture in Island County, Washington typifies agricultural development in coastal farmlands. The wet westside of the Cascade Mountains in the Northwest did not require irrigation works and the land was not suited to extensive pastoralism. This history, however, effectively demonstrates the process of intensifying agriculture and the complicated ecological ramifications of introducing market agriculture to a landscape where it had not previously been.

In one of the founding works of environmental history, *Land Use, Environment, and Social Change: The Shaping of Island County, Washington,* Richard White explored interconnected social and environmental changes in the last four decades of the nineteenth century. The first settlers possessed the best land and thus enjoyed the best economic prospects. Later arrivals found the remaining forestlands, marshlands, and tidelands to be difficult landscapes that required backbreaking work to claim for agriculture. As many as one-third chose tenancy on good lands rather than ownership on poor lands. The broader market in agricultural products immediately shaped these farmers who shifted several times in a few decades from grain-growing to sheep-raising back to grain-growing and on to intensive farming of various specialty crops. These activities in turn marked the land. Farmers cleared forests and burned brush to prepare the land for grazing. They raised timothy and clover to feed their animals, a strategy that marked the intensification of pastoralism as extensive rangelands were unavailable. The simple struggle to find land, prepare it, and locate profitable markets preoccupied nineteenth-century farmers, so they thought relatively little about the ecological changes they were creating (White 1992, 54–66).

Increasing populations required intensification of methods in an effort to stabilize ecological and economic practices. Local editors argued for "good farming," by which they meant reproducing northern European landscapes they believed to be stable. Editors and farmers assumed the fundamental stability of markets, production methods, and ecology. Such stability proved far more difficult to create than they expected. Market fluctuations led to shifting crops. More troubling and less controllable were environmental changes. Farmers battled exotic flora, especially the thistles and velvet grass that thrived in disturbed lands and served no useful purpose to farmers. To solve the problems, in 1883 the state legislature passed a law that made it a misdemeanor for farmers to let Canadian thistles go to seed and required road superintendents to eradicate it on roadways. Ironically, methods of eradicating it included chopping the thistles up, which likely led to its increase. In the end, the history of this time and place is best characterized by a series of successions. White explained them:

> "Tenants replaced freeholders; machines replaced Salish laborers; Chinese tenants replaced whites, and whites returned to replace Chinese. In the fields potatoes replaced wheat, sheep replaced potatoes, apples replaced sheep, and then men destroyed the orchards and planted wheat once again" (1992, 73–74).

Furthermore, market agriculture attempted to simplify and control the environment; however, exotic species and market forces, both originating outside Island County, challenged and compromised that control. Change was the only constant, and farmers seldom managed that change as well as they hoped or thought (White 1992, 67–76).

The story of western Oregon largely resembled that of Washington's Island County, only with accelerated transformations caused by stronger market influences. The same process of exotic species displacing and replacing native flora and fauna occurred, as did the displacement of Native cultures. The market in western Oregon's Willamette Valley proved even more dominant than in Washington because of its larger population. A belief in the valley's abundance and limitless economic possibilities fueled expansion, and Euro-American farmers rarely considered limits in their pursuit of profit (Bunting 1997; Robbins 1997).

If anything, dryland wheat farmers in the interior Northwest represented even greater connections to industrialized agriculture and market dependency. Since the 1860s, farmers fed nearby miners on a small scale. They also transported grain down the Snake and Columbia rivers to be shipped from Portland to Asian and other markets. Railroads made the process more efficient and promoted a significant expansion of wheat-growing in the Palouse country, which in turn tied them more closely to distant markets. By the late 1870s, for example, small-town newspapers reported San Francisco grain prices, illustrating the farmers' keen awareness that their financial fate rested with urban markets far outside the immediate region. At its foundation, though, economic expansion rested on nature. The famous Palouse loess, fertile and as deep as 250 feet, supported a productive agricultural base. By 1890, farmers replaced much of the open range of the interior, displacing pastoralists and replacing the native biota with cereal crops. It proved to be a remarkable economic success. Whitman County in Washington's Palouse produced the highest land values, the most improved farmland acreages, the largest farms, and the maximum valued agricultural crops in the state. Yet for all the productivity, dryland wheat farming came with noticeable ecological costs. When plows met the windblown Palouse hills, soil washed away. Intent on maximizing profits and acreage, farmers plowed

A short-line steam engine travels along the Columbia River in Oregon, ca. 1900. The railroad infrastructure created in the latter half of the nineteenth century accelerated the market economy's effects on the ecology of the western United States. (Library of Congress)

under more bunchgrass not imagining a day when the deep loess would deplete so much as to reduce yields significantly. Conserving soil would have to wait for another era (Duffin 2004, 197–200; Robbins 1997, 146–47).

When railroads arrived in the 1870s and 1880s, agricultural and ecological change became wedded to these economic links. To Oregonians, railroads represented improvement and provided a tool to complete the transformations already underway. Several historians have shown that nineteenth-century

Northwestern boosters believed industrializing the countryside through modern agriculture would merge a divine plan with human ingenuity to redeem nature. Technological change, then, did not spell the end of nature but the realization of its fuller purpose to serve human economies. The resulting consequences, both ecological and economic, later prompted reforms and reorganized industries in the early twentieth century (Bunting 1997, 104–119; Fiege 1999, 22–23; Robbins 1997).

Commercial Pastoralism in the Far West

Pastoralism in the Far West illustrates several ecological and social processes central to the region's environmental history. First, cattle- and sheep-raising contributed to the continued displacement of native species with imported species. Second, conflicts between cattle and sheep interests revealed pastoralism's own divisions. Third, as pastoralism developed, the intertwined processes of enclosure and intensification displaced Native peoples and reorganized ecological and economic processes. Throughout the Pacific West, evidence of these transitions abounded.

In California, the beef market arrived with the Gold Rush. This reoriented the long-standing ranchos' operations away from one organized to show wealth and social standing to one focused on profits. Quickly, rancheros organized cattle drives to mining country, relieving some of the pressure on Southern California grasslands. By the 1860s, cattle were increasing 600,000 annually, but only 400,000 were consumed. There were about 52 million acres for grazing that might have sustainably supported about 1.8 million cattle. But 3.5 million sheep, cattle, and horses were consuming the grasses. As these pressures mounted, disaster struck. Floods in the winter of 1861–1862, followed by the 1863–1864 drought, followed by grasshoppers decimated forage. The drought killed about 800,000 horses, cattle, and sheep. Some counties lost between two-thirds and three-quarters of their animals. The profligacy of California pastoralists created economic problems that compounded the devastating blows nature had inflicted. Land prices declined, and even with all the mortality, beef markets were still saturated. Rancheros unloaded their stock and land to remain afloat. The era of the rancho seemed gone (Isenberg 2005, 123–127).

Pastoralists adapted to these environmental and economic challenges in two ways. Some switched to sheep-raising. Because sheep required less labor and had a less restricted diet, costs were cheaper. In the 1860s and 1870s, production increased dramatically. So did the deleterious ecological effects of overgrazing. The other common adaptation was a switch from extensive

Immigrants camp with wagons and cattle at Eagle Lake (in the Sierra Nevada mountains) on August 14, 1859. Thousands of farmers and gold-seekers traveled overland with domestic animals, hastening the agricultural transformation of California following the Gold Rush of 1849. (Daniel A. Jenks/Library of Congress)

stock-raising to intensive stock-raising. One of Southern California's most successful ranchers, Abel Stearns, established a feeding ranch in northern California to which he drove cattle. This strategy moved toward intensive stock-raising, which eventually became a primary economic tactic. Pastoral intensification demanded enclosing the range, effected by an 1872 law from the California legislature. Enclosure displaced Native people, such as the Modoc tribe in northern California. Such developments reflected the interconnected environmental and social changes at work (Isenberg 2005, 128–62).

Although the interior Northwest did not have the same type of history as California's extensive cattle-ranching, pastoralism there soon boomed. Urban development in places such as Boise, Idaho and Walla Walla, Washington supported mining and prompted agricultural development. Cattle followed miners to the interior; sheep followed cattle; and ecological change accompanied pastoralism wherever it went. In the 1860s, livestock were the only product on the Columbia Plateau (Meinig 1995, 219). Rather quickly, the livestock industry increased from simply supplying miners to supporting a broader market, especially after transcontinental railroads arrived in the interior by the 1880s. Ranchers discovered the nutritious value of native bunchgrasses, such as bluebunch wheatgrass and Idaho fescue, that Native American horse herds had consumed for 150 years. Livestock-raising dramatically altered these grasslands because of intensifying pressures. By the 1880s, ranchers and travelers reported declining grasses, increasing weeds, and pesky thistles. Animals cropped native perennials so closely and so early in the season that the plants could not flower or seed; thus, they could not reproduce. Contributing to this were the races among herders to get their animals into rangelands first. Consuming grasses also meant reducing fuels and fires faded temporarily and brush, like sagebrush and juniper, invaded the grasslands. In addition, the disturbed soil allowed exotic weeds and annual grasses like cheatgrass to replace the more nutritious and hardy bunchgrasses. Paradoxically, this invasive cheatgrass provided ideal fuel for fires, as it colonized faster than native bunchgrasses. When fires struck, then, cheatgrass proliferated; when fires did not strike, brush invaded. Either way, range quality for grazing declined. Finally, in southeastern Oregon and other places where overgrazing prevailed, water sources frequently dried up, and forage near these scarce water sites felt severe pressures. Ranchers paid little attention to those ecological changes until they were well under way and nearly irreversible. To them, grass was simply another commodity to exploit through their animals (Langston 1995, 71–78; Robbins 1997, 142–75).

After cattle roamed through the open range, sheep arrived. Starting a sheep operation took less capital than cattle ranching. Sheep also increased more rapidly than cattle, because exporting wool did not require slaughtering the animals. Consequently, by 1890 on the Columbia Plateau, sheep numbers were four times as great as cattle; by 1900, they were eight times greater, and there were more than one million sheep. Cattle grazing transformed the grasslands in ways that hurt cattle but often helped sheep. Sheep ate the weeds that invaded alongside cattle, and sheep could crop grasses closer to the ground than cattle. Also, more so than cattle grazing, sheep raisers practiced transhumance, the seasonal movement of sheep between mountain grasslands and lowland ranges. Transhumance was often at odds with American sacred notions of private property. The rapid increase of sheep, the declining quality of the range, the constant migration of herds, and the likelihood that sheep-raisers and sheepherders were immigrants or American Indians all made the sheep industry frequent targets of cattle ranchers and farmers (Langston 1995, 71–78; McGregor 1982, 24–29; Robbins 1997, 163).

The Oregon Sheep Shooters Association furnishes the most dramatic evidence of the animosity between these two pastoral traditions. This group organized in the 1890s to keep sheep out of the public domain in the Cascade Mountains. One representative of the sheep shooters wrote to the *Oregonian*, the region's leading newspaper, explaining that when sheep herders violate range traditions, "our executive committee takes the matter in hand, and being men of high ideals as well as good shots by moonlight, they promptly enforce the edicts of the association. . . . Our annual report shows that we have slaughtered between 8,000 and 10,000 head during the last shooting season and we expect to increase this respectable showing during the next season." When woolgrowers in Oregon offered rewards for information leading to the identity of the sheep shooters, the association responded: "We have therefore warned them by publication of the danger of such action, as it might have to result in our organization having to proceed on the lines that dead men tell no tales. This is not to be considered as a threat to commit murder, as we do not justify such a thing except where the flock-owners resort to unjustifiable means of protecting their property" (quoted in Wilkinson 1992, 85–86). Hostility so explicit was rare, but it revealed the underlying cultural and environmental competitions within western pastoralism.

Both cattle- and sheep-raisers contended with factors beyond their control, even as they struggled with the ecological changes their pastoral mode created. Besides declining forage, predators, poisonous weeds, and livestock diseases threatened flocks. The climate at times proved even worse. The 1880s included several severe winters. Because few livestock raisers put up hay, they faced harsh winters with little to feed their animals. On the Columbia Plateau, two years of drought followed by extremely cold winter weather in 1889–1890 caused livestock losses between 60 and 90 percent. Such experiences encouraged stock-raisers to begin buying or harvesting hay and turned their extensive operations into more tightly managed intensive enterprises (McGregor 1982, 35–39).

Economic changes also influenced sheep and cattle operations. From the 1860s when large herds appeared in the interior Northwest to the 1880s when railroads connected the region to larger and distant markets, herds grew and grew while feeding on the public domain. Responding to market demands and depleted grasslands in California, large corporate cattle operations relocated to southeastern Oregon as did some family ranchers (Robbins 1997, 159). The presence of corporate ranches since the 1860s and the expanding rail networks points to agriculture's industrial connections from near its beginning. As a result, the fortunes of livestock-raisers depended on market demands, freight prices, and commodity values, not to mention the region's ecology. As William G. Robbins noted, Oregonian ranchers "confronted both an unpredictable market and an unpredictable climate" (1997, 162). The combination proved disastrous in the 1890s when a drought and depression corresponded. The national depression beginning in 1893 bankrupted many operations in the Northwest and much of the country. The federal government and corporate landowners responded to the disastrous 1890s with a series of reforms in land law and resource uses in the early twentieth century that marked the beginning of a new era.

California's Industrialized Agriculture

As with so much else in American environmental history, California agriculture was an exemplar, an exception, and an innovator. With roots in the rancho past, pastoralism in California was much larger than Northwestern counterparts and quickly adapted to an industrial ideal typified by the enormous company of Miller & Lux, the founders of which the historian David Igler aptly characterized as "industrial cowboys" (2001). The history of late nineteenth-century California agriculture reveals important intersections of agriculture—pastoral and hydraulic—legal and governmental innovations, markets, and environmental change. Pastoralism slowed the development of other agriculture even as it stimulated irrigation development in service of large herds. A variety of reforms and market changes eventually gave rise to irrigation farming. Simultaneously, then, enormous ranching operations, wheat farms of immense size, and smaller irrigated specialty farms developed. Because of supportive legal structures and market connections forged by mining booms and railroads, California agriculture expanded early and did so on large scales; thus, the environmental changes caused by industrialized agriculture portended continued change well into the twentieth century.

Two nineteenth-century German immigrant butchers, Henry Miller and Charles Lux, became the leaders of one of the century's greatest industrial enterprises. By careful manipulations, their firm acquired 1.25 million acres and more than 100,000 cattle in three states, and by 1900 the company was the only agricultural company to crack the nation's top 200 companies. It achieved such scale by applying industrial methods, such as vertical integration of production and marketing and horizontal consolidation of resources. More specifically, Miller & Lux gained rights to water and land over vast areas, concentrated in California's Central Valley (Igler 2001, 4–5).

Although one can read the history of Miller & Lux simply as an economic success, that would slight the centrality nature occupied in the narrative. From the beginning, nature's independence shaped Miller & Lux's operations. Fearful of the recurring droughts and floods that plagued the 1860s, Miller & Lux began acquiring land at critical places. Their choices were designed to meet two concerns—one, the company needed to create an economically viable land-based system; and two, it needed to stockpile enough resources so that it could withstand nature's vagaries. The firm focused on finding land close to urban markets (e.g., Rancho Buri Buri), near transportation nodes (e.g., Rancho Las Ánimas), and near the Central Valley's abundant grasslands (e.g., Rancho Sanjon de Santa Rita). California offered some advantages to accumulating large-scale lands, because Spanish and Mexican land grants were large, and by the 1850s, many were mired in debt and thus easy for well-capitalized Americans to acquire. Indeed, some have argued that California was "uniquely suited" for corporate agriculture and that model dominated early in the state's history and set subsequent land-use trends (Pincetl 1999, 9). About half of Miller & Lux's lands originated in land grants, and federal and state laws were flexible enough for large corporate bodies like Miller & Lux to acquire vast acreages, especially because the firm developed strong networks among local land agents, government officials, and San Francisco financiers. Other Californians did the same thing. Indeed, an 1872 California agency reported that 100 individuals held titles to nearly 5.5 million acres—lands larger than Massachusetts—and thirteen landowners in the San Joaquin Valley held holdings averaging 238,464 acres. With this strategic and growing land base, the company felt secure in its economic position and set out to reshape the hinterlands to maximize cattle production (Igler 2001, 35–71; Opie 1987, 135–139; Pincetl 1999, 4–9).

Miller & Lux understood that besides a broad land base for cattle production, it could generate even greater wealth by controlling water rights and developing irrigation. To that end and to further prepare for contingencies such as droughts, in 1878 Miller & Lux acquired the San Joaquin and King's River Canal & Irrigation Company (SJ&KRC&ICo) to ensure that it maintained its land and water rights (Igler 60–80). However, acquiring and maintaining water rights did not prove easy, as demonstrated in the struggle that resulted in the California State Supreme Court's ruling in *Lux v. Haggin* (1886). In California, riparian water rights developed along with prior appropriation leading to inevitable conflict. Unlike prior appropriators, all riparian owners possessed equal rights, providing they did not impair others. They did not need to use water to keep their rights. The courts and legislature tried to use both approaches and to view them as complementary rather than competitive. Gradually, state authorities began to allow some river withdrawals for irrigation, because irrigation proved necessary for successful farming. Such practices were untenable with riparianism. A burgeoning population caused conflict as riparian sites diminished and many firms and individuals sought monopolies of land and water (Hundley 2001, 85–87).

As irrigated agriculture developed, large ranches came to be seen as major obstacles to keeping irrigation flowing because they possessed riparian rights.

Furthermore, droughts accelerated the collision between riparian and prior rights. Meanwhile, the smallest farmers believed only the appropriation doctrine could help them. In the 1870s and 1880s, then, many Californians came to call for a change in the system of riparian rights (Hundley 2001, 87–93).

These abstract principles, agricultural trends, and environmental changes converged along the Kern River in the southern San Joaquin Valley's Tulare Basin. The place abounded in wetlands and swamps—wastelands to farmers' eyes. As historian David Igler has explained, farmers "viewed their task in terms of bringing order and productivity to the soil" (2001, 92–93). Historian Donald Worster characterized such work as nearly a religious impulse to nineteenth-century Californians: "The future of California required a strenuous work of creation, redeeming the land from its desolate condition, making it through farming what God had in mind for it" (Worster 1985, 97). So farmers and their employees reclaimed these lands by draining wetlands, simplifying ecosystems, and irrigating new fields. Miller & Lux began its reclamation project along the Buena Vista Slough, building canals and levees to engineer this landscape into a productive one that supported the company's industrialized agricultural enterprise. Meanwhile in the same basin, James Haggin and his Kern County Land Company acquired more than 400,000 acres and bought up most of the prior rights along the Kern River. In the meantime, Miller & Lux lost nearly 10,000 cattle in 1877 because the slough ran dry, in part because of drought and in part because of Haggin's upstream diversions. In response, Miller & Lux asserted riparian rights to the Buena Vista Slough to guard against future diversions as well as to allow the company to divert the slough for its own purposes. Before governmental investment in irrigation, only highly capitalized individuals or groups like these could afford such massive schemes to re-engineer nature (Hundley 2001, 94–96; Igler 2001, 92–103; Worster 1985, 103–104).

The lawsuit turned on the interesting question of what constituted a river. Miller & Lux claimed the Buena Vista Slough was a well-defined watercourse, for without such a definition they could not claim riparian rights. Haggin claimed the slough was indistinguishable from the surrounding swamplands. Ultimately, the California Supreme Court affirmed Miller & Lux's riparian rights. The court created what became known as the California doctrine. As historian Norris Hundley explained it, the California doctrine said that all private lands possessed riparian rights, but prior appropriation could supersede riparianism if an appropriator claimed the water before the riparian owner used the watercourse. Timing was everything, and this spelled doom for the small owners, who could not find available riparian areas left. In the end, the decision kept large landowners well in control and it reaffirmed that nature's highest

Kern Island canal head gate, near Bakersfield, in Kern County, California, ca. 1888. The control of flooding and irrigation in Kern County was imperative to industrialized agriculture's survival. (Carleton E. Watkins/Library of Congress)

duty was to serve human economies (Hundley 2001, 93–99; Igler 2001, 103–11; Pisani 1992, 34–35; Worster 1985, 105–08).

The decision had several results, including legal and political reforms and environmental changes. In the aftermath of *Lux v. Haggin*, it became apparent that monopoly, not the system of water rights, constituted the main problem. After all, both Miller & Lux and Haggin operated immense projects. A California state legislator, C.C. Wright, devised a new solution: the irrigation district. The Wright Act (1887) allowed communities, with a two-thirds majority, to form irrigation districts. These districts could tax all the local property and issue bonds to pay for land they condemned to provide irrigation for smaller landholdings. The irrigation district operated like a mini-legislature with its power to tax and condemn property. Moreover, it developed the first public irrigation projects in California to compete with the private systems developed by large owners like Haggin or Miller & Lux. Reformers expected, among other things, the tax burden to force speculators to sell. With the Wright Act, the water right rested in the district, not the individuals. Specialty crops increased and farm size declined slightly, suggesting some success with this political innovation. However, opposition to it meant court challenges and expenses, even as drought, depression, and poor farming techniques made the districts largely fail amid mounds of debt (Hundley 2001, 99–103; Opie 1987, 141–143; Worster 1985, 108–109).

Environmental legacies were less direct but more momentous. Throughout the larger Tulare Basin, reclamation efforts spelled ecological change and problems. For instance, Miller & Lux converted 1,200 square miles of wetlands, swamps, and small lakes into agricultural land. Without these, waterfowl lost habitat on their Pacific flyway, not to mention the fish or wildlife that depended on riparian areas. Irrigated fields of single crops replaced diverse riparian zones. Moreover, the agricultural methods increased soil alkalinity, and monocultural specialization made fields vulnerable to pests. This ecological transformation followed the nineteenth century's emphasis on farming, and it greatly simplified California's landscape in ways that ultimately threatened its sustainability (Hundley 2001, 98; Igler 2001, 111–121). Indeed, Igler concluded: "The environmental advantages gained through resource consolidation and landscape engineering relied on balancing the costs of ecological and social exploitation. Those material and ecological costs eventually became too high. The firm gradually collapsed upon exhausted soils, crumbling waterworks, and simplified ecosystems—products of the drive to engineer the landscape during the late nineteenth century" (2001, 178). Such a story was common throughout the Far West's industrializing agricultural landscape (Igler 2001, 175–178).

Mass production on the farms rapidly replaced self-sufficiency or small-scale market agriculture in California and spread throughout the West. Wheat fields were enormous, so large that teams plowed one field length a day and camped out until the next day to return the other direction. Industrial wheat dominated only temporarily, though, as soil fertility and prices declined. Growers then searched for higher-priced crops. Distribution technologies like railroads transformed consumption, making it much further from sites of production than before. Further exacerbating this distance was the urban locus of so much power that shaped the hinterlands. Such distances hid environmental and social costs from consumers, one of the most important shifts in American environmental history (Brechin 1999; Steinberg 2002, 177–178; Worster 1985, 99).

California led the West in irrigation. Governmental aid, legal innovation, and technologies combined to reorganize riverine ecologies and farmland. As much as irrigation engineers and farmers believed they controlled water for their own purposes, nature repeatedly demonstrated its ability to thwart human manipulation and plans. The irrigation ditches and fields merged human and natural elements. Finally, the irrigation systems emerging by 1900 promoted specialized agricultural products, especially in Southern California, and ultimately transformed markets and ecologies and redirected the region's environmental history. The irrigation projects by Miller & Lux or James Haggin represented important precedents.

INDUSTRIALIZING FISHERIES

Economic development and multiple migrations to the Pacific West changed Native fisheries significantly. Fishery production and habitat changes sharply reduced fish populations and hampered their reproduction. The sustainable relationship between human and fish populations suffered. Meanwhile, migrants to the region brought with them increasingly efficient technologies that took from the fisheries with amazing speed and in vast quantities. These technologies, just as in mining, required substantial capital. To recoup the investment, companies increased output to maximize profit. Such activities severely strained nature's ability to survive the human onslaught toward select species. Within this larger fishery decline, smaller pockets of less intensely industrial fishing practices prevailed among Chinese immigrants and American Indians. These groups certainly participated in the developing commercial economy and responded to the ecological changes at work in the fisheries. As vulnerable members of society, though, they also faced discrimination in their practices. These various factors combined to create complex changes in fishing that reflected economic growth, ecological change, and cultural contests at the end of the nineteenth-century Far West.

External Economic Factors

The new fishing regime itself affected what fish were taken and how they were harvested, but other economic activities intensely affected the Far West's fisheries and perhaps spelled the greatest change. A survey of the effects of other industries helps explain the ecological context and vulnerabilities within which the industrial fishery operated. The fur trade was the first economic activity to affect salmon runs in the Northwest. Although beaver dams could obstruct upstream migration, they generally could be breached by freshets. Moreover, the beaver ponds provided excellent habitat for juvenile salmon; thus, the fur trade's end because of overtrapping and scarcity meant fewer refuges for young salmon (Taylor 1999, 45–47). These changes were minor, however, compared with the changes mining wrought.

Not long after Marshall's discovery of gold, periodic strikes throughout the Northwest sustained various booms for nearly two more decades. Mining, especially by hydraulic methods, negatively affected fisheries. In the process of gaining gold, miners also extracted gravel, sand, and limestone from streams, removing important elements of salmon habitat. Hydraulic mining technology, exported from California, required abundant water and extensive manipulation of it. Miners built dams, ditches, flumes, and pipes to route water to high-powered nozzles directed at hillsides to wash away all but the gold. This supporting apparatus was immense; one ditch, the Eldorado Ditch, ran 136 miles long when completed in 1873. These hydraulic systems hurt fish; in the descriptive words of environmental historian Joseph E. Taylor III, fish "were sucked into ditches and blown through nozzles" (1999, 51). Diversions took salmon from their spawning streams, lowered water levels, and raised water temperatures. Meanwhile, debris occasionally blocked upstream migrations. In Idaho, the Boise River lost its salmon runs in 1865 to such a blockage, while Idaho's Bruneau River and Oregon's Grande Ronde were lost to salmon at various points in 1892. Finally, the mining camps and cities that grew up placed pressure on water and salmon habitat because of consumption and waste disposal. Such a combination of factors greatly pressured fishery sustainability (Taylor 1999, 47–51, 57–59).

Meanwhile, agricultural development further degraded the fishery. Agriculture in the Northwest increased impressively. For example, between 1850 and 1870, dairy cattle increased fivefold; wheat production more than tenfold; and sheep increased more than twenty times. Livestock concentrated in river valleys, and by staying there, they consumed forage prodigiously and prevented effective reseeding. Their increasing numbers also compacted the soil, decreasing its capacity to retain water and thus increasing runoff and erosion. Moreover, livestock waste poured into the streams, harming salmon habitat. Like mining diversions, irrigation lowered water levels in streams and raised its temperatures, while salmon swam into irrigation canals without screens. Then, farmers and miners overappropriated water, exacerbating the already low water levels. In one dramatic example, the Yakima River in 1906 dropped from an average flow of 3,900 second-feet to a mere 105. Finally, the milling operations to support wheat growing and wool-growing industries put dams in streams without fish ladders. These agricultural impacts added another particularly widespread factor contributing to salmon declines (Taylor 1999, 51–55).

Just as mining encouraged agriculture, it spawned a significant logging industry. At the end of the nineteenth century, Oregon loggers cut nearly a billion board feet per year. Loggers initially cut in riparian areas and the nearby hills, destabilizing the soil, promoting erosion, and reducing cover that kept streams cool. The debris left behind from logging was vulnerable to frequent fire, and these fires were hotter and more damaging than fires in most anthropogenic fire regimes. Each burn made it more difficult for the soil to retain water, so erosion increased. Other logging methods hurt habitat. To get logs to mills, loggers floated them down rivers. They frequently jammed, which blocked fish passage. The common practice of using dynamite to break apart log jams, besides not being terribly effective at unlocking jammed logs, predictably harmed streambed habitat. Also, loggers built splash dams to regulate flow so that they could use giant flushes to push logs downstream. These splash dams not only blocked fish passage, but when flushed, the increased water volume also scoured the streambeds and destroyed habitat. By hurting and blocking upstream habitats, such activities also concentrated salmon spawning downstream. At the final step of processing, sawmills contributed to problems. They used water power, and the dams did their usual damage. Moreover, mills' sawdust "covered streambeds, smothered plants, and consumed oxygen in the water. Incubating eggs and alevins literally suffocated" (Taylor 1999, 57). Logging represented the final economic activity that harmed salmon runs to a high degree (Taylor 1999, 55–57). All of these effects were essentially external to the fishing industry itself, while deeply affecting it.

Internal Economic Factors

When commercial fishing and processing arrived in the Pacific West, it vastly changed the economic and ecological nature of fishing. At first, New Englanders came and developed the salmon canning industry. By 1853, they located a ready market in Australia. Hume, Hapgood, and Company formed in 1864 and packed salmon for three years in California before the runs became less profitable and they moved north to the Columbia River. Historian Arthur F. McEvoy likened this mobile canning industry to swidden farmers, who cycled through various fields, slashing and burning before moving on. Indeed, McEvoy explains that the decline was the fisherman's problem in classic form. Fishers would not leave fish in the water for fear a competitor would harvest it. Consequently, too few salmon remained to spawn. "Key to the salmon fishery's self-destruction," McEvoy wrote, "was the fact that, to the market, a fish was a commodity with a price set by the specifically human, short-term functions of supply and demand" (1986, 72). Such market thinking did not value the biological functions and value of fish. From an ecological perspective, salmon work and gain nutrients from the environment, while fishers transfer the results of salmon's work to humans for consumption. By using market logic and ignoring ecological function, fishers and canners worked together and devastated one stream after another (McEvoy 1986, 69–73).

The dynamics McEvoy described for California largely reproduced themselves in the Northwest, including some of the same canneries. The production of Northwestern canneries was impressive and increased quickly. For example, in 1866, the single cannery on the Columbia River packed 272,000 pounds of salmon; by 1875, there were fourteen canneries that packed 25 million pounds; by 1884, it was up to 42 million pounds. Throughout the 1870s and 1880s, Columbia River salmon catches were routinely more than 30 million pounds (Taylor 1999, 63, 122). Although the total numbers were comparable to those of the indigenous fishery, the fishery's broader ecological context was much different and thus less sustainable.

Moreover, harvesting operated under much different principles. Canneries concentrated the harvest in time and space, reducing the time fishers took salmon from nine to four months and focusing largely on the lower Columbia only. Technological changes introduced in the 1870s (e.g., fishwheels and fish traps) also increased the pressure. Indeed, few factors limited harvesting. Taylor explained that

> only the technical limitations of harvest, preservation, transportation, and markets constrained the industrial fishery. Capitalism freed fishers from the governing influence of ceremonies and taboos, and fishmongers harnessed ships and, after 1883, railroads to create a global market for salmon. Quite suddenly the East Coast, Great Britain, and Europe consumed more Pacific salmon than northwesterners (1999, 64).

These shifting economic and technological relationships changed the ecological strain salmon confronted. Market forces dominated. When canneries established quotas to allow them enough time to process caught fish, fishers high-graded their catches. That is, after meeting the quota, fishers kept catching fish and replaced small ones with larger ones. Thus, reported figures of canned salmon underestimate the total fish caught. The market structured the industry this way. Without government regulation, canners produced more and cut costs as they tried to stay competitive, while fishers caught more to compensate for lower wages (Taylor 1999, 64). Although fishing matured as an industrial practice, it declined in the late 1880s in large part because of canners' and fishers' success.

But human economic activity cannot be blamed entirely for salmon declines. Nature also contributed. Climatic conditions mattered significantly when an El Niño arrived in 1877 and again in the 1890s. El Niños warm Oregon winters and make them wetter, while summers are drier. The warmer ocean water limits many of the organisms salmon feed on, and it moves some fish north to salmon territory at the time when young salmon are reaching the ocean and feeds on them. As a result, "Runs plummet, size decreases, and fecundity declines, but the effects are temporary" (Taylor 1999, 106). Contemporaries blamed fishing, but climate may have been a controlling factor for these episodic declines. Similarly, repeated floods in the 1870s and 1880s, in part climatic and in part resulting from environmental changes in western watersheds, damaged spawning grounds. Conversely, droughts also harmed spawning, as streams went dry and salmon could not reach them (Taylor 1999, 65–66). Thus, nonhuman factors certainly exacerbated the more typically cited problems of market-induced environmental changes.

In California, too, changing predation patterns and new sources of pollution altered fisheries. For example, the abalone fishery could sufficiently support a commercial harvest only because sea otters had declined to negligible populations because of the fur trade. Chinese immigrants dominated the abalone fishery with methods transplanted directly from China. So effective were those methods that they largely stripped abalone from the coast by the 1880s. Ironically, that may have helped other fisheries. An abundance of abalone meant heavy grazing of kelp, abalone's main source of sustenance. Kelp is important habitat for other fish. So, getting rid of abalone allowed kelp to grow, which would help other fish. Chinese harvesting may have helped some near-shore fisheries rebound in the 1890s (McEvoy 1986, 75–76, 81–82). Meanwhile, industrial and municipal pollution damaged fisheries. Eutrophication smothered marine life in the 1890s as a byproduct of the new fruit-packing industry present along San Francisco Bay. In addition, increased siltation because of mining and urban development led to shrinking marshlands in the delta. Besides this loss of habitat, the tides were smaller, which lowered the nutrients and energy that supported the estuary's biological functions (McEvoy 1986, 86). Thus, a series of ecological changes significantly influenced the fisheries' economic viability.

Compounding the dynamic nature of the fisheries was the intersection of ecology and economy with various immigrant and Native cultures. In California, for example, the 1880 census indicated that 92 percent of fishers were immigrants, one-third from China. Fishers remained loyal to both the industry and the methods and markets of their cultural group. These immigrant fishing industries were diverse and occupied only specialized economic niches. Each one was virtually isolated from the others, using unique methods and selling largely in ethnic markets. Each immigrant fishery seemed to successively fall on hard times, some because of ecological changes and some because of legalized discrimination. For instance, a series of laws effectively prevented Chinese from participating in the commercial fishery. In some cases, too, violence accompanied discriminatory laws. Ultimately, the smaller immigrant groups were unable to compete with the larger organizations that were developing by the turn of the twentieth century (McEvoy 1986, 65-92).

At the same time, American Indians in northern California and in some parts of the Pacific Northwest continued to fish, sometimes commercially. The Karok, Hupa, and Yurok of California withstood Euro-Americans much better than other California tribes. They were effective enough to have a strong part of the commercial market just past the turn of the twentieth century (McEvoy 1986, 51–62). Further north, Euro-Americans deployed state power in the 1850s and 1860s to largely exclude American Indians from their traditional fishing grounds, despite the fact that Native communities maintained their right to fish in the treaties they signed in the 1850s. It proved more difficult, but Natives did still fish. Some groups, in fact, became active in the commercial fishing and canning industries (Taylor 1999, 134–137). The Lummi tribe in northwestern Washington lived in an ideal location to capture returning Fraser River salmon and to work in the canneries developing in Puget Sound. The Lummi took advantage of this fortuitous situation, but they quickly found themselves excluded from the industrial fishery for two reasons. First, Chinese laborers worked in canneries at lower wages and in conditions more favorable to cannery owners. Second, and more important, new fishing technology proved too expensive for most Lummi to purchase. Fishing technology mechanized to save labor and to take greater quantities. Gillnets and fish traps, especially, displaced Native fishers and captured much more fish for the rapidly industrializing enterprise. The story of the Lummi, characterized by displacement by expensive technology, represented a common story for Native fishers of the Northwest (Boxberger 2000, 35–60).

By 1900, fishing had been thoroughly industrialized as an economic sector. In combination with the ecological influences of various other industrial sectors, commercial fishing and canning devastated fish populations and disrupted cultural subsistence patterns. The tremendous decline in salmon runs, recognized in the mid-to-late nineteenth century, tended to be blamed ubiquitously on "over-fishing." Such a censure focused on fishers and misunderstood the much broader social and ecological contexts. Enormous cultural and economic changes hurt the habitat of salmon and other fish while fishing continued at the same rate. Of course, disaster for the fish and the industry followed.

A Hupa fisherman fishes from a platform ca. 1923. The Hupa remained active subsistence and commercial fishers into the twentieth century, longer than many Native American tribes. (Edward S. Curtis/Library of Congress)

INDUSTRIALIZING FORESTS

As with so much else, the Gold Rush stimulated the exploitation of forests. This proceeded locally (i.e., adjacent to the mines) and regionally (e.g., California's redwood forests and the Northwest's Douglas fir forests). Miners demanded wood for construction and fuel. Forests located adjacent to the mines quickly fell to the axe, facilitated by miners who frequently found work as timber cutters as the easy placer deposits played out. This transition of miner to logger encouraged interest in developing the Far West's timber resources commercially. Markets were readily available in mining camps and growing cities such as San Francisco and Portland. Moreover, shipments across the Pacific promised profits. Investment in industrial logging began slowly and only improved as new technologies proved capable of generating substantial profits. Just as in the Gold Rush and salmon canning, greater investment prompted intensified exploitation to recoup costs. The cycle damaged forests and markets. Furthermore, forest ecology affected far more than simply trees. Water quality and quantity, along with soil conditions and wildlife habitat, were central to forests' ecological function. By the turn of the twentieth century, the Far West's forests were so decimated that a consensus emerged that forest conservation must become a national priority.

California's redwood forests furnish a first example. The California coast redwood *(Sequoia sempervirens)* is the world's largest tree. One tree yielded enough wood to build a large building. Moreover, redwoods resist fire, insects, and decay—beneficial ecological and economic qualities. The rainy climate on the coast allows male pollen and female seed cones from different branches to pollinate easily and fall to the ground to resprout profusely. Redwoods' size, endurance, and prolific reproductive capacity make them amazing organisms, but they are only found in a narrow strip along the coast because the necessary soil and climate requirements exist only there (Isenberg 2005, 79–81).

Late nineteenth-century entrepreneurs thought timber was nearly inexhaustible and set out to profit from cutting the massive redwoods. By 1900, a third of the coastal forests had been cut (Isenberg 2005, 77). The Homestead Act (1862), the Timber and Stone Act (1878), and other generous land laws facilitated mill owners in acquiring vast timber holdings legitimately and fraudulently. Mill owners might pay their employees to acquire such land and sell it at a fraction of its value, for example. The federal government also allowed cutting on the public domain, essentially providing a free product to mills. Experienced lumbermen from Canada and the American Midwest relocated in the 1850s and built northern California's first commercial sawmills. For these men accustomed to midwestern pine forests, the redwoods' size was astounding and

Loggers pose with a felled sequoia in Humboldt County, California, ca. 1905. The sequoias of this stand are now part of Humboldt Redwoods State Park, which was established in 1921. (Library of Congress)

offered enormous yields. However, the very size that made the redwoods so attractive slowed development because normal saws could not cut the trees. Furthermore, once cut, they were too big to move to the mills. And then sawmills could not process them. Thus, the industry needed technological innovation, and that required capital. But capitalists wanted assurances of profitability (Isenberg 2005, 81–82).

The evolution of logging technology in California's coastal redwood forests followed a similar course as those in other forested regions throughout the Far West. Forest industries began with little more than animal and water power, human muscles, and axes or crosscut saws. These technologies and labor sources created significant limits. To float logs to mills, forests had to be cut near large enough streams and rivers. Of course, human and animal strength could only do so much. The first innovation came in the 1850s with steam power in the mills; steam engines generated more power than the existing waterwheels. Steam power in California necessarily depended on wood, as coal was not readily available, creating an ironic cycle. As historian Andrew C. Isenberg noted: "[U]sing wood, usually Douglas fir, as fuel to power steam engines, loggers and mill workers increased their capacity to consume still more trees" (2005, 84). Such a relationship demonstrates the sometimes tortured logic of markets when considered from broad environmental perspectives.

Lumber production increased dramatically with the next two big technological changes: the steam donkey and the railroad. Both technologies expanded the range of timber cutting, allowing the industry to move farther from rivers and accelerating deforestation. Donkey engines developed in 1882 by John Dolbeer in California and replaced animal power, and they furthered forest industrialization. Basically, the steam donkey was a steam-powered winch, or winches, set on skids from which long cables could be hooked onto a log and yarded the logs to a central point for further transportation or to skid roads. The steam donkey halved costs by reducing labor needs and costs. It also allowed work to continue in hot summer months and in cold, wet months, because they made logging operations independent of animal needs. Finally, steam donkeys permitted loggers to cut trees that had previously been too massive to move. A lumberman on the Washington coast raved about steam donkeys: "When one considers [that] . . . they require no stable and no feed, that all expense stops when the whistle blows, no oxen killed and no teams to winter, no ground too wet, no hill too steep, it is easy to see they are a revolution in logging" (quoted in Robbins 1997, 214). It truly was a revolution and rapidly dispersed throughout the Far West (Bunting 1997, 147; Isenberg 2005, 86–87; Rajala 1998, 14–17; Robbins 1997, 213–215; White 1992, 96).

Logging railroads furthered industrial incursions in the forests. Railroads ended loggers' reliance on floating logs to mills. Combined with the steam donkeys, they greatly increased production capacities in coastal forests. In Washington, Oregon, and California, production increased from 600 million board feet in 1879 before the invention of the steam donkey and before most railroads arrived in the woods to 2.6 billion board feet in 1889 and 2.9 billion board feet in 1899. With this increased production and railroad connections, loggers shipped these forest products to markets in the American Midwest. This process extended Far Western lumber products east for the first time, as well as west to Pacific markets that had been tapped since the 1850s. Like the steam donkey, the railroad liberated logging operations from the seasonal and geographic obstacles that once limited them. Finally, railroads were usually controlled by outside investors, so local control declined as Western forests often served midwestern capitalists. Railroads improved the quantity of timber production, but they also subjected forest ecologies throughout the Far West to profit imperatives from outside the immediate region (Isenberg 2005, 87; Robbins 1997, 215–216; White 1992, 85–86; Williams 1995, 228–232).

Because technological innovation was not free, entrepreneurs demanded more cutting to meet profit expectations. Many historians have explained that these new technologies impelled investors to step up production to meet market expectations. Unfortunately, this logic hurt profitability because markets were soon flooded with timber. Even before technological innovations in the California redwood forests, in 1871 lumber companies created a cartel, the Redwood Lumber Association, to try and stabilize the market and increase profits. They set annual minimum prices and maximum production for their mills. However, producers still could not control the San Francisco market and its wild fluctuations. A depression in 1873 broke the cartel as prices dropped dramatically. By the 1880s a new cartel had again emerged to fix prices and reduce supply to the urban markets. As prices rose, though, individual mills left the cartel to try to maximize production and profit. The cartel lost its authority, and mills simply increased production, further destabilizing the market. Thus, mill owners plundered the forest. Similar stories were repeated in Northwestern forests. For example, certain tax policies in nineteenth-century Washington encouraged unrestrained cutting, as landowners had to pay taxes on the assessed value of standing timber. Thus, economic policies and practices throughout the Far West promoted deforestation and discouraged conservation (Isenberg 2005, 88, 94–97; White 1992, 97–98).

The ecological consequences of all this were severe and devastating. Logging the Western forests was wasteful. Loggers left the lower ten feet in stumps and much of the upper portions were also left. One contemporary expert estimated 28 percent of each tree was wasted. Perhaps up to 20 percent of coastal redwoods shattered when felled, making them useless for market purposes. Then, at the mill, millers did not use the entire log and simply burned much of the waste, perhaps as much as half the wood. Much of this waste remained in debris piles that were prone to damaging fires and that harmed forest regeneration. Also, the forest that did come back was often not composed of the same species. However, with trees removed, browsing animals increased to feed on the seedlings, shrubs, and forbs that grew. Browsers kept the next generation of some forests in check, which displaced animal species that demanded shade or closed canopies like martens or fishers. Deforestation caused the worst habitat changes for rivers. Trees stabilized soils and their removal increased erosion, hurting aquatic habitat for fish and changing the depth and

In 1918, the largest spruce sawmill of the era was in Fort Vancouver, Washington. All available fur resources had been expended by the mid-nineteenth century, so the Hudson's Bay Company turned to agriculture and logging. The U.S. Army took over Fort Vancouver in 1860. (Library of Congress)

channels of some streams. Cutting trees also exposed soil and water to increased sunlight, which raised temperatures often to levels too high for fish and other organisms. Deforestation, then, did much to change forest habitats besides simply removing trees (Isenberg 2005, 89, 97–98; White 1992, 88–91).

The technology also caused ecological harm. Steam donkeys nearly required clearcutting, for the donkey engine needed a clear ground to pull the logs over distances. Thus, loggers removed virtually all trees, when before mechanization they took only the best and left much standing timber. Without standing timber, reseeding and reforestation demanded much greater effort, an endeavor usually not undertaken because of costs in time and money. Furthermore, clearcutting drastically simplified ecosystems, which reduced the species living in forests and weakened the forest's biodiversity and stability (Bunting 1997, 148–149; Isenberg 2005, 88; White 1992, 107–108).

By 1900, a first round of industrial logging had swept through much of the Pacific West. Animal- and human-powered logging cut millions of board feet near rivers and the coast. When railroads and steam donkeys spread through the region, deforestation proceeded apace. West-side forests in the Northwest and California's redwoods shrank before the axe and saw. Interior forests, by comparison, were lightly touched by industrial logging, in part because railroads were somewhat slow in reaching into the drier interior pine forests. A federal source reported that by 1900 only 33,700 acres of merchantable timber had been cut from a total of 1,450,420 acres on the Cascades' eastern slopes (Langston 1995, 80–82; Robbins 1997, 230–232). This would change in the twentieth century, as various locations in Oregon, Idaho, and elsewhere became sites of industrial logging with its concomitant ecological costs. In its wake, observers worried about community and environmental sustainability, but those concerns were small compared with those who looked for new forests to open with rails and new profits to make.

Overall, this late-nineteenth-century process historian William G. Robbins has called "industrializing the woodlands" severely disrupted the region's ecology (Robbins 1997, 205–237). Just as in the Gold Rush, technological changes induced investment that demanded increased production. Ironically and troubling to many, expanded production glutted markets and profits declined. Rooted behind all of this was a vision that lost the forest for the trees (Isenberg 2005, 98). That is, entrepreneurs saw board feet, not a forest ecosystem, just as other industrialists only saw gold, bushels of wheat, crates of fruit, or canned salmon. Nineteenth-century resource economies constantly faced such problems and yet few argued for technological restraint or anything approaching ecosystem management. Fueled by technological changes and profit motives, capitalist imperatives changed forests, in some cases irrevocably.

CONCLUSION

A forty-niner who lived to 1900 would have experienced a dizzying pace of environmental change. That forty-niner would have arrived in a Far Western landscape undergoing ecological shocks caused by European contact. The subsequent environmental history, though, would make the initial changes seem comparatively minor. Forty-niners and those who followed them repopulated a region recently decimated by disease. With this new population came new economic and political beliefs and practices that changed the Far West's human relationship to the natural world. Barely present in the early nineteenth century, capitalism thoroughly dominated by century's end. Responding to the comparative lack of labor and capital in the region, capitalists exploited nature to fuel economic expansion. This often meant technical innovations that devastated the region's resources by reducing them to commodities whose only value was their market value. Ecological functions, if recognized, were seldom appreciated or protected. The Pacific West provided abundant resources that allowed this exploitation—rich gold deposits, fertile soils, plentiful fisheries, and superb redwood and Douglas-fir forests. Exploiting resources became easier, too, as railroads connected the Far West to American markets and shipping across the Pacific became more reliable. Capitalist logic and new technologies reduced minerals and fish, forests and soil simply to potentially profitable commodities. A forty-niner would have witnessed this change and may have been among those who, by 1900, recognized an impoverished environment and called for reforms in far westerners' relationship to the region's ecosystems.

REFERENCES

Aiken, Katherine G. *Idaho's Bunker Hill: The Rise and Fall of a Great Mining Company, 1885–1981.* Norman: University of Oklahoma Press, 2005.

Arrington, Leonard J. *History of Idaho: Volume I.* Moscow and Boise: University of Idaho Press and Idaho State Historical Society, 1994.

Beesley, David. *Crow's Range: An Environmental History of the Sierra Nevada.* Reno: University of Nevada Press, 2004.

Boxberger, Daniel L. *To Fish in Common: The Ethnohistory of Lummi Indian Salmon Fishing.* Lincoln: University of Nebraska, 1989. Reprint, Seattle: University of Washington Press, 2000.

Brechin, Gray. *Imperial San Francisco: Urban Power, Earthly Ruin.* Berkeley: University of California Press, 1999.

Bryant, Keith, Jr. "Entering the Global Economy." In *The Oxford History of the American West,* edited by Clyde A. Milner II, Carol A. O'Connor, and Martha A. Sandweiss, 195–235. New York: Oxord University Press, 1994.

Bunting, Robert. *The Pacific Raincoast: Environment and Culture in an American Eden, 1778–1900.* Lawrence: University Press of Kansas, 1997.

Duffin, Andrew. "Remaking the Palouse: Farming, Capitalism, and Environmental Change, 1825–1914." *Pacific Northwest Quarterly* 95 (Fall 2004): 194–204.

Fiege, Mark. *Irrigated Eden: The Making of an Agricultural Landscape in the American West.* Seattle: University of Washington Press, 1999.

Hundley, Norris, Jr. *The Great Thirst: Californians and Water: A History.* Revised ed. Berkeley: University of California Press, 2001.

Hurst, James Willard. *Law and the Conditions of Freedom in the Nineteenth-Century United States.* Madison: University of Wisconsin Press, 1956.

Igler, David. *Industrial Cowboys: Miller & Lux and the Transformation of the Far West, 1850–1920.* Berkeley: University of California Press, 2001.

Isenberg, Andrew C. *Mining California: An Ecological History*. New York: Hill and Wang, 2005.

Kelley, Robert. "Mining on Trial." In *Green Versus Gold: Sources in California's Environmental History*, edited by Carolyn Merchant, 120–125. Washington, DC: Island Press, 1998.

Langston, Nancy. *Forest Dreams, Forest Nightmares: The Paradox of Old Growth in the Inland West*. Seattle: University of Washington Press, 1995.

McEvoy, Arthur F. *The Fisherman's Problem: Ecology and Law in the California Fisheries, 1850–1980*. New York: Cambridge University Press, 1986.

McGregor, Alexander Campbell. *Counting Sheep: From Open Range to Agribusiness on the Columbia Plateau*. Seattle: University of Washington Press, 1982.

Meinig, D. W. *The Great Columbia Plain: A Historical Geography, 1805–1910*. 1968. Reprint, Seattle: University of Washington Press, 1995.

Milner, Clyde A., II. "National Initiatives." In *The Oxford History of the American West*, edited by Clyde A. Milner II, Carol A. O'Connor, and Martha A. Sandweiss, 155–193. New York: Oxford University Press, 1994.

Opie, John. *The Law of the Land: Two Hundred Years of American Farmland Policy*. Lincoln: University of Nebraska Press, 1987.

Opie, John. *Nature's Nation: An Environmental History of the United States*. Fort Worth, TX: Harcourt Brace College Publishers, 1998.

Pincetl, Stephanie S. *Transforming California: A Political History of Land Use and Development*. Baltimore, MD: The Johns Hopkins University Press, 1999.

Pisani, Donald J. *To Reclaim a Divided West: Water, Law, and Public Policy, 1848–1902*. Albuquerque: University of New Mexico Press, 1992.

Rajala, Richard A. *Clearcutting the Pacific Rain Forest: Production, Science, and Regulation*. Vancouver: UBC Press, 1998.

Robbins, William G. *Landscapes of Promise: The Oregon Story, 1800–1940*. Seattle: University of Washington Press, 1997.

Rohe, Randall. "Mining's Impact on the Land." In *Green Versus Gold: Sources in California's Environmental History*, edited by Carolyn Merchant, 125–135. Washington, DC: Island Press, 1998.

Schwantes, Carlos A. *In Mountain Shadows: A History of Idaho*. Lincoln: University of Nebraska Press, 1996.

Steinberg, Ted. *Down to Earth: Nature's Role in American History*. New York: Oxford University Press, 2002.

Taylor, Joseph E., III. *Making Salmon: An Environmental History of the Northwest Fisheries Crisis*. Seattle: University of Washington Press, 1999.

White, Richard. *"It's Your Misfortune and None of My Own": A New History of the American West.* Norman: University of Oklahoma Press, 1991.

White, Richard. *Land Use, Environment, and Social Change: The Shaping of Island County, Washington.* Seattle: University of Washington Press, 1980. Reprint, Seattle: University of Washington, 1992.

Williams, Michael. "The Last Lumber Frontier?" *The Mountainous West: Explorations in Historical Geography,* edited by William Wyckoff and Lary M. Dilsaver, 224–250. Lincoln: University of Nebraska Press, 1995.

Wilkinson, Charles F. *Crossing the Next Meridian: Land, Water, and the Future of the West.* Washington, DC: Island Press, 1992.

Worster, Donald. *Rivers of Empire: Water, Aridity, and the Growth of the American West.* New York: Oxford University Press, 1985.

Worster, Donald. *Under Western Skies: Nature and History in the American West.* New York: Oxford University Press, 1992.

5

RESPONDING TO INDUSTRIALIZED ENVIRONMENTS

Gifford Pinchot stood tallest among the conservationists. Not only did he enjoy a close relationship with President Theodore Roosevelt and head the U.S. Forest Service, but Pinchot also publicized the conservation cause in ways to rally the public. He likened conservation to a battle but saw it, too, as a moral issue. "The central thing for which Conservation stands is to make this country the best possible place to live in, both for us and for our descendants," he wrote in *The Fight for Conservation.* "It stands against the waste of the natural resources which cannot be renewed, such as coal and iron; it stands for the perpetuation of the resources which can be renewed, such as the food-producing soils and the forests; and most of all it stands for an equal opportunity for every American citizen to get his fair share of benefit from these resources, both now and hereafter" (1967, 79). Such idealistic prose belied the reality of conservation. Although conservation did attempt to limit waste and plan for renewable growth, it frequently failed and at times failed spectacularly. Nonetheless, conservation inaugurated a new day in the relationship of Americans and westerners with nature. As Pinchot declared in his conclusion, "Conservation has captured the Nation. . . . It has taken firm hold on our national moral sense, and when an issue does that it has won" (1967, 132–133). In this respect, Pinchot was correct. Conservation captured Americans' attention, nowhere more so than in the Far West. Although results seldom met the grand praise of Pinchot, conservation's great cheerleader, never again would issues concerning natural resources hide from public view.

CHARACTERIZING THE CONSERVATION MOVEMENT

The Progressive Era conservation movement transformed North America's political and natural landscape. Many historians have praised the movement and its accomplishments, although recently scholars have also criticized some conservation policies that excluded and harmed certain populations and their

Gifford Pinchot, the first chief of the United States Forest Service and governor of Pennsylvania, ca. 1940. (Library of Congress)

subsistence strategies. Moreover, new scholarship has highlighted the scientific shortcomings of many conservation programs. Despite these problems, the conservation movement represented a fundamental change in Americans' relationship with the natural world by placing a debate about appropriate use and management of natural resources prominently on national and regional agendas. The Far West helped lead this transformation. In part because the massive ecological changes of the previous decades prompted some of the greatest concern and in part as a site of numerous archetypal conflicts, this region's history symbolizes the dynamic ways Americans in the early twentieth century related to nature anew.

Like most social or political movements, conservation was diverse and remains difficult to summarize. Indeed, historian Donald Pisani aptly explained this challenge of characterization: conservation "stood for the denial of monopoly, democratic control over the allocation of natural resources, the prevention of waste, the protection of natural beauty, and the affirmation of the rights of future generations" (Pisani 1999, 123). Clearly, conservation encompassed multiple strategies and targets, making it a sprawling, almost uncontainable movement.

Generally, conservation was a government response to limit, or more accurately, to shape and control economic activities that seriously threatened natural resource sustainability. It was not about stopping the use or economic exploitation of nature; indeed, most conservationists advocated various methods to ensure full development of natural resources. Instead, conservationists sought to implement what historian Samuel P. Hays fittingly called the "gospel of efficiency" (1975). As the most prominent American conservationist Gifford Pinchot famously put it: conservation meant "the greatest good to the greatest number for the longest time" (Pinchot 1967, 48). To implement this vision, reformers established or enlarged a number of government agencies to promote and regulate the relationship of private industry to nature. Hays argued that conservationists desired decision making by experts in positions of centralized authority. Furthermore, he explained that the conservation movement did not represent an expansion of the democratic process insofar as experts were frequently appointed to their positions and instituted their changes without input from local populations. According to Hays, the decisions "came from a limited group of people, with a particular set of goals, who played a special role in society" (1975, 4). This strategy would create, the conservationists maintained, a more efficient and rational use of resources, allowing for long-term planning and full use of resources without significant diminishment of environmental qualities. To be sure, conservation embodied an optimistic, some might say naive, belief about the ability of experts to manage resources sustainably. And indeed, conservationists' plans seldom worked out as they anticipated (Hays 1975, 1–4).

Conservation developments within the Pacific West context elucidate the national story. From the initial efforts rooted in the late nineteenth century, conservation through the first half of the twentieth century developed in myriad ways. By the 1940s and despite occasional recalcitrance and constant problems, far westerners faced a conservation structure that governed much of their relationship to the natural world. In addition, new economic and techno-logical arrangements, such as urbanization, modified westerners' experiences with the natural environment. The several decades after 1900 thoroughly reor-ganized the region's ecology and the institutional structure surrounding nature. In short, another ecological revolution occurred, one driven by structural changes compelled by the late-nineteenth-century industrial assault on nature.

FISHERIES, GOVERNMENT SCIENCE, AND EMERGING CONSERVATION

The earliest efforts at federal conservation came in fisheries. The late nine-teenth century found intense pressures on fish populations in the Pacific West's river and coastal fisheries. Industrialized fishing itself, of course, harvested tons of fish every year, and changes in land use degraded the remaining fish habitat. Thus, stopping the rapid depletion of fisheries proved a difficult and multi-faceted task, as it demanded significant adjustments in several economic sec-tors. In 1880, the California Commissioners of Fisheries aptly declared, "Neither the fish, the public, nor the future of the business appears to have many friends. Any restrictions upon unlimited fishing and unlimited canning, while a fish can be found in the river, is looked upon as a personal injury, in-flicted by a meddlesome and tyrannical government" (quoted in McEvoy 1986, 93). Despite the complaint, that the state even had a commission on fisheries indicates the recognition that regulation was necessary.

Legal restrictions may have been the way to slow or alter fishery produc-tion patterns, but as we have seen, the law in the nineteenth century focused mostly on economic expansion, not restriction. Fish and wildlife were particu-larly vulnerable because they were common resources owned by no individual. While some ethnic groups developed their own regulation patterns, govern-ments gradually became interested in regulating the use of common resources to ameliorate the worst consequences of economic development (McEvoy 1986, 93–100). In the case of West Coast fisheries, regulating fishers or other sectors of the economy proved far less attractive a political option than managing the fish themselves. The solution became fish culture, or artificial propagation (Taylor 1999, 68–98). Government scientists developed a great interest in maintaining the fisheries and believed the technical solution of scientific management through artificial propagation could reverse the declining yields and thus satisfy the demands of the fishing and canning industries. However, this approach severed the fishery from its ecological context, and in so doing, was bound to fail by divorcing fish management from fish ecology. This example demonstrates that conservation often had more to do with political and economic contests than environmental sustainability.

Fishery conservation in the West began with scientific questions but evolved rather quickly away from those roots. In the mid-nineteenth century, concern over declining salmon and shad runs on the East Coast stimulated elite interest in fish culture. Although advocates such as George Perkins Marsh understood that broad ecological factors contributed to fishery losses, reform-minded scientists desired practical solutions that would minimize the need to restrict commercial fishing, canning, or other economic practices. As the West Coast industry faced a similar decline, officials predictably moved in the same direction—toward a fish culture solution. In 1870, Congress created the U.S. Fish Commission (USFC), while states formed their own commissions. These bureaucracies aimed to study the problem and so possessed broadly scientific missions. With respected scientist Spencer Fullerton Baird heading it, the USFC enjoyed immediate prestige and acted quickly. Baird hired Livingston Stone, a fish-culture evangelist, to establish a hatchery on California's McCloud River. Although enthusiastic, Stone misunderstood fishery science; for instance, he thought the Pacific salmon were *Salmo*, not *Oncorhynchus*. This fundamental problem symbolized the immaturity of the science. Moreover, the fact that the USFC did not care about the ignorance showed that the agency focused on production, not scientific research. Increasingly, government efforts applied science to commercial questions, while ignoring biological inquiries. Thus, hatchery production crowded out any other avenue to conservation, and ironically, no evidence showed hatchery production improving or even stabilizing salmon runs. Despite these significant and elemental shortcomings, hatcheries satisfied various economic constituencies and thus proved a political success (McEvoy 1986, 100–108; Taylor 1999, 68–98).

Meanwhile, states worked with and against the federal efforts. Indeed, the USFC relied on state regulation to coordinate with federal fish culture programs. However, states preferred not to regulate because that required taxes and pitting one group against another. In Oregon, furthermore, divisions within the state complicated efforts, as urban constituencies favored state intervention while rural populations did not. Despite the state schism, the Oregon legislature finally devised new regulations to control pollution, close seasons temporarily, and restrict certain types of gear. Predictably, the laws were widely flouted, and salmon runs continued to decline (Taylor 1999, 99–105).

Some changes in the state regulatory structure promised change. Although fishery regulation faced opposition from economic interests, legal innovations allowed California to assert greater regulatory power. An 1893 law gave the

Ichthyologist and first commissioner of the U.S. Fish Commission (USFC) Spencer Fullerton Baird. His efforts at fish culture rank him as one of the first conservationists to work on the West Coast. (Library of Congress)

state power to regulate the market for wildlife. This assertion of title constituted, according to legal and environmental historian Arthur McEvoy, "probably the most significant advance in the law of wildlife before 1900" (1986, 115). In addition, in *People v. Truckee Lumber Company* (1897) the California Supreme Court decided that the state enjoyed power to regulate water uses that damaged the opportunity to fish. Now, California could demand that industries explain how their use of water affected fisheries. Even though fishing constituted a minor player within the state economy, fish themselves are among the most susceptible species to ecological damage, serving as excellent barometers of worsening environmental conditions. Together, these legal changes bolstered the state's regulatory power and suggested the state recognized, however incompletely, that it possessed an interest in maintaining some ecological integrity over the state's fisheries, even if that included sacrificing some economic growth in other sectors. Although the stage was set for expanding regulations, California experienced little success (McEvoy 1986, 114–119).

Meanwhile, as runs plummeted, fishers turned on one another over access to the dwindling resource. In the early 1880s, fishers began fighting with one another. Historian Joseph E. Taylor described the situation as "devolving into a Spencerian struggle more concerned with identifying and excluding the politically weakest members of the fishery than with addressing the fundamental problems of overfishing" and broader ecological damage (1999, 133). Although the rhetoric at the time sounded a familiar conservationist tone about saving fish for "the people," campaigns simply rationalized why one group was entitled to more fish than another. Whites fought against Indians; citizens struggled against immigrants; fishwheel operators challenged gillnetters. As was true of many conservation battles that followed, experts' optimistic claims of imminent solutions belied the complexity of biological systems, environmental change, and economic influences (Taylor 1999, 133–165).

The first several decades of the twentieth century brought new problems to the fisheries and to conservation efforts. Commercial fishing used more technology allowing greater and more distant exploitation of fishery resources; urban constituencies gained power to shape conservation programs; and hydropower and irrigation advocates built destructive dams across streams. All of these developments affected fish ecology, as did the ever-dynamic climate.

Technological and economic developments that affected fish in many ways also produced urbanization. The economic expansion that fueled urban growth depended on several industries that harmed salmon ecology. Mining, lumbering, and farming all compromised salmon habitat, as they had done since their commencement in the nineteenth century. Now, too, industrial processing fouled many Northwestern streams; in fact, the Willamette River's oxygen levels dropped so low as to constitute a "virtual dead zone" (Taylor 1999, 179).

Urbanites also responded negatively to the increased taking of salmon, as they developed a penchant for angling. Following a set of elite social rules, anglers rejected the industrial fishery and embraced instead a code that defended their class and cultural interests. Thus, working-class immigrant fishers and cannery workers or Native fishers all fell short of the idealized angler who fished for pleasure, not economic necessity, and who did so in large part to reconnect with the seemingly natural world that was disappearing under an increasing urban footprint. These urban anglers acquired substantial political power and used it to shape fisheries management. One of their first actions was to encourage transplanting so-called sport fish, such as trout. Often these transplants preyed on juvenile salmon or eggs, but the exotic fish fulfilled anglers' vision of a better nature. As recreational fishing overtook industrial fishing in power within political and social realms, anglers used the Progressive reforms of the initiative and referendum to circumvent Oregon's rural-dominated state legislature for fisheries management. Unable to achieve all their reforms legislatively, urban interests launched a series of statewide initiatives that sought to close certain streams from commercial fishing, as well as limit the types of gear fishers, especially commercial fishers, could use. Not all initiatives passed, but it became clear that urban anglers possessed greater political power than commercial fishers and could do much to shape conservation politics (Taylor 1999, 166–202). This trend of urban power continued and accelerated through the rest of the century.

A combination of additional factors significantly modified California fisheries. The biggest technological change was fossil fuel–driven boats. Abundant fossil fuels in Southern California aided this technological transformation and hastened wider fishery exploitation. Between the turn of the twentieth century and 1925, California's harvest increased ten times with relatively the same number of fishers. This upsurge grew from using fossil fuels that allowed fishing further offshore. Thus, fishing space expanded. Also, the California fishery diversified during this era. Both sardines and tuna became new species marketed for a variety of products, including fishmeal fertilizer. Indeed, to profit, many fishers now relied on the by-product industry. Not surprisingly, this expansion demanded higher capital investments and greater specialization, which required ever-greater profits for backers to recoup their investments. Seemingly, no one thought about long-term sustainability for the fish or the fishing, canning, and by-products industries. Combined with the characteristically shifting climate, fish populations suffered under economic and ecological pressures. Most disastrously, during World War II, fishers caught sardines relentlessly as the War Food Administration increased the established catch by one-third to meet wartime demands, even as experts knew the sardine fishery had been declining already for nearly a decade. This allowable total doubled the believed maximum sustained yield, which caused a near-total collapse of the sardine fishery. Clearly, the economy's capacity to harvest and the regulators' willingness to relax rules far outpaced nature's ability to recover (McEvoy 1986, 123–166).

By the 1940s, after decades of weakening fisheries and nearly as long trying to solve that deterioration, fishery experts did not speak with a single voice. Indeed, both Arthur McEvoy's and Joseph Taylor's studies of West Coast fisheries demonstrate how rife with politics the scientific efforts to conserve fisheries were. Scientific fisheries research demonstrated clear evidence of declining fisheries in the 1920s, however, the political will to slow or stop harvesting remained small and essentially powerless. Similarly, regulatory efforts at state and federal levels yielded minimal results. Meanwhile, efforts at propagating fish continued apace but did so against a backdrop of increasing scientific skepticism. Scientists voiced several concerns. For example, hatchery fish limited the natural diversity of salmon species and runs. Indeed, fish culture set salmon on new evolutionary pathways in ways that did not benefit fish in the wild. Higher incidents of disease outbreaks, less genetic diversity, and new schooling behavior that weakened their protection from predation were just a few of the problems scientists discovered in hatchery-raised fish. Still, managers supported continued hatchery production as a way to mitigate the continuing problems economic expansion caused. Of course, it was a palatable political solution, just not an ecological one. In the fisheries case and others, conservation measures could not stem the tide against powerful and countervailing economic forces (McEvoy 1986, 156–184; Taylor 1999, 203–225).

The years surrounding the turn of the twentieth century produced new developments in the environmental history of West Coast fisheries. Ecological pressures on fish populations continued to flow from economic intensification within the fishing industry and other sectors. However relentless the decline seemed, these pressures did elicit some of the first conservation measures. With great confidence in scientific solutions, conservationists believed they could solve the crisis by simply producing more fish. This politically acceptable solution did not require them to sharply reduce anyone's economic activities. But it did not work as planned. The other direction conservation took turned one set of fishers against another. This approach also failed to address fundamental problems of unsustainable harvests and ecological changes. The interaction of these multiple factors created complex problems, but conservationists, and especially the public, wanted simple solutions. In many ways, this experience generated a template for other ecological problems: experts researched and proposed solutions while those who depended on the industry proposed different solutions, leaving a confused morass of policies that failed.

CRUSADE FOR RECLAMATION AND THE ILLUSION OF CONTROL

Unmistakably, one of the biggest factors decimating fish habitat inland came from the irrigation and hydroelectric projects of the early twentieth century. These projects enjoyed widespread support, as they tapped into the desire to promote both rural and urban economic development and to transform nature to serve capitalism. Although American Indians had irrigated small portions of this region and mid-nineteenth-century Euro-Americans had developed some more extensive irrigation works with private capital, from the 1890s on, the federal government furnished capital, expertise, and control over the West's irrigation and hydroelectric projects. The hydraulic West is full of successes and problems, reform visions and corrupted results, alternating control and lack thereof.

The hydraulic West's institutional framework developed from a mixture of environmental circumstances and reformers' ideals. Except for western Washington and western Oregon, most of the Pacific West required irrigation for agriculture. The aridity and unpredictability of the climate demanded water storage and manipulation for successful farming. From an ideological perspective, Americans had long believed in the superiority of an agrarian economy and culture. Thus, finding a way to make the desert bloom like a rose, as advocates often said, became a cultural, economic, and political priority. Central to this vision was the belief that small, family farms would promote democracy. Also, westerners' faith in fully using nature meant putting river water to work and not wasting it by letting it flow to the ocean without it first watering a farmer's field or turning a hydroelectric turbine (Hundley 2001; Reisner 1993; Worster 1985).

Thus, armed with necessity and vision, politicians, boosters, and farmers pursued government policies that would provide avenues for irrigation development. In 1877, Congress passed the Desert Land Act, which provided 640 acres of land for an extremely cheap price to any settler who promised to bring water to the property within three years. This acreage was far too large to manage as a workable irrigated farm, however, and the law primarily benefited cattle ranchers and land monopolists who used it fraudulently to acquire vast landholdings (Pisani 1992, 89–90; Worster 1985, 103). Even had it not been used improperly, individual farmers were unlikely to be able to muster the financial and technical resources necessary to construct substantial irrigation works. The federal government tried to improve the situation with another reform, the 1894 Carey Act. This law granted desert states a million acres if they promised to reclaim them, and it guaranteed construction firms a profit. Most states did not bother implementing the Carey Act, and those that did failed almost everywhere, despite modest success in Idaho (Pisani 1992, 251–265; Worster 1985, 157). Despite the shortcomings, it put the federal government partially in the irrigation business. These efforts started some moderately successful reclamation projects, but the federal government substantially increased its commitment to reclamation development with the 1902 National Reclamation Act, also known as the Newlands Act.

The Reclamation Act was among the most important environmental initiatives of the era. The law emerged out of myriad concerns of the 1890s, including a demographic shift that favored urban and industrial interests much to the chagrin of westerners and other Americans who lauded rural values and economies, as well as concern over the lack of assimilation among immigrants. The Newlands Act thus faced a tall order: it needed to support American nationalism and revitalize the rural West while delivering water over vast distances to farmers' fields. To accomplish this, the U.S. Reclamation Service (created in 1902 and changed to the Bureau of Reclamation in 1923) would promote small, irrigated farms throughout the nation's arid regions. Funds gathered from sale of public lands would be reserved to finance the construction of dams, canals, and other requisite infrastructure for reclamation projects. (Farmers agreed to repay the costs later, a requirement that the federal government postponed and eventually forgave.) Further, because of the desire to promote family-based farms, the act limited the acreages of those farms to 160 acres, although the secretary of the interior, who administered the program, could establish even lower limits. Farmers must also reside at or near the irrigated land. Ideally, these restrictions would prevent speculation and absentee ownership. Frequently, though, officials either ignored those measures or exempted them. For instance, California's Central Valley Project and reclamation in the Imperial Valley both enjoyed exemptions based on preexisting acreages that were often several times larger than the established maximums. A lack of political will to curtail the substantial economic and political influence of agribusinesses ruled the day. Accordingly, although the program inserted the national government into reclamation more than it had previously, it still deferred much to local power. Ultimately, the Newlands Act promised to rearrange social and ecological systems to serve an agrarian vision of the West, a vision that countered the industrializing direction of much of the nation and region (Hundley 2001, 115–120, 223–226, 262–266; Pisani 1992, 273–325; Worster 1985, 160–163).

Two case studies from southern Idaho demonstrate flaws in these reclamation schemes. Carey Act provisions funded the Twin Falls South Side Project. Ira Perrine was the booster who arranged investment and promoted Twin Falls, Idaho, as a new community vibrant because of reclamation. Perrine and the other principal promoters ran the Twin Falls Land and Water Company and were uninterested in reforming society. Instead, making money, largely through speculation, captured their interest and energy, thus demonstrating that reclamation development did not always follow reformers' dreams. As a business venture, Twin Falls real estate was a smart investment. Per-acre prices increased from around $50 in 1905 to nearly $400 by 1919. The Twin Falls experience demonstrated how business interests shaped many reclamation projects (Pisani 2002, 65–77).

In nearby Rupert, Idaho, 50 miles east of Twin Falls, the reclamation experience followed a different trajectory. Here, the Reclamation Service established the Minidoka Project, in part to counter Carey Act developments nearby. Minidoka's costs were smaller than those of the private investments at the Twin Falls South Side Project. One reason for lower costs was that the Reclamation Service only constructed canals to within a mile and a half of farms, forcing farmers to build the rest. Farmers had flocked to the Minidoka project ahead of water. This calculated risk harmed them, as mortgages often came due before water arrived. Meanwhile, the government tried to promote town development, although Rupert lagged far behind Twin Falls in urban amenities and growth. It seemed clear that government reclamation could not meet its myriad goals (Pisani 2002, 77–89).

Despite these representative shortcomings, reclamation projects—private, federal, or combination—dramatically altered the Far West's landscape, but the changes seldom accomplished what advocates planned. By 1920, southern Idaho boasted 13,000 miles of ditches and canals through 18,000 farms to water 2 million acres. This apparent success belied the great challenges that remained. Reclamation advocates and engineers believed their work would control nature and fulfill a providential design. Dams, canals, and storage reservoirs would make droughts a memory, would eliminate waste, and would produce only marketable crops. This supposition of control, of course, was overdrawn. Droughts still came, frustrating irrigators' best-laid plans. Seepage, leakage, and evaporation meant much water was not used on the fields. On one project, in fact, 96,000 acre-feet out of 314,000 acre-feet diverted was lost primarily from evaporation; another recorded losses of 2,000 acre-feet daily. And weed species infiltrated the canals and fields. Nonnative fish and plants spread in and along canals, while weeds made their way into farmers' fields. But, of course, the biggest change was a welcome one—the crops of alfalfa, hay, and potatoes that replaced the native sagebrush. Reclamation, thus, both failed and succeeded in living up to its promises (Fiege 1999, 11–80, 143–170).

Environmental issues also made arid agriculture a challenge. Farming in these deserts often led to conflicts and sometimes innovative solutions that took their cues from the natural world. In southern Idaho, conflicts over the nature of water demonstrate how ecological conditions clashed with institutions like private property and other political abstractions. According to historian Mark Fiege, Idahoans discovered ways to cooperate in the face of these struggles in part because of their willingness to respond to ecological conditions rather than solely trying to change them. The first conflict occurred when water crossed private property lines. Early irrigators frequently overapplied their water. This created several problems. For instance, it waterlogged land and compromised drainage systems. Moreover, it raised groundwater levels so high that occasionally crops drowned by their roots. Perhaps most difficult, this "extra" water crossed property lines, forcing farmers to confront what Fiege termed the "hydrological commons" (1999, 35). Eventually, irrigators banded together to create a drainage district with the power to levy taxes to create drainage systems. As a result, the law adapted to environmental realities—something uncommon in U.S. history (Fiege 1999, 34–35).

On a larger scale, Idaho irrigators pursued other types of cooperation amidst conflict. Stories of violent confrontations between irrigators are scattered throughout the history of the hydraulic West. Indeed, in southern Idaho one Joe Koury died as the result of a blow from a shovel at the hands of his neighbor William Grover, Jr., who appeared and challenged Koury, who seemed to be taking Grover's water. Despite such tensions, which seldom manifested so violently, the hundreds of farmers who relied on Snake River water and numerous reclamation projects devised ways to cooperate. An example emerges from the 1910s and 1920s when Upper Snake River Valley irrigators competed with downstream farmers in the Minidoka and Twin Falls projects. The latter group of farmers enjoyed rights to water stored behind the upriver Jackson Dam, while farmers in the Upper Valley possessed rights to natural flow in the Snake. Complicating the matter, the spatial arrangements of rights differed, for natural flow users enjoyed their rights at the headgates of canals, while stored water users acquired their rights at the reservoir. Moreover, the different water rights came in different measurements with natural flow measured in cubic feet per second—a mobile measure—and stored water measured in acre-feet—a static measure. With inferior rights because of prior appropriation, the upstream users had to let water flow past them sometimes, even though their crops were withering during years when natural flow was insufficient. Differentiating between water in the river bed that had been stored in the reservoir or that was natural flow made little intuitive sense and led to conflict, resentment, and confusion. In 1923, the Snake River Committee of Nine formed as a nonbinding extralegal organization that would divide water between stored water and natural flow users. This committee, with

One of several dammed sections along the Snake River in Idaho. (Corel)

representatives from all concerned areas, generally cooperated and secured arrangements that met most users' needs. The most common cooperative action was a temporary water transfer where stored users lent water to other farms. Fiege explained that "a moral economy of water underlay the transfers: the practice rested on the basic belief that it was wrong to let a neighbor or nearby farmer fail for lack of water. Temporary transfers showed that irrigators could both uphold the individualist, competitive system of prior appropriation and periodically agree to set that system aside in the interest of helping each other" (1999, 114). The Committee of Nine symbolized the local accommodations that occasionally emerged within larger frameworks of national reclamation programs (Fiege 1999, 97–112).

By the 1930s, numerous reclamation projects converted deserts or dry rangelands into farms—some prosperous, some not. These farms depended on federal support, as well as local capital, and they all used scientific and engineering expertise to harness the power of rivers, fitting well within typical conservation prerogatives. Although the majority of these farms were small (an important goal of Progressive conservationists), numerous large farms contained the most economic value in the hydraulic West. These vast business operations accelerated the industrialization of agriculture begun earlier. They, too, would benefit the most from new reclamation projects from the 1930s to 1970s.

FORESTS AND CONSERVATION

If fishery conservation offered an example of early efforts at scientific management, and if irrigation provided an example of compromised efforts at reform, forest conservation was arguably the most famous and extensive effort. Part of the reason for its prominence, both at the time and in subsequent scholarly accounts, is Gifford Pinchot. The first chief forester of the U.S. Forest Service, Pinchot claimed to have coined the term "conservation" and was a tireless promoter—of himself and the broader conservation cause (Pinchot 1947; Pinchot 1967). Pinchot and forest conservation efforts responded to broad and diverse national concerns related to forests. First, the lumber industry had most recently decimated the Great Lakes forests, furthering a cut-and-run strategy that was wasteful economically and disastrous ecologically (Cronon 1991, 148–206; M. Williams 1995, 25). Second, foresters and public officials worried that the profligate cutting was rapidly directing the United States toward a timber famine in the near future, a sure portent of economic calamity (Hirt 1994, 28; Langston 1995, 85). Third, forests served as important watershed protection for many urban areas, and recent floods encouraged many to think seriously about forest protection. So, protecting new reclamation projects from excessive erosion influenced forest conservation (Pyne 1997, 185). As these concerns coalesced at the end of the nineteenth century, they spawned a new era of forest history. In the Far West, especially the Pacific Northwest, forest policy played a significant and evolving environmental and economic role. Similar to other conservation campaigns, the best-laid plans set out by foresters, however, seldom turned out as expected.

Federal concern with western forests received legislative support in 1891, evolving over the next several decades to strongly influence western forests. In 1891, President Benjamin Harrison signed a land law reform bill that included a little-considered section allowing presidents to reserve public domain from entry and settlement by individuals. These forest reserves, later termed national forests, represented a shift in American land policy. To this point, the overriding goal of federal land policy had been to get land into private hands. What purpose these new forest reserves served was unclear at the time. By 1905, though, a new federal agency—the U.S. Forest Service (USFS)—had emerged within the Department of Agriculture with the charge to manage these lands. By 1911, there were about 76 national forests with more than 80 million acres in Washington, Oregon, Idaho, and California. With Pinchot leading it, the USFS tried to manage the public forests entrusted to it following the precepts of scientific forestry. These guidelines included treating trees as a crop (hence the agency's placement in the Department of Agriculture) and eliminating wasteful practices—both economic and ecological practices deemed inefficient. The USFS hoped to ensure forests would be productive for human economies long into the future. The record was mixed (Cronon 1994, 607–610).

Consistent with Pinchot's mission to control forests was his desire to control fires. However understandable, such a commitment was not inevitable. Indeed, in California a contemporary controversy arose over light burning. Many Euro-American settlers recognized the historic role anthropogenic fire played in shaping forests and other ecosystems, and they mimicked Native American practices. Several groups in California found light burning useful. Some timber owners used it to reduce potentially dangerous fuel loads. Ranchers used it to encourage forage growth and prevent brush encroachment. Farmers burned to clear land and prepare fallow land for future planting. All these groups publicized their efforts and saw fire as essential to maintaining their ecological regimes. This pioneer practice, however, was anathema to most professionally trained foresters. Despite its utility, foresters saw deliberate burning as dangerous and wasteful (Pyne 1997, 100–122).

The nail in the coffin of light burning came in the Northern Rockies. Beginning in spring 1910, a drought descended on the interior Northwest. Small fires erupted throughout the early summer, culminating in the Big Blowup on August 20. Only the fall rains in September finally extinguished all the fires. In the process, some 5 million acres of national forest lands burned that year. Idaho and Montana received the brunt—3 million acres burned in those states. These figures do not account for the acres and acres of private lands decimated. Small mountain towns were evacuated; although some were ultimately spared, others were destroyed. Smoke from the 1910 fires carried soot to Greenland. Almost 8 million board feet of commercial timber burned, and erosion dirtied rivers. But such statistics do not fully reflect the impact of these fires (Pyne 1997, 239–259; Pyne 2001).

The Big Blowup's institutional effect on the Forest Service was monumental. Still a young agency, the USFS was struggling to gain public support and to define itself. As it turned out, foresters forged in the 1910 fires dominated the agency for decades and created Forest Service fire policy. That policy decidedly did not include anthropogenic fire management. Referencing the 1910 fires and the contemporaneous light-burning controversy, fire historian Stephen J. Pyne

Fire-lookout building atop Santiago Peak in California's Cleveland National Forest, ca. 1925. Fire lookouts were used throughout the West to spot and suppress fires. (Library of Congress)

explained: "No one who had lived through the holocausts in the Northern Rockies could seriously accept the proposition that fire should be liberally distributed throughout the woods" (1997, 251). Instead of intentionally setting fires, the USFS pursued a policy of total suppression. The story the Forest Service interpreted from the Big Blowup was the need for a national suppression policy with the agency in charge. The Weeks Act, passed the following year, represented the political aftermath of the 1910 fires. Although a key part of the Weeks Act authorized the USFS to purchase forest lands in the East, the legislation also promoted cooperative fire control agreements between the Forest Service and the states, effectively funding and extending USFS fire policies beyond national forest boundaries. In the aftermath of the fires, the USFS developed a network of trails and fire lookouts to ensure early detection and quick suppression. So, as much as anything, fire-fighting legitimated the conservation role of the fledging Forest Service (Pyne 1997, 242–264; Pyne 2001; Wilkinson 1992, 132).

As the USFS pursued its national policies, it frequently ran up against local forest ecosystems and found their best-made conservation plans wanting. One case study of the Forest Service's early years might represent the agency's larger experience and impact. In northeastern Oregon, the Blue Mountains contained several national forests administered by earnest foresters who came, in the words of historian Nancy Langston, "to save the forest from the scourges of industrial logging, fire, and decay" (1995, 86). With these good intentions, federal foresters arrived at the turn of the twentieth century to survey and classify the forests. Many locals resented this federal intrusion and feared for economic losses if, as expected, foresters restricted logging, grazing, and mining. Local residents based their fears on the undefined mission of the Forest Service in its earliest years. They need not have worried too much though; the Forest Service wanted to ensure forest production, albeit with government supervision (Langston 1995, 86–97).

The Forest Service promised to bring scientific understanding to the natural processes of forests and then redirect them to make the forest produce for human welfare with machinelike efficiency. Unfortunately, foresters assumed that forests functioned in certain ways, ways that were frequently at odds with ecological reality. Moreover, in assuming economic production to be the highest use of forests, they misread much of the forest and managed it in ways that ultimately compromised the forests' ability to sustain themselves. Their view and management of old-growth forests especially reveal their attitudes.

To the founding generation of American foresters, old growth represented a wasteful and inefficient nature. These forests were decaying; the trees no longer displayed vigorous growth. Decaying trees, however, play an essential function to forest ecosystems by providing habitat to all sorts of insects and the birds that preyed on them, as well as providing essential ingredients to building soil. Yet foresters saw nothing but waste and worked to eliminate the old growth (not to mention the insects that relied on it) so that a new, more vigorous, and in their minds, *healthy* forest might take its place. Moreover, foresters believed they could manage a younger forest on a regular, self-sustaining schedule. They would produce a crop of trees. Or so the theory went (Langston 1995, 148–156).

What happened in the Blue Mountains (and elsewhere) was more complicated. To start their crop of efficiently growing trees, foresters had to sell old-growth ponderosa pine forests. In the 1910s and 1920s, the Forest Service pursued large-scale timber sales, usually by developing cooperative arrangements with large corporations. (Ironically, the USFS claimed part of its authority to manage Blue Mountain forests by blaming outside economic interests for cutting and running, which worsened both the ecological and economic states of the region.) The USFS calculated timber sales expecting that the forest they were designing would be able to withstand annual harvests equal to annual growth. They called this model sustained-yield forestry. A fundamental problem with it rested on foresters' uncertainty about annual growth rates. A second problem arose when superiors in the Forest Service, desiring higher rates of sales, would simply increase the growth estimates. So, if foresters in the Blue Mountains determined (or guessed) that the local ponderosa pine forest produced a growth of 0.5 percent, superiors in the regional office might decide that it was actually 2 percent, thus allowing four times the cut local foresters had recommended. Thus, as in fisheries management, politics more than biology determined sustained-yield forestry, despite foresters' claims to be using science to achieve their management strategy (Langston 1995, 157–174).

Moreover, by focusing entirely on producing a timber crop in the future, Forest Service planners ignored the broader ecological roles of the forests, such as watershed protection and wildlife habitat. This neglect was ironic, for protecting water supplies constituted one of the principal reasons for reserving land in national forests. How to best protect the ecological function of watersheds remained contested within the USFS, because foresters, engineers, and others thought about water in distinct ways. Wildlife was no different. Locals demanded range resources for grazing cattle and sheep. Thus, the Forest Service managed the forest to promote domestic grazing. Administrators perceived wildlife as competitors to livestock. However, many valued wildlife for hunting, and in the 1920s agency biologists attempted to determine how best to promote hunting. Consequently, the USFS imported game and eliminated predators, especially coyotes. The absence of predators led to a skyrocketing rodent population that threatened new range growth. Thus, competition between deer and livestock animated conflicts in the 1930s between grazing and wildlife managers. By managing forests for production, USFS personnel seldom viewed the forest ecologically with numerous connections among each of these various "resources." This restricted worldview limited managers' abilities to understand the land (Langston 1995, 91, 142–144, 234–235, 240–246).

Similarly, because the USFS in the early twentieth century remained imbued with the ecologist Frederic Clements's ideas of ecological succession and climax, they tried to exclude disturbances that would hinder succession. From this perspective, coupled with their economic desires and agency policy, the USFS worked to exclude fire from the Blue Mountains, which set off a number of unexpected ecological consequences. For example, sagebrush invaded grasslands, shrinking available grazing habitat. Fire exclusion allowed some young pine forests to gain a foothold, a result the Forest Service favored. But then again, it also meant that pine forests often crowded with new fir trees that previously would have been checked by periodic fires, a result the Forest Service did not favor. Although foresters recognized that fire played an ecological role, they saw some fires as damaging, and thus all fires needed suppression. Such a policy lacked nuance and led to a series of much more damaging fires later in the century (Langston 1995, 132–134, 247–263). Management values and complications such as these dominated the forested landscapes of the Far West and reshaped those forests significantly.

The national depression of the 1930s also affected forests in the Far West in myriad ways. The economic exigency hit timber communities hard. In part, this economic decline was rooted in profligate economic behavior. That is, before the depression, the timber industry faced intense competition and few restrictions, so they cut forests rapidly. Overproduction plagued the industry, devastating the human and ecological communities in many Western forest lands. By the 1930s in the Blue Mountains, the USFS had written large sales agreements that prompted the construction of several large mills that required more trees than could be harvested sustainably. The agency felt obligated to assist mills and local communities, even as foresters began to acknowledge that their predictions about forest growth were far outpaced by local mill production demands. The Depression presented an opportunity for reflection and reform. Accordingly, foresters and planners suggested sustained-yield timber production plans that reduced annual timber production to encourage long-term economic and community stability. The moment for real reform seemed to pass before it started, though, when World War II began and full production became the order of the day (Langston 1995, 189–200; Robbins 1987, 187–196; Robbins 2004, 147–148).

While the Depression continued, federal responses to the economic disaster prompted other changes in these Western forests. President Franklin D. Roosevelt possessed a deep interest in conservation and forestry. Consequently, among the numerous New Deal programs initiated to mitigate Depression conditions was the Civilian Conservation Corps (CCC). Staffed by young men from around the nation, the CCC did much work in the nation's forests. They planted trees, worked to stop erosion, and built trails, campgrounds, and roads. This labor profoundly reshaped forest ecosystems and paved the way for subsequent changes.

One enormous change allowed by this surplus labor came in firefighting policy. It had long been the policy of the Forest Service to suppress fires, and the 1930s saw a series of large fires reinforce the agency's suppression penchant. The 1933 Tillamook fire along the Oregon Coast, for instance, consumed billions of board feet and about a quarter-million acres of prime Douglas fir, western red cedar, and white fir forest. Observers estimated a loss of nearly half a billion dollars in merchantable timber, not to mention the profound habitat changes (Wells 1999, 7–12). (The Tillamook region faced three more devastating fires over the next two decades.) Similarly, fires in the Selway region of the Northern Rockies along the Idaho and Montana border renewed important fire policy discussions. One veteran forester and experienced firefighter, Elers Koch, responded with an essay lamenting the tendency of the USFS fire policy to destroy unique wilderness areas, especially through roads created for fire protection. He believed some fires should be allowed to burn themselves out. Koch's was a minority opinion, and instead Forest Service in 1935 instituted the 10 a.m. Policy, a policy that largely remained in place until the 1970s. This policy did not represent a big shift in thinking. It simply stated the goal to contain fires by 10 a.m. the following morning. Its originality lay only in that it was a national policy to blanket all national forests. And with CCC labor available, it seemed possible. Forest supervisors guided the CCC as it built miles and miles of roads and a fire lookout system to aid in early detection. The national policy derived from Far Western experiences and showed the increasing influence of the region's national importance in formulating resource solutions (Pyne 1997, 278–287).

The early decades of the twentieth century saw forest policy develop rapidly. Foresters hoped to standardize timber production and forest protection. Like most Progressive conservationists, they believed in the power of expert management to make nature fit their goals. They worked to end forest fires, to stop waste, to eliminate decadent old growth, to promote a new and vigorous crop of trees, and to match production with consumption goals to stabilize timber communities. In time, all of these policies came to be seen as socially or ecologically suspect. Moreover, these policies seldom achieved the desired results; instead, they created something entirely new, demonstrating the contingency of nature that environmental historians have long explored.

PRESERVATION MOVEMENT

Forests, of course, did not simply produce timber or provide wildlife habitat. They became sites for many national parks, and with that designation came frequent conflict over the best use of resources. America's first national park, Yellowstone, was reserved in 1872. A number of other national parks received designation in the 1890s and early twentieth century, many of them in the Pacific West. By 1916, Congress created the National Park Service (NPS) to administer these areas. These parks represented some core cultural ideas about nature, including the notion that monumental North American landscapes symbolized national greatness. Indeed, this scenic nationalism informed many early national park advocates (Runte 1997). In the late nineteenth and early twentieth centuries, preservationists worked on the behalf of many such places in the Far West, activities that became important to the landscape and that broadened notions of nature's value. Although many Americans found it worthwhile to preserve spectacular landscapes like Yosemite Valley or Mount Rainier in parks, such protection also sparked numerous political and economic controversies.

National parks have a long history in the Far West. In 1864, Congress granted the state of California pieces of Yosemite Valley and part of the Sierra redwoods for scenic preservation. In 1890, the federal government took over those lands and more in what became Yosemite, Sequoia, and General Grant national parks. These were the nation's second set of parks. After these reservations, preservationists soon secured national park status for Washington's Mount Rainier (1899) and Oregon's Crater Lake (1902). Next, President Theodore Roosevelt signed the Antiquities Act (1906) giving the president the authority to remove from public domain areas of scientific or historical significance. Roosevelt used this power to make Lassen Peak, California (1908), and Mount Olympus, Washington (1909), national monuments. All of these parks and monuments, according to historian Alfred Runte, were basically worthless for traditional Western industries. This worthlessness limited the objections to taking such places out of the public domain and precluding private economic development. Except for watershed protection and some limited grazing, reserved park lands promised comparatively little economic return (Runte 1997, 48–73).

Tourists drive a horse-drawn wagon through the Wawona Tunnel Tree at Mariposa Grove, California, ca. 1905. Mariposa Grove is one of three major giant-sequoia groves in Yosemite National Park. (Library of Congress)

Despite the appearance of worthlessness, various development interests soon found parks desirable. These conflicts led to local and national debates over the purpose of these lands. The best known example of this conflict occurred in Yosemite National Park over Hetch Hetchy Valley. Historians have long used this example to show the conflict between preservationists and conservationists. Following Roderick Nash's lead, many historians have interpreted the conflict as one of wilderness versus development (Nash 1982, 161–181). However, as Robert Righter has recently demonstrated, the battle at the time was not so much about use versus nonuse, but about competing ideas of use.

That is, one group wished to develop the valley for tourism, while the other wanted it for municipal water and hydroelectric growth—both sides wanted development of one kind or another. Moreover, an essential aspect to the debate centered on public versus private ownership for San Francisco's water supply (Righter 2005). The story of Hetch Hetchy, only briefly outlined here, illustrates the first modern environmental conflict in the region's political history, and it was one that took on national proportions and spawned countless legacies within and beyond the region.

San Francisco's urban growth and need for fresh water caused city leaders to look for a supply. The private water supplier upon which the city had relied, Spring Valley Water Works, faced charges of high prices and inefficiencies. To solve this problem and in the Progressive spirit of the day, city voters passed a city charter in 1900 requiring municipal water ownership. The city believed it had found an ideal water source in the Tuolumne River that flowed through Hetch Hetchy Valley, a beautiful valley north of the more famous Yosemite Valley but still within the national park boundaries. That the potential dam and reservoir site rested in public lands meant the federal government would need to approve the project; that the site lay within a national park meant an especially bitter fight ensued (Hundley 2001, 171–173; Nash 1982, 161–181; Righter 2005).

City leaders and government conservationists led the faction favoring a dam in Hetch Hetchy Valley. In the mid-1890s, having never visited the valley, San Francisco Mayor James Phelan became convinced (and never wavered) that Hetch Hetchy furnished the superior site for a new municipal water supply. Subsequently, in 1901, Phelan applied to the Department of the Interior for reservoir rights. Early that year, Congress passed a Right-of-Way Act authorizing transmission lines, canals, or other infrastructure through California parks and forests. Despite these legislative preparations, Secretary of the Interior Ethan Hitchcock denied the city's request in 1903, basing his decision on the idea that reservoirs and national parks were inconsistent uses. The city both appealed and looked elsewhere for reservoir sites. After the 1906 San Francisco earthquake and ensuing fires, city leaders falsely suggested that a more secure water supply would have helped extinguish the fires more quickly. This strategy played on national sympathy and renewed the campaign for a Hetch Hetchy dam. This time President Theodore Roosevelt, Chief Forester Gifford Pinchot, and new Secretary of the Interior James Garfield seemed to offer San Francisco greater hope. Although Roosevelt occasionally expressed preservationist values, he ultimately embraced the type of utilitarianism Pinchot personified. To these federal officials, to leave Hetch Hetchy Valley undammed seemed wasteful, especially as it clearly weakened San Francisco. Garfield approved the city's petition for development in May 1908. While these conservationists asserted the appropriateness of Hetch Hetchy as the best source for San Francisco's water needs, others offered different perspectives (Brechin 1999, 99–102; Righter 2005, 23–95).

John Muir became the most vocal critic of San Francisco's plans. Long an explorer of the Yosemite region, Muir helped to found in 1892 the Sierra Club, one of the nation's earliest outdoor organizations, and presided over the club. Muir valued Hetch Hetchy as a place of beauty suffused with spiritual values. Indeed, he viewed nature as full of God's glory, and thus any wanton destruction of nature, such as a dam in Hetch Hetchy Valley, represented sacrilegious defilement of God's natural sanctuaries. Armed with such perspectives, Muir and allies such as Robert Underwood Johnson, editor of *Century Magazine,* waged a national campaign to bring attention to their cause. Supporters published articles in national magazines and wrote letters to officials in Congress and the executive branch. Marshaling romantic notions of nature, they charged prospective dam-builders not only with sacrilege but also with base greed. For instance, Muir concluded an essay, originally published in *Sierra Club Bulletin* and later revised and included in his book *The Yosemite,* with a charged and oft-quoted passage combining those criticisms:

> These temple destroyers, devotees of ravaging commercialism, seem to have a perfect contempt for Nature, and, instead of lifting their eyes to the God of the Mountains, lift them to the Almighty Dollar.
>
> Dam Hetch Hetchy! As well dam for water-tanks the people's cathedrals and churches, for no holier temple has ever been consecrated by the heart of man (Muir 1988, 196–197).

Such sentiments resonated with a growing number of Americans who saw national parks not only as sacrosanct but also as places for tourists. Indeed, these preservationists sought not to leave Hetch Hetchy entirely alone but to promote it for scenic tourism, even suggesting a road through the valley. Not wishing to seem cold-hearted to legitimate needs for water, Muir and others suggested alternative sites for the city's reservoir outside the national park (Nash 1982, 161–181; Righter 2005, 66–116).

The two sides clearly were at odds. However, as a political issue, the status quo did not last long. Although Secretary Garfield had approved San Francisco's plans, a new administration arrived in 1909 that replaced Roosevelt with William Howard Taft and Garfield with Richard Ballinger. Internal politics also soon ousted Pinchot as the head of the Forest Service. To persuade the new political powers, Muir accompanied both President Taft and Secretary Ballinger on different trips to Yosemite National Park, and he reported that the politicians seemed favorable to overturning Garfield's earlier order. Indeed, Ballinger required San Francisco to "show cause" why Garfield's order should stand and set a hearing for May 1910. After much expert testimony and typical bureaucratic delays, an established board of engineers favored the dam project, in part because the questions they asked and answered focused on technical, not aesthetic or environmental, concerns. By this time in 1913, though, Woodrow Wilson's administration presided over the nation. Not helping the preservationists, Wilson's new secretary of the interior, Franklin K. Lane, enjoyed long ties to San Francisco as a city attorney who worked with Phelan to assert the city's water rights in Hetch Hetchy. The die seemed cast, but these officials placed the ultimate decision in the hands of Congress. Despite the continued national preservationist campaign, which included tens of thousands of pamphlets and booklets and countless letters to senators and representatives, Congress sided with San Francisco. Wilson signed the bill on 19 December 1913, acknowledging its opponents as well-meaning but misguided (Righter 2005, 81–133).

The Hetch Hetchy story has become an important parable to environmentalists, although it has often become a simplistic caricature of the episode. Environmental historians have drawn many lessons from the political battle. One way to understand Hetch Hetchy is to see it as two competing visions of the common good with one side arguing for preserving the valley undammed for aesthetic and cultural reasons and the other side favoring the dam to promote urban growth. Another aspect seldom considered is the fact that much of San Francisco's effort was based on acquiring a public water supply to end the city's dependence on a private monopoly, as well as gaining a public source of hydroelectric power. San Francisco's efforts, thus, represented two elements of Progressive conservation: efficient use of resources and public water and power. For the preservationists, though, debates over public or private power missed the larger point, which was rooted in the need to maintain parks for scenery. For them, as well, they learned the value of political mobilization, especially on a national level. And they discovered the value of asserting development, at least tourist development. Finally, Hetch Hetchy showed Americans that some places attracted national attention and demanded national protection (Nash 1982, 161–181; Righter 2005, 3–10, 191–215).

The eventual loss by preservationists at Hetch Hetchy encouraged the creation of the National Park Service (NPS) in 1916. According to the legislation creating the Park Service, the agency's purpose was "to conserve the scenery and the natural and historic objects and the wild life therein and to provide for the enjoyment of the same in such manner and by such means as will leave them unimpaired for the enjoyment of future generations" (quoted in Runte 1997, 104). Such a mission entailed a difficult balancing act. Indeed, inherent contradictions seem built into the statement. How was one to promote enjoyment and leave parks unimpaired? Debates over policies designed to promote tourism and maintain that ecological integrity became increasingly rancorous as the twentieth century pressed forward, but the roots rested in these competing demands for parks.

Early park management demonstrated some of these problems. One of the more interesting aspects to this conundrum is found in Yosemite's human history. Nineteenth-century visitors to the Yosemite Valley found it inhabited and generally wrote derisively about the Yosemite Indians. Some early tourists associated Indians with wilderness and so were pleased to have seen native peoples living and working in the region, although this satisfaction was often marred by dehumanizing or patronizing attitudes. Once the NPS took over managing Yosemite, federal managers expressed unease about the presence of people, permanently or seasonally, in a nominally natural area. Some Indians were incorporated into the park economy as guides or as merchants of their material culture. The Park Service often perpetuated stereotypes and frequently restricted traditional subsistence activities within Yosemite. Federal officials built a new Indian village in 1930 where they relocated the remaining Yosemite Indians who were employed in park work. This greater confinement occurred at a time when Park Service policy began to value ecological processes, and allowing a group to actively live in the park and draw from its natural bounty no longer conformed to the vision managers possessed of national parks. People working in what was supposed to be a scenic nature preserve struck managers and visitors as inappropriate. By 1969, all residents and structures were gone. Although their long presence within park boundaries was exceptional, Natives' presence in and eventual removal from this park landscape underscores the basic fact that most national parks had been inhabited and used for generations, so creating national parks helped dispossess these cultures. Indeed, Native and sometimes Euro-American uses strongly shaped the natural scene the Park Service sought to preserve, an imprint most actively ignored (Spence 1999, 101–132).

The park's natural landscape, it turned out, bore strong human imprints from the earliest years of the park. Evidence clearly suggests that the grassy valley floor partially resulted from frequent anthropogenic burning. In turn, this plentiful forage attracted many ranchers to Yosemite and nearby valleys for grazing. Indeed, Muir famously fumed about sheep, or "hoofed locusts," overgrazing in the high meadows near Yosemite. Subsequently, vigilant fire protection changed the Yosemite landscape. Early managers, too, wanted streams and lakes full of fish, so they stocked parks with native and nonnative fish. Similarly, preferred wildlife, such as deer, enjoyed protection as the NPS operated effective predator control programs that reduced wolves and mountain lions (Sellars 1997, 23–26). Besides these measures, NPS personnel shaped visitors' experience by promoting particular visions of what parkland ought to resemble, often drawing from English landscape traditions (Olwig 1995). To this end, landscape architects created scenes that seemingly removed humans and their impacts, as well as making the park accessible for visitors while conforming to tourists' preexisting notions of national parks. Yosemite was not unique in these respects; indeed, its history marks it simply as a quintessential national park working to integrate wilderness values and tourist development inherent in the Park Service mission.

As the NPS initially confronted humanized landscapes and created ostensibly natural ones instead, automobiles further challenged the circumstances. Mount Rainer National Park became the first park to allow automobiles in the 1910s. NPS managers tried to harmonize nature with modern machinery, in the process creating what historian David Louter has termed "windshield wilderness." That is, the Park Service sought to make roads the avenue whereby visitors viewed and appreciated wilderness. It was an ironic vision given that prominent wilderness advocates such as Bob Marshall and Robert Sterling Yard defined wilderness as roadless. The Park Service remained wedded to the automobile, though, producing canned tours from park to park, along highways through a western circuit that included several of the Far West's prominent parks. Amazingly, in the 1930s, about half of all the nation's automobiles passed through national park gates, underscoring the close relationship between these paragons of nature and technology. However, in that same decade and somewhat as a reaction to such trends, a series of parks were established meant to serve as biological reserves. Olympic National Park, created in 1938, was one example, featuring vast areas of ecological interest and not simply a single, monumental feature such as nearby Mount Rainier. Leading government figures, such as Secretary of the Interior Harold L. Ickes, wanted Olympic National Park to be roadless. Nevertheless, Park Service planners ultimately relied more on roads to present the park, albeit a single road that concentrated motorized visitation. This approach brought a road into the park but not far, just to the edge of the wilderness. By bringing visitors to the edge, NPS roads made the park visible by windshield. Similarly, the challenges of blending the modern with the supposed primeval remained at the heart of many preservation struggles through later eras (Louter 2005).

Although national parks received the greatest attention, preservationists also defended areas that did not become national parks. In Idaho, for example, a state that still has no national parks, early preservationists attempted unsuccessfully to create national parks in the Sawtooth Range and at Craters of the

Moon. The Sawtooths were a beautiful mountain range, but federal managers did not believe the scenery to be spectacular or accessible enough to warrant inclusion in the national park system. Moreover, locals and the Forest Service, which administered the area, believed there was much wealth to be made there. All of these—standards of spectacle, accessibility, and potential economic development—marked it as wanting. Meanwhile, Craters of the Moon represented a geologic wonderland of strange craters, vents, and cinders. Such a place would not support agricultural or mineral development, passing the worthlessness test. But it could not match other sites for impressive scenery. Although it became a national monument in the 1920s, it barely received 5,000 visitors in peak years in that decade. These failed efforts reveal some Idahoans' desire to benefit from park status, as well as the difficulty of obtaining such designation (Neil 2005, 35–42).

Idaho preservationists did enjoy other successes, though. Heyburn State Park, which Congress carved out of the Coeur d'Alene Indian Reservation in 1908, became the Pacific Northwest's first state park. The following year the state legislature created new state parks in part to prevent commercial overdevelopment. Near Twin Falls, southern Idaho's Shoshone Falls, a waterfall taller than Niagara, marked one example, although subsequent reclamation projects reduced the water to a trickle. Following a somewhat different rationale, many national forests contained vast tracts of roadless wilderness areas that some Idahoans wanted preserved. Indeed, by 1930, forest supervisors of the Idaho, Payette, and Salmon national forests created the boundaries of the Idaho Primitive Area, borders that were later enlarged. This area of central Idaho encompassed far more than a million acres and was to be preserved in so-called frontier conditions. Advocates touted the area's lack of modernity as a salve to contemporary life, along with its qualities of solitude and its function as a game preserve. These examples of less well-known landscapes help to flesh out the broad efforts and successes of preservationists in the first half of the twentieth century (Neil 2005, 25–34, 45–61).

The preservation movement's beginnings lay in the East and in the mid-nineteenth century. By the early twentieth century, though, that movement had been tempered in searing political battles and had developed a large following and enjoyed important achievements. Having the national park system administered by a federal agency suggested a coming of age for American preservationists. Although not the only place where preservation battles played out, the Far West included some of the most prominent early national parks and most contentious political struggles. While the conservation movement faced complexities in managing ecosystems for multiple environmental and economic values, preservationists faced their own challenges in balancing tourists and protection of scenery and wildlife values, a challenge often ignored or confronted much later. As modernity reached into the mountains, preservationists faced increasingly difficult choices.

URBAN ENVIRONMENTAL CHANGE

Preservation interests grew in large part because of the Pacific West's growing urban population. In the early twentieth century cities grew in importance economically, culturally, and politically. They also generated significant ecological changes associated with their economic transformations. New industrial technologies produced new pollutants that fouled the region's air and water. Meanwhile, urban leaders set out to reconfigure most of the major river systems that flowed through cities, and these truly astounding engineering feats reorganized urban ecologies unlike anything that had come before.

In Portland, a city at the junction of the Willamette and Columbia rivers, Progressive conservationists pursued what historian William G. Robbins called "industrial modeling" within and beyond the metropolitan area (1997, 241). The important shipping interests in Portland promoted plans for a canal where the Columbia River pushed through the Cascade Range. The Celilo Canal, finished in 1915, made Columbia River shipping more predictable but did not constitute a major change in economic or ecological terms, though it portended much greater river changes. More substantially, like most rivers, the Willamette proved unpredictable. Long used by valley farmers for irrigation and transportation, as well as for a dumping site, the river inspired several engineering schemes by the 1930s. Concerned about predictability and navigability, engineers sought to remake the Willamette River with a series of dams. Conservationists hoped, too, that the newly systematized river would be better able to flush the accumulating pollutants along the river bottom. Federal engineers from the Army Corps of Engineers, local politicians, and many Oregonians heralded the proposals that circulated. Politicians ensured that the Willamette Valley Project passed Congress in 1938, a measure that initiated wholesale changes to the river system. Besides improving flood control, navigability, irrigation, and generating hydroelectricity, the project employed many Depression-era Oregonians. According to Robbins, this engineering work "reconfigured . . . waterways into instrumentalized, mechanized units of production. The multiple dams converted those rivers from seasonally fluctuating, free-flowing streams into functional segments designed to augment an expanding industrial economy" (1997, 292). The Willamette River, then, served Oregon's evolving industrialization. Of course, salmon and other fish suffered, and industry continued to dump its waste into this environment. Tragically, greater river regulation—a key component to these conservation efforts—did not lead to a healthier ecology (Robbins 1997, 241–244, 283–293).

Seattle illustrates even better the era's penchant for engineering rivers and associated ecological and social changes. Like several West Coast cities, Seattle is hemmed in by waters. Lakes Washington and Union, the Duwamish and Black rivers, and Puget Sound all bounded or flowed through parts of the city. Several Native American communities used these lakes and rivers for sustenance and transportation. As Euro-Americans arrived in the mid-nineteenth century, they began using these waterways. Competition over these hydrological spaces exacerbated cultural conflict. By the beginning of the twentieth century, too, new economic development in Seattle increasingly changed the population's use of these waters. To press these watersheds more fully into service of the emerging industrial order, Seattle boosters set out to reengineer nature (Klingle 2005; Mighetto 2005; Thrush 2006).

The engineering work in Seattle represented another example of Progressive conservation ideology at work. Those who envisioned a massive reordering of urban water expected such work to promote a more efficient, ordered, and controlled landscape. This mastery seemed necessary and lacking because routine floods—to which Natives had adapted—wreaked havoc on the more sedentary, less adaptable Euro-Americans and their property. Floods from the Cascade foothills overflowed riverbanks, stranding salmon and inundating lowland real estate development (Mighetto 2005, 47–48). Tidal bores could push water the other direction, reportedly raising the spacious Lake Washington fifteen feet in a single day, seriously threatening shoreline development (Klingle 2005, 22). Predictably, engineers believed rerouting rivers would ensure ecological predictability and allow secure industrial development, especially in the tidelands and along the shorelines.

Their flood-control dreams were monumental. They included filling in the tidelands (which eliminated many important wetlands), straightening and channelizing the Duwamish's many oxbows (which reduced the river by almost 10 miles and harmed salmon habitat), diverting many tributaries to the Duwamish (which severely reduced the river's flow), lowering Lake Washington between ten and twenty feet (which vastly reduced wetlands and altered shoreline ecosystems), and building the famed Hiram M. Chittenden Locks and the Ship Canal from Puget Sound to Lake Washington (which completely changed the lake's outlet point and accelerated industrial development). All of these water projects were in addition to the massive regrading of the city's hills that took the earth that made up the hills and placed it in city's tidal flats. In eight years, the Seattle General Construction Company filled in the tidelands with 24 million cubic yards of silt. Clearly, Seattle after the turn of the twentieth century was in a fit of engineering fury. These efforts were part of a vision of finishing nature, improving it through human efforts. Indeed, a city promoter explained that the Ship Canal would conclude "nature's . . . uncompleted purpose" of connecting landlocked lakes Washington and Union with Puget Sound (quoted in Mighetto 2005, 54). To Progressives, these actions represented significant progress to a burgeoning Western city (Klingle 2005, 23–25; Mighetto 2005; Thrush 2006, 101).

To the ecosystems within and around Seattle and those who depended on them, it could hardly be called progress. First, salmon lost in myriad ways. Former spawning grounds disappeared under fill or were rerouted entirely as the Black River was closed off from its natural outlet. The locks trapped salmon, making them easy prey to seagulls and sea lions. Industrial development along lakeshores and riverbanks reduced shade and increased erosion, consequences that hurt salmon habitat. Second, pollution created by sawmills and untreated sewage contaminated the city's water sources. Lake Washington, especially, experienced problems as engineering reduced its outflow and shrank its overall area, concentrating the lake's pollution. Third, Native peoples experienced severe displacement as reworked rivers and disappearing salmon sharply reduced their subsistence and commercial activities among Seattle's watersheds. This dispossession rooted Seattle's urban environment in inequality based in part on race. Land hunger by Euro-Americans combined with ecological changes to make Seattle an unwelcome place for many Native people. Despite all of this, Indians resided in and visited Seattle while making use of the declining marine resources that had long sustained them, albeit with frequent conflicts. The Progressive Era in Seattle demonstrated the power of engineers to alter nature in accordance with particular visions of modernity, visions that did not embrace nature on ecological terms or all cultural groups equally (Klingle 2005, 26–27; Mighetto 2005, 55; Thrush 2006).

If Seattle illustrates one set of ecological transitions in early twentieth-century cities, Los Angeles represents a host of other issues. The second case study in this volume illustrates the voracious appetite Los Angeles developed for scarce water after 1900, as well as demonstrating the city's subsequently problematic and paradoxical relationship with flooding. The city also symbolized a relatively new environmental relationship associated with another resource for which Angelenos and other Americans possessed an insatiable appetite: oil.

It has become commonplace to satirize Californians' love affair with the automobile and, by extension, oil. The bond between Californians and petroleum developed in the early twentieth century when petroleum discoveries in the Los Angeles Basin fueled oil development. By the 1920s, the state's residents led the nation in gasoline dependency. Petroleum not only modernized transportation, but it also provided heat. Instead of burning large coal deposits like much of the nation, California burned oil, which constituted 80 percent of their fuel needs compared with the national average of 20 percent. California could afford this reliance because of ever-increasing production. With the abundant deposits and despite the reckless use of this nonrenewable fuel source, the state enjoyed energy independence through World War II (Williams 1997, 115–117, 126–130).

However, numerous problems developed. First, political debates arose over the methods used for extracting oil. The initial discoveries of offshore oil deposits and the lack of clear legal title encouraged fast exploitation and resulted in economically disastrous market gluts, not to mention waste and premature depletion of finite resources. Meanwhile, beach-going Angelenos protested the developments as unsightly. Throughout the 1920s, Californians largely rejected drilling from piers, and in 1928 the state legislature banned it. This was one of more than fifty bills the legislature considered to control oil production in the decade. The oil industry won a referendum in 1936 allowing them to slant drill from the shore while earmarking half of all royalties for beach and park development. This sort of trade-off illustrated a common strategy of simply shifting, rather than changing, the offensive practice, as well as placating opponents with recreational developments. Moreover, this recreational and consumer-based opposition to development, in addition to local working-class complaints, portended many environmental conflicts in the decades to come (Elkind 2005a, 8–11; Quam-Wickham 1998, 189–203).

Next, in the midst of the 1930s Depression and rising concerns of national security, the federal government asserted its authority over tideland oil. Senator Gerald P. Nye of North Dakota and Secretary of the Interior Harold Ickes viewed federal oversight of offshore oil as necessary for military preparedness. Smaller, independent oil producers also favored the idea, believing federal control would offer them a way to break into the lucrative market heretofore dominated by major oil producers. In the meantime, both federal and state governments asserted title to offshore oil, but such competition subsided during the war crisis in the 1940s. However, the war did lead to oil shortages, and Ickes's poor management reduced public faith in federal control. Nevertheless, in *United States v. California* (1947) the Supreme Court declared that the federal government possessed title. This period of federal, state, and corporate challenges illustrates the multiple economic and political demands oil brought to California (Elkind 2005a, 11–13).

Besides jurisdictional issues and concerns about rampant oil development, petroleum mattered most, perhaps, because of multiplying automobiles.

Population in California generally and in the southern regions specifically grew after 1900 at enormous rates; between 1920 and 1924, for instance, the population of Los Angeles increased by 100,000 per year. These residents desired low-density settlement to contrast with crowded eastern and midwestern cities. Consequently, suburban growth organized Angelenos' urban space, and that ultimately depended on shaping the landscape for automobiles. Ironically, in the 1920s, Angelenos relied on one of "the best public transport systems in the world, with more than a thousand miles of fast electric interurban and street railways," and carrying more than 100 million passengers annually (Frost 2001, 366). However, in the 1920s, automobile ownership growth outpaced population growth 451 percent to 155, and California's figure doubled the national average. Newcomers clearly preferred individual automobility to public transportation, and the city allowed the public transportation system to atrophy with a large push by the automobile industry. In exchange, the city accommodated this switch to automobiles by building roads and other necessities, such as service stations. New residents with their automobiles that guzzled petroleum profoundly reshaped California's urban environment (Frost 2001, 362–367; Williams 1997, 153–157).

The impact of cars on the urban landscape was enormous. First, the built environment changed. Suburbs spread; gravel and dirt roads were paved; service stations and motor courts were built; and restaurants, attractions, and billboards appeared on roadsides. These physical changes had ecological and aesthetic effects. For instance, paved roads increased rapid runoff for water and exacerbated flooding. Also, decentralized suburban development extended the consumption of rural spaces for residences. As these neighborhoods became located farther and farther from places of employment and services, Southern Californians drove farther and farther spewing more pollution into the atmosphere (Williams 1997, 156–158, 167). Thus, air pollution became the most recognizable environmental problem in Los Angeles. By World War II, smog frequently blanketed the Los Angeles Basin. Increased automobile use was one culprit, but intensified industrial (including petroleum) development in wartime contributed to declining air quality. The city's location in a basin exacerbated air problems because of frequent temperature inversions (Elkind 2005b, 38–44). Urbanization in Southern California led to undesirable effects. Of course, Los Angeles, and other cities, did not have to choose to accommodate the automobile-suburban complex so readily. Alternatives existed, and leaders made deliberate choices without considering the ecological toll those decisions wrought. However, once put on this path, alternatives became fewer and choices more restricted (Frost 2001, 363).

A dense layer of smog sits among the skyscrapers of Los Angeles, California, and the surrounding urban sprawl. (PhotoDisc/Getty Images)

These few examples hint at the multiple ways cities changed the Far West's environmental trajectory. Engineers had a penchant for reshaping nature and and new technologies had an appetite for nonrenewable resources, and the much-expanded environmental needs meant that cities launched the West Coast in new ecological directions. In the second half of the twentieth century, cities increased both their impacts and their awareness of potential problems. That recognition began in the early twentieth century but change took much longer.

CONCLUSION

When the twentieth century dawned, the Far West had already experienced a series of monumental environmental upheavals. Indeed, the closing decades of the nineteenth century had produced industrialized systems that massively rearranged existing ecologies. To respond to these radical new directions, westerners turned to government reforms. These new approaches combined a faith in science and expertise with an optimism that conservationists could manage nature. Indeed, they believed they could improve nature whether that was by producing salmon, regulating forest growth, straightening rivers, or framing nature's spectacular vistas. Conservationists imagined a nature they controlled. What they achieved typically failed to live up to their expectations. Moreover, they discovered that their solutions frequently caused as many complications as the initial problems did. In the following decades, conservation continued to work toward adjusting the relationship of humans to the environment. Added to these preexisting patterns came a massive infusion of new dynamics that accelerated environmental change even more and simultaneously restrained it with a full-fledged environmental movement.

REFERENCES

Brechin, Gray. *Imperial San Francisco: Urban Power, Earthly Ruin.* Berkeley: University of California Press, 1999.

Cronon, William. *Nature's Metropolis: Chicago and the Great West.* New York: W. W. Norton, 1991.

Cronon, William. "Landscapes of Abundance and Scarcity." In *The Oxford History of the American West,* edited by Clyde A. Milner II, Carol A. O'Connor, and Martha A. Sandweiss, 603–637. New York: Oxford University Press, 1994.

Elkind, Sarah S. "Black Gold and the Beach: Offshore Oil, Beaches and Federal Power in Southern California." *Journal of the West* 44 (Winter 2005a): 8–17.

Elkind, Sarah S. "Los Angeles's Nature: Urban Environmental Politics in the Twentieth Century." In *City, Country, Empire: Landscapes in Environmental History,* edited by Jeffry M. Diefendorf and Kurk Dorsey, 38–51. Pittsburgh, PA: University of Pittsburgh Press, 2005b.

Fiege, Mark. *Irrigated Eden: The Making of an Agricultural Landscape in the American West.* Seattle: University of Washington Press, 1999.

Frost, Lionel. "The History of American Cities and Suburbs: An Outsider's View." *Journal of Urban History* 27 (March 2001): 362–376.

Hays, Samuel P. *Conservation and the Gospel of Efficiency: The Progressive Conservation Movement, 1890–1920.* 1959. Reprint, New York: Atheneum, 1975.

Hirt, Paul W. *A Conspiracy of Optimism: Management of National Forests Since World War Two.* Lincoln: University of Nebraska Press, 1994.

Hundley, Norris, Jr. *The Great Thirst: Californians and Water: A History.* Revised ed. Berkeley: University of California Press, 2001.

Klingle, Matthew W. "Fluid Dynamics: Water, Power, and the Reengineering of Seattle's Duwamish River." *Journal of the West* 44 (Summer 2005): 22–29.

Langston, Nancy. *Forest Dreams, Forest Nightmares: The Paradox of Old Growth in the Inland West.* Seattle: University of Washington Press, 1995.

Louter, David B. "Wilderness on Display: Shifting Ideals of Cars and National Parks." *Journal of the West* 44 (Fall 2005): 29–38.

McEvoy, Arthur F. *The Fisherman's Problem: Ecology and Law in the California Fisheries, 1850–1980.* New York: Cambridge University Press, 1986.

Mighetto, Lisa. "The Strange Fate of the Black River: How Urban Engineering Shaped Lake Washington and the Duwamish River Watershed." *Journal of the West* 44 (Fall 2005): 47–57.

Muir, John. *The Yosemite.* Boston: Houghton Mifflin, 1914. Reprint, San Francisco: Sierra Club Books, 1988.

Nash, Roderick. *Wilderness and the American Mind,* 3rd edition. New Haven, CT: Yale University Press, 1982.

Neil, J. M. *To the White Clouds: Idaho's Conservation Saga, 1900–1970.* Pullman: Washington State University Press, 2005.

Olwig, Kenneth R. "Reinventing Common Nature: Yosemite and Mount Rushmore—A Meandering Tale of a Double Nature." In *Uncommon Ground: Toward Reinventing Nature,* edited by William Cronon, 379–408. New York: W. W. Norton, 1995.

Pinchot, Gifford. *Breaking New Ground.* New York: Harcourt Brace, 1947.

Pinchot, Gifford. *The Fight for Conservation.* Introduction by Gerald D. Nash. 1910. Reprint, Seattle: University of Washington Press, 1967.

Pisani, Donald J. *To Reclaim a Divided West: Water, Law, and Public Policy, 1848–1902.* Albuquerque: University of New Mexico Press, 1992.

Pisani, Donald J. "The Many Faces of Conservation: Natural Resources and the American State, 1900–1940." In *Taking Stock: American Government in the Twentieth Century,* edited by Morton Keller and R. Shep Melnick, 123–155. New York: Woodrow Wilson Center Press and Cambridge University Press, 1999.

Pisani, Donald J. *Water and American Government: The Reclamation Bureau, National Water Policy, and the West, 1902–1935.* Berkeley: University of California, 2002.

Pyne, Stephen J. *Fire in America: A Cultural History of Wildland and Rural Fire.* 1982. Reprint, Seattle: University of Washington Press, 1997.

Pyne, Stephen J. *Year of the Fires: The Story of the Great Fires of 1910.* New York: Viking, 2001.

Quam-Wickham, Nancy. "'Cities Sacrificed on the Altar of Oil': Popular Opposition to Oil Development in 1920s Los Angeles." *Environmental History* 3 (April 1998): 189–209.

Reisner, Marc. *Cadillac Desert: The American West and Its Disappearing Water.* Revised and updated ed. New York: Penguin Books, 1993.

Righter, Robert W. *The Battle Over Hetch Hetchy: America's Most Controversial Dam and the Birth of Modern Environmentalism.* New York: Oxford University Press, 2005.

Robbins, William G. "Lumber Production and Community Stability: A View from the Pacific Northwest." *Journal of Forest History* 31 (October 1987): 187–196.

Robbins, William G. *Landscapes of Promise: The Oregon Story, 1800–1940.* Seattle: University of Washington Press, 1997.

Robbins, William G. *Landscapes of Conflict: The Oregon Story, 1940–2000.* Seattle: University of Washington Press, 2004.

Runte, Alfred. *National Parks: The American Experience,* 3rd edition. Lincoln: University of Nebraska Press, 1997.

Sellars, Richard West. *Preserving Nature in the National Parks: A History.* New Haven, CT: Yale University Press, 1997.

Spence, Mark David. *Dispossessing the Wilderness: Indian Removal and the Making of the National Parks.* New York: Oxford University Press, 1999.

Taylor, Joesph E., III. *Making Salmon: An Environmental History of the Northwest Fisheries Crisis.* Seattle: University of Washington Press, 1999.

Thrush, Coll. "City of the Changers: Indigenous People and the Transformation of Seattle's Watersheds." *Pacific Historical Review* 75 (February 2006): 89–117.

Wells, Gail. *The Tillamook: A Created Forest Comes of Age.* Corvallis: Oregon State University Press, 1999.

Wilkinson, Charles F. *Crossing the Next Meridian: Land, Water, and the Future of the West.* Washington, DC: Island Press, 1992.

Williams, James C. *Energy and the Making of Modern California.* Akron, OH: University of Akron Press, 1997.

Williams, Michael. "The Last Lumber Frontier?" *The Mountainous West: Explorations in Historical Geography,* edited by William Wyckoff and Lary M. Dilsaver, 224–250. Lincoln: University of Nebraska Press, 1995.

Worster, Donald. *Rivers of Empire: Water, Aridity, and the Growth of the American West.* New York: Oxford University Press, 1985.

6

MODERN ENVIRONMENTAL CONFLICTS

In the mid-twentieth century, Americans had high hopes. Those living in the Far West were no different. One of those westerners, Richard Neuberger, a journalist-turned-politician from Oregon, captured those large ambitions in his book of the Northwest, *Our Promised Land* (1938). He praised the Bonneville Dam, then being built across the Columbia River, for its role in delivering cheap hydroelectricity to thousands of northwesterners that would promote industry and improve the lives of rural residents. Neuberger betrayed some ambivalence, however, when he acknowledged that the fate of salmon was unclear: "The food of the Indians, the future of an industry, the sport of thousands of anglers, and the usefulness of the most elaborate and costly fish equipment ever built will be at stake when the salmon runs of the near future reach the massive barrier at Bonneville." Even the head of the U.S. Fish Commission "admits he is not certain what will happen, but states he is hopeful the giant stairways and lifts will meet the crisis" (Neuberger 1938, 139). Like so many others, Neuberger expressed his hope that the engineering solutions would work. If not, he seemed to suggest, then a larger, more important cause would still be served. In these attitudes, Neuberger represented most of his regional contemporaries. They possessed grand visions for transforming nature even further to serve human betterment.

And they expected it to all work out for the environment. In the post–World War II era, such confidence met its match as environmental change proceeded rapidly, so rapidly that those directing the change could not control it. Technological advances during the war changed the relationship of all humans with the natural world, as atomic bombs rearranged the basic building blocks of atoms. New chemicals became widespread, and Americans used them on their farms and forests. The era of big dams harnessed rivers' power for rural and urban economies through hydroelectricity and irrigation. Meanwhile, an expanding urban and suburban population developed greater interest in recreation oriented around natural amenities, an interest that often sparked efforts for wilderness preservation. In this period, thus, a common pattern emerged.

Buttresses—part of the Bonneville Lock and Dam on the Columbia River—near Bonneville, Oregon, ca. 1937. Dams like this one promised economic growth but came with environmental costs, such as declining salmon runs. (Library of Congress)

Resource management led to ecological problems that demanded legislative responses. Frequently, these proposed solutions solved little. Thus, a typical pattern resulted whereby a resource was developed, followed by recognition of environmental problems, followed by proposed solutions, and usually followed by continued concerns. Hence, successive conflicts describe this period's environmental history more than anything else.

THE REVISION OF MODERN AGRICULTURE

Far westerners devised numerous ways to modernize farming. Technological changes did the most to transform modern agriculture. Larger dams and reclamation projects brought new areas under cultivation, hastening several environmental problems in the arid region. Perhaps more powerfully, fertilizers and pesticides made agriculture dependent on chemical inputs heretofore unheard of. Farmers hoped to make these changes in the name of efficiency, control, and profit. Subsequent ecological problems and the public's decreasing faith in the suitability of chemicals challenged the new directions of modern agriculture. But proposed solutions were never easy or uncontested.

Furthermore, postwar agricultural prosperity depended on a significant institutional apparatus. Farmers enlisted state support, especially from land-grant institutions that, according to historian William G. Robbins, "fostered an ideology of progress and a belief that seemingly endless increases in agricultural productivity were possible" and desirable (2004, 83). Reminiscent of Progressive Era conservationists, postwar promoters emphasized technology and efficiency for the fields.

Chemicals were central to this effort. At the end of World War II, many chemicals evolved from military to civilian use. The U.S. Department of Agriculture and university scientists formed a close partnership with farmers and chemical companies, an association that focused on technological and chemical solutions that precluded alternative strategies for reducing pests. For example, integrated pest management, a strategy that relied on natural predators and not chemicals, had enjoyed a period of enthusiasm in the 1950s among California scientists as an alternative to DDT and other chemical pesticides (Perkins 1998, 418–423). While all agreed that DDT was toxic, officials claimed that it would not harm humans. Although crop yields increased with pesticides, so did chemical residues in soil and plants. Further, increasingly chemical-resistant insects plagued farmers. When the public expressed concern, the legislature responded with nominal controls but provided little enforcement. Complaints and alarm increased, but the official response focused on education, voluntary controls, and individual responsibility for protection (Robbins 2004, 82–85, 94–101, 114–130).

Chemicals altered farming practices, but federal irrigation projects drastically rearranged existing ecologies. These projects shaped the Pacific West's landscape for a century, but from the late 1930s through the 1960s, larger projects with farther-reaching environmental, economic, and social consequences initiated a series of complicated changes. Advocates for these reclamation projects recognized their revolutionary impact, but they strongly emphasized only positive effects. Federal and local officials, not to mention the farmers themselves, hailed the water provided to the dry fields and the cheap hydroelectric power generated by turbines at many of the dams. Many contemporaries viewed reclamation projects as blending nature and artifice in appropriate, even beautiful ways (White 1995). Some Native people, however, recognized that dams would spell the end of what was left of the salmon fishery. From the promise heralded at the dawn of this renewed and ambitious reclamation regime came complications that remain unresolved, and are perhaps unresolvable, without major changes in economic and environmental values.

Lake Mead, the largest engineered lake in the United States, was created by the damming of the Colorado River in the 1930s with the Hoover Dam (originally called the Boulder Dam). (iStockPhoto.com)

The reclamation experience in California's vast agricultural lands demonstrates these and other environmental issues at a magnified level. Although agriculture existed in California without irrigation and although irrigation projects already existed, three major reclamation projects in the mid-twentieth century transformed California agriculture and made the state's farmers more dependent than ever on hydraulic manipulations. The Boulder Canyon Project acquired and moved water from the Colorado River to the Imperial Valley; the Central Valley Project reconfigured the waterscape in the Central Valley to harness and store water from the northern valley's winter rains. Both of these projects were federal and primarily benefited agribusiness, although Los Angeles and its environs also profited from Colorado River water. The final reclamation scheme, the State Water Project, avoided some federal restrictions and involved a massive water transfer from northern to Southern California and amounted to the most significant state-funded water project ever. Cumulatively, this major reclamation work watered deserts and semiarid lands, converting them into a cornucopia of agricultural productivity and raising a host of complicated ecological and political questions (Hundley 2001, 203–302; Walker 2004, 175–181; Worster 1985, 194–212, 240–256, 290–295).

These projects achieved much. Irrigated acreage increased from 1.5 million acres in 1899 to 4.2 million acres in 1919, the era of initial investment in California reclamation. But not until after World War II did irrigation become the general agricultural practice of the state. Before that, farming focused overwhelmingly in Southern California, and this irrigation came almost entirely from wells and pumps, taking groundwater at rates far faster than natural replenishment (Walker 2004, 127–128, 174–176). Thus, the new reclamation works altered the geography of farming and expanded water development.

The first major California project centered on moving Colorado River water. The Boulder Canyon Project delivered 4.4 million acre-feet of the Colorado River to California, despite the fact that the state contributed almost nothing to the river's flow. In 1942, the project's All-American Canal brought water to the Imperial Valley at subsidized rates and irrigated nearly half a million acres, primarily to farms averaging 300 to 700 acres and some as large as 3,000. Thirty years later one could find holdings of 9,000 acres. This large size subverted the Reclamation Act's acreage limitation, but the valley landowners received exemption—evidence of agribusiness' political clout (Hundley 2001, 221–226). This sort of disproportionate development prompted historian Donald Worster to write, "Water had indeed made this desert bloom, and the crop was oligarchy" (1985, 206). At the same time the Imperial Valley began receiving Colorado River water, Los Angeles completed a 242-mile aqueduct to the river, although cities in the Metropolitan Water District did not begin using the water until the 1960s. Still, according to Norris Hundley, this water "had a profound psychological effect on city leaders and planners. It obliterated any sense of restraint about Los Angeles's capacity to absorb ever more people and industries" (2001, 231). As it had throughout the century, faith in engineered solutions continued unabated (Hundley 2001, 230–233).

Arguably, the Central Valley Project accomplished even more. Valley farmers had been pumping groundwater onto its best lands for decades. Indeed, in the 1940s, 35,000 wells extracted 6 million acre-feet yearly in the San Joaquin Valley (Hundley 2001, 277). The Central Valley Project put more land into crops, aided in flood protection, and prevented saltwater incursions in the delta. Hundley presented the project's basic figures: "When completed the Central Valley Project consisted of 20 dams and reservoirs and 500 miles of canals for managing 9 million acre-feet of water supporting 2.5 million city dwellers,

3 million acres of farmland, and a vast array of fish and other wildlife. In addition, it generated 5.6 billion kilowatt hours of electricity annually at eleven power plants for some 2 million people, receiving $34 million in power sales to help pay for the project" (2001, 257–258). Although much of the Central Valley Project had been proposed as a state project, the Bureau of Reclamation took over and put in place its acreage restriction of 160 acres. As in the Imperial Valley, though, enforcement did not follow. Political changes in Washington, DC, found waxing and waning interest in enforcing the limits. At one point, the head of the Bureau of Reclamation, Michael Straus, advised agriculturalists how to meet technical compliance without actually fulfilling the spirit of the law. For instance, corporate farms could obtain 160 acres for each shareholder or a grower could deed land to each of his or her relatives. This technical compliance, still the rule, allowed a system to develop where large farms dominated the landscape and squeezed smallholders out. Still, agribusiness feared changing political winds and desired a state plan that would ensure the bureau's 160-acre rule would never apply (Hundley 2001, 234–276).

The State Water Project constituted that ambitious plan. It grew out of growers' concerns about potential enforcement of acreage limits and groundwater pumping. This siphoning resulted in declining water tables (three to five times deeper in the 1950s than in the 1930s) and land subsidence. The program's roots lay in the 1940s, but not until the 1950s and Governor Edmund "Pat" Brown's astounding political ambition did the plan move forward. The program basically transferred water from the northern portion of the state to the southern. A series of subsidies lowered the cost of water for growers for several decades after the project's completion in 1972. Like the others, this project benefited agribusiness disproportionately. For instance, cheap and abundant water led to glutted markets that large operations could absorb far easier than small farmers could. Moreover, it put often marginal land into production. Perhaps because of these problems, Californians ended their consensus over large hydraulic projects after this project was completed. From the 1970s on, critics of various aspects of reclamation and agribusiness challenged business as usual (Hundley 2001, 276–302; Walker 2004, 177–179).

Indeed, critics of California agriculture have found much to indict in the state's efforts to remake nature. Historian Donald Worster has argued controversially that projects such as these have created a hydraulic society that is "coercive, monolithic, and hierarchical . . . [and] ruled by a power elite based on the ownership of capital and expertise" (1985, 7). In twentieth-century California, this elite has been basically the government—state and federal—in cooperation with agribusiness. Moreover, Worster maintained that control of water has been a primary way of controlling workers, too. Historians such as Norris Hundley and Donald Pisani have challenged Worster's characterization, seeing a more diverse region where small local owners enjoy more influence than Worster allowed. Although the almost conspiratorial tone of Worster's analysis lacks the subtlety of Hundley or Pisani, Worster's environmental critique remains more trenchant. He points out the fairly predictable results of the Far West's massive hydraulic engineering: oversubscription of available water; declining water quality, especially because of increased chemicals and salinization; and the loss of free-flowing rivers. Among these problems, the rise of the chemical presence may be the newest and most revolutionary and problematic (Hundley 2001; Pisani 2002; Worster 1985, 311–326).

Rising chemical use in the postwar period reflected the rise of what geographer Richard A. Walker called petrofarming. Farmers increasingly relied on chemical fertilizers and pesticides to boost production and fight various pests. They had long added inputs into their fields, but now these were increasingly inorganic chemicals. In California between 1940 and 1980, fertilizer use multiplied twenty times. Yields greatly increased. With chemicals and irrigation, farmers ended time-honored practices of fallowing fields and rotating crops. Thus, fields never recovered from cropping, as fertilizers substituted for rest and the biological quality of soil became decreasingly relevant. Meanwhile, this system of monocropping and never fallowing fields made them vulnerable to pest and disease outbreaks, problems that farmers hoped even more chemicals would solve. Consequently, California led the nation in pesticide use. In the 1970s, the state spent $500 million to spread more than 100 million pounds of pesticides, about 5 percent of the entire world's pesticide application. Yet pesticide use caused a number of problems. Many killed all insects, including natural predators. Moreover, insects evolved new resistance to many of these poisons. Thus, the need for newer and stronger pesticides constantly pushed agribusiness into greater chemical dependence. Finally, chemicals leached into water, polluting rivers and groundwater. This new reliance on chemicals promoted the chemical industry's power within the agribusiness complex (Walker 2004, 181–189).

This new agricultural-chemical complex created a series of human problems, too. In the late 1960s, the United Farm Workers' Organizing Committee (UFWOC) used the pesticides issue to help organize their union. The campaign highlighted the problematic relationship between economic and political weakness and environmental and health vulnerabilities often called environmental racism. Historically, California farm laborers have comprised various marginalized ethnic and racial groups, beginning with American Indians and proceeding through Chinese, Japanese, Filipino, Punjabis, Okies, and Mexicans (Walker 2004, 66–75). Workers of Mexican origin dominated the postwar era and thus bore the brunt of the interface between petrofarming's ecological and health problems. When the UFWOC raised concerns about ill health effects from pesticide exposure, regulatory officials disbelieved them. This official doubt stood against significant, but still undercounted, evidence of disease attributed to pesticides. From 1954 to 1967, California officially received an annual average of far more than 600 cases of occupational diseases related to the petrofarming complex, including poisoning, lung and skin problems, and chemical burns. The UFWOC used such problems to argue for worker safety provisions in the union contracts they hoped to make with growers. The organization also filed a series of lawsuits to improve how pesticides were used and to ban some of the more dangerous chemicals, especially DDT. Perhaps most famously, concerns about pesticides played a role in the United Farm Workers' national boycott of grapes. The boycott included advertisements that warned consumers about the dangers of pesticide residue on fruit, as well as its danger to farmworkers. In the midst of negotiations that had just broken down, the union movement's leader, Cesar Chavez, wrote a caustic letter to growers:

> As producers of food you must be aware of your social responsibility to the workers and consumers to insure their health and safety. Surely, California table grape growers are not so interested in profit from the sales of grapes that they would willfully do harm to unsuspecting workers and consumers. . . .
> Again, this grape season the indiscriminate use of DDT and another wholesale slug from the arsenal of economic poisons dumped on the grapes—enough poison to destroy the balance of nature for generations to come. The most efficient, inexpensive, and expedient way to produce the damn grape seems to be the only touchstone of morality in your industry. Our city water, our streams and rivers, our ocean, our atmosphere and environment, our fish and wildlife—indeed, our very health and well-being as consumers and workers seem to have no place in the grape market (quoted in Pulido 1996, 110–111, 111–112).

Such an image of the grape industry immorally poisoning nature, not to mention harming workers, made an impact as some growers soon broached negotiations anew. Pesticides, thus, were a powerful publicity tool for farmworkers. When contracts were agreed upon in 1970, health and safety conditions protected workers from abuse by pesticides. Indeed, scholar Laura Pulido credits the UFWOC for being the first group to crack open the pesticides issue in California agriculture, eventually allowing for greater regulation and better protection (Pulido 1996, 57–124).

Another critique of petrofarming arose when some western farmers adopted organic agriculture. Modern organic agriculture grew from various roots. Geographer Julie Guthman identified four social movements from which the organics movement derived strength: "alternative production technologies, the health and pure food crusades, the 1960s counterculture, and modern environmentalism" (2004, 4). Broadly, organic agriculture challenged the prevailing use of chemicals, the highly specialized large farms that characterized most of modern agriculture, and the exploitive social and environmental practices of agribusiness. It promised a revolutionary critique of modern production by adopting a chemical-free agriculture that mimicked nature's rhythms and reduced exploitation by relying on small-scale family or cooperative farms. At least that was the radical vision; the reality turned out somewhat differently (Guthman 2004).

What occurred, at least in California, undermined this broad reform vision. Among the most important discouraging factors were land values. California's agriculture had evolved into both large enterprises that encompassed massive acreages of forage crops and smaller specialty farms with high crop values. These two trends made land values unusually high, and organic farmers were forced to interact within that same market. It put them in an unenviable position for enacting a truly revolutionary agriculture, as they would need to profit enough to afford the exorbitant rents. Paradoxically, though, organic products came to earn high prices in urban markets, making them an attractive option for specialty growers who constantly searched for innovative ways to make money. As agribusiness entered the organics market, usually as mixed growers who still grew most of their crops conventionally, they threatened and then obliterated the organics movement as a movement to fundamentally reform agriculture. One way this happened was through the development of organic standards. Essentially a gatekeeping mechanism, organic standards defined what were acceptable inputs in the agricultural regime. But this strategy made "organic" only a narrow, technical definition about inputs, not broader ecological, social, and economic goals. Indeed, the *California Certified Organic Farmers Certification Handbook* made the point clearly:

> The Material List [of acceptable and unacceptable inputs] is emphatically not a recipe for organic farming; a grower who relies primarily on highly soluble mined fertilizers for fertility management and botanical insecticides for pest control may be "organic" within the letter of the law, but cannot be viewed as truly farming organically. They are merely replacing a synthetic treadmill with a botanical one (quoted in Guthman 2004, 120).

Thus, although it was important to create some sort of criterion, the standards do not represent a comprehensive measure for the broadest organic vision. Moreover, the standards quickly reduced the ability of smaller farmers to enter the organic market. Today, those advocates who see the potential for a greater change in agriculture want to deepen the meaning of "organic" so that it represents more than simply conforming to a forbidden input list. Whether reform is achieved within organic agriculture remains to be seen, but continuing conflict seems certain (Guthman 2004).

Despite its long history in the region, agriculture in the Far West faced significant changes in the second half of the twentieth century. Agricultural processes intensified the use of land, especially with water and chemicals. Resulting changes caused declining water quality, human and land health concerns, and a backlash against conventional agriculture. In many ways this trend reflected broader dynamics in the modern era, for in forestry and urban development far westerners also faced an intensification followed by reactions and attempts to reform standard operations. And just as the organic solution proved incomplete, other reforms did not always live up to their promises.

FOREST DEBATES

One of the most contentious environmental practices in the modern era focused on forests, and the Pacific Northwest's forests stood tall in these debates. After World War II, an unprecedented appetite for wood led to increased timber production, especially in national forests. Because of the public nature of these forests and the immense production totals, forest practices received tremendous public scrutiny all the while balancing the environmental and economic priorities that challenged the modern forest industry. However, these problems were not apparent to most far westerners as the post–World War II era began; instead, optimism characterized their belief in the Forest Service's ability to manage forests and in the timber industry's ability to produce responsibly. Very quickly, that faith proved to be misplaced.

Forestry Practices

The approach to forestry was familiar. Indeed, historian William G. Robbins noted striking similarities between agriculture and forestry: "The quest for efficiency, the attempt to achieve economies of scale, and cooperation between the private sector and public agencies to advance agricultural production also extended to modern forest management" (2004, 143). Although the public never quite saw timber and agriculture as the same, they operated with a similar belief in science and technical solutions and a strong reliance on chemicals. Armed with these practical tools and predilections, postwar foresters demanded more out of the national forests than their predecessors had. Aptly, historian Paul W. Hirt (1994) claimed that the Forest Service operated under "a conspiracy of optimism," an ideology that promised with greater technology, better expertise, and more investment federal foresters would solve environmental problems and promote high levels of timber production even as evidence mounted against such a confidence. These convictions shaped a new era in the timber industry that quickly led to several environmental, economic, social, and political problems. This mismanagement, though, was nothing new having enjoyed a long gestation.

During the 1930s, the federal government embarked on several programs of resource planning to stabilize communities and their economy, but they ultimately fell by the wayside. For their part, federal foresters wanted to adopt sustained-yield programs through cooperative agreements with public and private agencies and lands, as there was a strong sense among some that cutting was proceeding too rapidly for either the economy in timber towns or ecology in the woods. Cooperative agreements would stabilize communities by promising lower levels of production over longer periods; however, the competitive economic model of the United States complicated this faith as anything resembling centralized planning lost appeal during the Cold War era. Moreover, because the federal government tended to prefer to negotiate with better-capitalized companies, cooperative arrangements tended to leave out small mills. Then, timber interests called for federal timber to supply those mills, an approach that followed neither the free market nor ecological principles (Robbins 2004, 147–156).

Meanwhile such practices set timbered regions on a tragic track. For example, in central Oregon at the dawn of World War II, lumber companies were cutting at three times the pace of sustainability. Clearly, this was not a plan for permanence. Two big sawmills in Bend, Oregon, insatiably cut into the regional ponderosa forests. Although timber company spokespersons assured everyone that they were practicing conservation, the harvesting stood at demonstrably and obviously unsustainable levels. Meanwhile, mills could only continue producing if they gained access to new federal timber. To gain access to this timber, companies demanded roads, something more important now that trucks rather than trains transported most of the logs. This new mobility, along with gasoline-powered chainsaws, accelerated and spread the ecological damage, such as increased landslides and siltation caused by roads. Although they contemplated reform for the timber industry during the depression, federal foresters largely capitulated to industry demands for rapid forest "conversion," a euphemism usually for clear-cutting that would convert old-growth forests into new-growth forests. Clearly, marketplace priorities and production goals overwhelmed any decision based on ecosystems. Thus, high production figures hid real costs in degraded land, water, and communities. Foresters

Trucks transport ponderosa pine from the Malheur National Forest in Oregon, July 1942. Trucks quickened the pace and spread of logging after the 1940s. (Russell Lee/Library of Congress)

optimistically embraced intensive forestry as a solution to problems (Hirt 1994; Langston 1995, 264–269; Robbins 2004, 177–212).

In many ways, intensive forestry simply repackaged the maximizing and efficiency ethos of Progressive Era conservation. Washington's Gifford Pinchot National Forest exemplifies the approach to intensifying forest uses and the resulting negative consequences. In this forest in Washington's southern Cascade Range, just north and east of Portland, the USFS pursued a policy consistent with management throughout the Pacific Northwest. The policy promised maximum timber production alongside broad environmental protection. Foresters expected to be able to control nature with their own expertise along with technological and capital infusions. The rush to maximize production through intensive management came after World War II when private forestlands (already largely cut) could no longer meet public demands. Believing in their power to maintain both economic and ecological values, managers of the

Gifford Pinchot National Forest "facilitated logging on 10 to 20 square miles of forest land *per year* [emphasis in the original] since the 1960s, sending to local mills annually 400 to 560 million board feet of lumber (enough to build 40,000 to 56,000 three-bedroom homes each year), and generating gross receipts of $17–$20 million dollars per year from its timber sales," according to historian Paul W. Hirt (1999, 210). To achieve these impressive production figures, in the 1980s the Forest Service authorized significant road construction and repairs, amounting to 80 to 90 new miles and 40 to 50 reconstructed miles annually. In the areas of the forest dedicated to intensive timber production, there were about 3 miles of road per square mile of forest. Such development was immense, indeed, and it also fragmented ecological systems, hastened unsustainable timber harvesting, and accelerated erosion (Hirt 1999).

Moreover, such practices directly undermined established policy. Forest Service policy dictated only sustained-yield harvest. However, managers frequently revised this number upward. One district in the forest set a limit of harvesting 34 million board feet in 1940, but the USFS revised that number to 55 million board feet in 1953. Multiply such decisions across the entire forest or the entire region and one discovers massively inflated allowable cuts. And by the 1960s and 1970s, the Forest Service did not even stay within these exaggerated allowable cuts with the actual cut between 400 million and 560 million board feet annually exceeding the established allowable cut that was established at 200 million board feet in 1949 and revised in the early 1960s to 381 million board feet. In the late 1980s and early 1990s, these deliberate decisions to ignore their own figures—and to do so simply to increase production quotas—led to hard times for Forest Service managers who could no longer ignore the mounting evidence of serious ecological damage or the growing protest among the public (Hirt 1999).

In the inland Northwest, fire suppression illustrated different predicaments of intensive forestry. With some dissenting opinions, the Forest Service practiced fire suppression according to both their management values and scientific evidence that suggested that suppression would improve forest growth. This practice overwrote fire suppression on a rich palimpsest of both natural and anthropogenic fire regimes. In addition, insect infestations attacked the undergrowth of many timber stands adding even more fuel to potential fires. Intensive management eliminated fire, as best it could, to prevent waste and that helped promote fir growth in the pine understories that helped support such insects as the spruce budworm and the Douglas-fir tussock moth. In the mid-1990s, fires in the Blue Mountains of Oregon burned almost 300,000 acres. In recent decades, more frequent and more intense fires have become annual occurrences, sparking a debate about the proper role of fire in forest management.

Managers now have become interested in determining how to include anthropogenic burning in management plans. Reinserting human fire regimes has proven difficult, however, as the forests have changed irrevocably in the intervening years since a frequent anthropogenic regime was in place. With denser fuels, forests bear much greater chances for escape or escalation from small-scale to devastating fires. Thus, anthropogenic burning today does not and cannot resemble past regimes. Tending to these changed ecological and political circumstances is a main challenge of modern resource managers, as they confront a shifting political context (Langston 1995, 3, 36–37, 247–263).

Forest-Policy Changes

The backdrop to intensive forestry was a number of policies that at first facilitated it, then challenged it. Timber interests long acted as though sustained yield meant maximum production and multiple use meant timber first and any other values far behind. Historically, the USFS had largely adhered to this view, although by the 1950s some in the agency challenged that perspective. However, the last half of the 1950s found Congress and the Forest Service in an extended debate over the direction of USFS policy, a debate that promised great impacts for the Far West. The ultimate result of these debates, the Multiple Use-Sustained Yield Act of 1960 (MUSY), mostly just articulated USFS policy as practiced; however, it became the first time a law purposefully put outdoor recreation alongside the more traditional range, timber, watershed, and wildlife and fish management priorities, a slight but notable change. MUSY allowed for some areas of national forests to be used in ways that did not maximize their financial potential (Hirt 1994, 171–192; Steen 2004, 278–307; Wilkinson 1992, 137).

The 1960s saw a flourish of challenges to MUSY and dissatisfaction by nearly all sides. Increased environmental activism stirred the placid waters of forest management, and by decade's end a new law promised major changes to the Forest Service, not to mention many other resource agencies. Although not focused strictly on timber, the National Environmental Policy Act (NEPA), signed by President Richard Nixon on 1 January 1970, affected national forests in the Far West significantly. Among other things, NEPA required federal agencies to write environmental impact statements before undertaking activities that would significantly affect the natural environment. Thus, before the Forest Service planned a timber sale or built a road, it was now required to evaluate the proposal, lay out the likely ecological costs, establish alternatives, and open the environmental impact statement to public debate. All of this was required before the agency proceeded. NEPA constituted a landmark in environmental legislation, clearly the high-water mark of environmental legislation of Congress. The law forced agencies like the Forest Service (but also the Bureau of Reclamation, National Park Service, and many others) to consider ecological values more holistically than ever before when creating management plans. The law did not require an agency to choose the least environmentally disruptive option, but in making multiple options clear and then opening it up for public debate, NEPA significantly altered the dynamics involved in decision making. Moreover, it brought the discussion to the public, giving environmentalists their strongest weapon to fight against what they perceived to be poor resource management. Since 1970, environmentalists have used NEPA in myriad ways to stop countless development plans.

Then, in 1976, Congress passed the National Forest Management Act (NFMA), a noteworthy law and the last in a series of reforms during the active previous decade. Initiated as a response to a lawsuit over management in the Monongahela National Forest in West Virginia, the NFMA created a far-reaching solution with ramifications far beyond the Monongahela. The NFMA offered some small restrictions on clear-cutting, a practice that the public increasingly opposed vigorously. The law also required the Forest Service to maintain viable wildlife populations; in fact, the NFMA has been used more than the Endangered Species Act to force the USFS to improve wildlife management. In the spirit of NEPA, too, the NFMA demanded more comprehensive planning and more options for public input. Most importantly, though, the NFMA defined sustained yield as maintaining "'non-declining even-flow': a forest's output of timber must be capable of being sustained perpetually without declines" (Hirt 1994, 263). Unfortunately, but predictably, this law included a large loophole that allowed unsustainable harvest practices to continue under certain circumstances. In practical effect, the status quo was maintained with only slight modifications (Hirt 1994, 260–271; Steen 2004, xv).

Finally, in 1992, the USFS Chief, Dale Robertson, announced a new policy: ecosystem management, also known as New Perspectives. It was a broad statement and as such meant that ecosystem management could mean different things to different people. Many traditional-minded foresters and timber industry representatives interpreted ecosystem management as permission to practice ever more intensive forestry. Environmentalists, on the other hand, wished to use the insights of conservation biology to manage forests so that natural processes dominated. To be certain, ecosystem management contained enough vagueness to ensure confusion and continued debates over specific management practices. Conflict over managing habitat for endangered fish and wildlife animated much of the Forest Service's activity in the Far West since the 1970s, accelerating in the 1980s and 1990s as federal courts upheld numerous lawsuits and halted old-growth logging in the Pacific Northwest (Hirt 1994, 287–289; Langston 1995, 269–270; Steen 2004, xxix-xxxi).

Forest Conflicts

Indeed, these changing laws and policies all evolved within the context of a changing environment and public values that increasingly gave voice to environmental concerns. As much as anything in the Pacific Northwest, concerns about forests invigorated the environmental movement. Chemicals, clear-cuts, and roads were perhaps the three most contentious public issues. (Endangered species protection also engendered much debate, but that is explored in this book's final case study.)

As in agriculture, chemical use in forestry increased after World War II. Foresters used chemical herbicides to control insect and fungi outbreaks and to promote the growth of shade-intolerant Douglas fir. Industry officials likened DDT, the favored pesticide, to antibiotics in promoting health. Foresters justified the spraying as necessary to prevent moths from spreading. Typically, tussock moths would come and last for a couple seasons until a virus would mostly eradicate the population. In eastern Oregon, these recurring insect infestations were likely related to the replacement of ponderosa pines with firs, which were weaker in withstanding insect attacks. DDT, industry officials argued, provided benefits far outweighing long- or short-term costs. Thus, chemicals became a mainstay for the forest industry (Robbins 2004, 188–197).

However, as it had for agriculture, chemical use in forestry disconcerted the public. Despite assurances from Oregon forestry officials that the chemicals were safe, several public protests occurred. For instance, in western Oregon, controversies erupted over using 2,4,5-T, a notorious chemical used in creating Agent Orange and suspected of causing cancer, birth defects, and miscarriages. Spraying in southern Oregon's Siuslaw National Forest caused a public controversy and was compounded by the fact that the Siuslaw was producing more timber than any other forest in the nation. Besides legitimate health concerns, the public was angry that the national forest's officials did not distribute the environmental impact statement widely and thus stunted public debate. Ultimately, Citizens Against Toxic Sprays (CATS) and other organizations filed a lawsuit to stop the spraying and succeeded in slowing the practice. Actions like that of CATS forced changes in spraying practices, including forcing more public acknowledgment of the defoliant spraying and informing local landowners. The issue reached a head when nine women near Alsea, Oregon, an area that had been sprayed, had miscarriages. One of the women tracked the miscarriages and the sprayings to show a correlation, although she carefully never claimed cause and effect. She mailed her findings widely and further investigations revealed more instances of spontaneous abortions. Critics argued that the methods used informally by the local woman and more formally by the U.S. Environmental Protection Agency were flawed. Oregon State University scientists vocally criticized the studies, exposing the close ties between the university, the chemical companies, and the timber industry. The bad publicity finally led to the end of 2,4,5-T use. The pesticide controversies demonstrated the initial ubiquity of chemical use after the war, as well as the power of citizen activism to alter practices in a new political and environmental context (Robbins 2004, 194–205).

Clear-cutting was another widely used forest practice that was revised because of criticism from several quarters. Foresters saw clear-cutting as a practical tool and argued that it mimicked such ecological processes as windstorms and fires. The analogy was simplistic, false, and self-serving (Langston 1995, 290). Moreover, because the goal of forestry was wood production, they used clear-cutting to open the way for second growth, maintaining again that this constituted the most efficient silvicultural method. A series of localized protests against clear-cutting from all around the nation coalesced into a series of Congressional hearings led by Senator Frank Church of Idaho. Ultimately, political and scientific experts recommended that the Forest Service needed to be more careful with their clear-cutting practices. Even this mild condemnation angered Northwest timber industry leaders who argued that the Forest Service was managing more for recreation than for timber and in the process wasted trees and undermined true multiple-use management. Clearly, vested interests wanted the status quo maintained (Hirt 1994, 245–251; Robbins 2004, 181–184).

Many groups weighed in on the clear-cutting controversy. As the timber products industry maintained the necessity—economic and ecological—of clear- cutting, reform-minded forest scientists such as Jerry Franklin entered the debate and suggested a variety of possible approaches besides and in addition to clear-cutting. Franklin believed some areas that were primarily productive timberlands were being appropriately harvested with clear-cutting. Other areas should not be felled. In other words, timber managers ought to develop more comprehensive plans and allow greater flexibility. Meanwhile, politicians avoided fundamental questions and often simply urged managers to include clear-cutting in less obvious places so that the practice could be continued. Oregon Governor Tom McCall, for instance, always saw timber in an economic, not an environmental context, and sided with the timber industry. He wanted scenery preserved, but not larger ecological integrity. For some of the public and environmentalists, clear-cuts provided a visual weapon against the practice and could even persuade conservative legislators, as clear-cuts left such obvious ecological scars on the land. By the 1970s, there was a strong movement to ban the practice entirely. Eventually, the National Forest Management Act in 1976 ensured that some clear-cutting could remain in the sustained-yield program. Nevertheless, by the late 1970s, clear-cutting became a much reduced silvicultural practice (Robbins 2004, 184–188).

A last issue related to forestry that sparked public protest concerned roads. The roots to antiroad sentiment rested in the wilderness movement that germinated in the interwar period and was revitalized in the 1950s. To many wilderness advocates, roads symbolized an invasive modernity that was inappropriate to wilderness (Sutter 2002). To be sure, many protested roads because of ecological issues, too. Roads were notorious for creating landslides, fragmenting habitat, and exacerbating erosion, not to mention simply providing easier access to harvest more trees (Hirt 1994). For these reasons, environmentalists protested the constant construction of forest roads. They seemed ubiquitous, and after the Wilderness Act (1964) prohibited roads in designated wilderness areas, roads became centerpieces in numerous environmental conflicts. For example, U.S. Supreme Court Justice William O. Douglas was a Pacific northwesterner and avid hiker who fought publicly and privately from the 1950s through the 1970s for keeping roads out of several Northwest national forests, efforts that ultimately prevailed in what is now known as the William O. Douglas Wilderness Area (Sowards 2006). Roads became arguably the most symbolic target of environmental protest in the modern era.

In time, concerns developed about the inability to continue high production. Many mill closures from the late 1970s and into the 1980s wreaked economic havoc in Western communities. Still, the 1980s found the Forest Service under pressure to push up production that agency officials knew was unsustainable. Politicians such as Senator Mark Hatfield of Oregon argued that environmental protections were ruining a functioning Northwest economy. Despite such claims, most mill closures were unrelated to or only marginally connected to environmental laws. Instead, changing technology was a far more likely factor. Meanwhile, critics of road-building and logging practices on fish and wildlife populations became louder. And then came the northern spotted owl (*Strix occidentalis caurina*). Forest job losses were large, but the decline in the industry began before the spotted owl issue and were largely the result of technological changes in the woods and mills. Thus, as the twentieth century ended, the Forest Service and the timber industry faced increasingly difficult ecological and political problems, challenges that have yet to be worked out (Robbins 2004, 205–212).

Forests have particularly strong meaning in the Far West, especially in the Northwest, as they constituted a significant part of regional identity and economy. The postwar era experienced a rapid assault on the region's national forests because of housing demands and the inability of private forests to continue producing. Relying on their faith in science, technology, and expertise, foresters set out to reorder Northwestern forests to meet the economy's perceived needs. In the process, the forests changed markedly. Unfortunately, those changes frequently harmed forest ecology with clear-cuts and roads encouraging erosion and decreasing biodiversity. Foresters' substantial faith in their own abilities was found wanting by a public increasingly interested in intact forest ecosystems and by politicians hoping to gain environmental votes. Indeed, many of the legislative achievements of the modern environmental era focused on improving the Forest Service's management of ecosystems. This transformation of the public's forests, from being administered first as tree farms and then increasingly like ecosystems, represents an important, albeit incomplete, change in the Far West's environmental history and sensibilities.

ENERGY AND PROTESTS

Questions related to energy also fueled many environmental and political controversies in the postwar era. These issues ranged from concern over declining salmon runs in the Northwest caused by hydroelectric turbines to grassroots protests against nuclear power plants. The growth of cities was a significant reason for the rise of this power politics. They demanded electricity for both the increasing populations and the new technologies that sucked electricity from urban and suburban wall sockets. Indeed, following the pathways of power through ecosystems reveals surprisingly much about the modern Pacific West. These environmental questions are among the newest, as regional debates about water and forests and agriculture had been around for a century. Modern energy developments, though, introduced some new dynamics to Far Western environmental history, although the general themes of their environmental consequences were familiar.

The New Deal helped prepare the expanding energy infrastructure. Giant hydroelectric dams, such as the Bonneville and Grand Coulee on the Columbia River, promised cheap electricity along with improvements for navigation and floods and eventually some irrigation. With such benefits, residents and regional politicians broadly supported their construction. Many, like Senator Richard Neuberger from Oregon, believed no harm would befall salmon or that engineers could solve any problems. There were many critics, though—some conservationists, fisheries scientists, commercial and sport fishers, and American Indians. In 1946, the U.S. Fish and Wildlife Service even proposed, unsuccessfully, a decade-long moratorium on dam building, so it could study the effects on fish before further damming. Although many groups considered salmon, the fish were never deemed important enough to slow hydro-development or to displace the technological optimism that prevailed just as it had in agriculture and forestry. In combination with other ecological factors, dams constituted an insurmountable problem to anadromous fish and those who depended on them (Goble 1999, 248; Robbins 2004, 47–76).

Fundamentally, the dams remade the Far West's river systems and the life they supported. At a basic level, they changed the energy of rivers by tapping it and transforming it in to marketable electricity (White 1995, 59–88). Facilitating this revolution, from the 1930s to the 1980s approximately 75 dams were built within many of the Northwest's watersheds, although not all of them generated electricity. All of them did, however, change the hydrological nature of the region, and these dams modified the rivers to make salmon survival increasingly untenable. For instance, the Grand Coulee Dam by itself blocked a thousand miles of spawning grounds, as it was too large for any fish passage system (Taylor 1999, 225–228, 245). The many smaller dams created smaller individual effects but a greater cumulative one. Additionally, the storage reservoirs slowed the river, changed the seasonal volume of river flows, increased water temperatures, and multiplied opportunities for predation. All of these worked against salmon survival. Moreover, turbines caused a mortality rate of up to 15 percent on the downstream migration of smolts. Because some Idaho smolts needed to travel through eight dams before reaching the Pacific Ocean, chances for survival were minimal (Goble 1999, 247). The billions of dollars spent on hatchery programs has made it appear that politicians and scientists were solving the problems, but salmon suffered from habitat losses, more dams, the preference for hydroelectric power, and the failure of fish culture to make salmon that survived in a healthy state (Taylor 1999, 237–257). From the salmon's perspective, hydroelectric development has been an unmitigated ecological catastrophe. Thus, it is important to remember the optimism that animated postwar developers and to recognize that the challenges to this hydro order came slowly.

The U.S. Bureau of Reclamation and the Army Corps of Engineers, along with private power companies, went on a building frenzy after World War II, leaving few streams free-flowing. But such a construction spree could not go on forever. The history of Hells Canyon illustrates the politics of dams and power and how the tide slowly changed. Set on the Oregon-Idaho border and formed by the Snake River, Hells Canyon is the continent's deepest gorge and provides numerous excellent dam sites. Throughout the 1950s, regional politicians like Oregon Senators Richard Neuberger and Wayne Morse and Idaho Senator Frank Church and Representative Gracie Pfost tried to get a large dam in Hells Canyon and to make sure it would be a federal dam, thus promising cheaper public power. This group of politicians did not initially question the dam's effect on salmon or scenic values; they simply wanted it to be out of the hands of private power companies and believed unspecified technological solutions might save salmon runs. The large dam never was built. Instead, four smaller ones blocked the Snake (Robbins 2004, 215–225).

The issue reasserted itself in the 1960s when another proposal came to the Federal Power Commission (FPC) to build a private power dam in Hells Canyon at a place called High Mountain Sheep. Trying to stop that private development in favor of a public project, Secretary of the Interior Stewart Udall slowed the process and the case reached the U.S. Supreme Court in 1967 where northwesterner Justice William O. Douglas wrote the majority opinion. Douglas, a noted conservationist, questioned "whether any dam should be constructed," moving the debate beyond simply public versus private power development. "The ecology of a river is different from the ecology of a reservoir built behind a dam," Douglas argued, using "ecology" for the first time in Supreme Court history (*Udall v. FPC* [1967]). The Court remanded the case to the FPC for further study, which effectively closed the proposal by raising the questions the Court did. The case stood out for its conclusion that surprised those who had never questioned continued dam construction. Moreover, the Court's decision clearly represented a turning point where environmental values and ecology might make a difference in management and legal decisions (Sowards 2002, 165).

Similar turning points also affected nuclear power plants. Seen as a solution to the region's increasing power demands, nuclear power development proceeded slowly but deliberately in the postwar era. By 1970, California had almost 100 nuclear power plants, but the public began to question further nuclear projects. A protest in northern California illustrated how environmental protest connected to other political concerns. Pacific Gas and Electric wanted to build a nuclear power plant at Bodega Bay, about 50 miles north of San Francisco. Pacific Gas and Electric appealed to local political and business leaders. However, officials ignored 1,300 petitions from local residents and did not allow a public hearing, a prospect many 1960s activists found anathema to participatory democracy. Protesters cited the environment, scenic beauty, safety, and conspiracy against democracy. Significantly, antinuclear activists also cited the danger of radioactivity moving through the food web to create some fear locally, an argument all the more powerful given Rachel Carson's deployment of it in her then-recent book *Silent Spring*. Ultimately, a geologist's testimony that a fault ran through the building site scuttled the project. Thus, earthquakes and attendant safety concerns led California to adopt relatively stringent nuclear safety standards that slowed further development, although objections over seismic uncertainty did not stop all nuclear development as the approval of a plant at Diablo Canyon attested. Still, in 1976, the Nuclear Regulatory Commission ordered one California nuclear plant at Humboldt Bay closed because of the discovery of active faults. Furthermore, the Bodega Bay victory also emboldened the antinuclear movement; subsequently, nuclear power development proceeded more slowly (Wellock 1998; Williams 1997, 305–307).

Hanford, Washington, offers the most unique case of nuclear energy questions. Federal officials selected this lightly populated area of south-central Washington to produce plutonium to build an atomic bomb. Designed to aid the U.S. effort in World War II, this project rearranged the basic elements of the universe into powerful weapons and, by the mid-1960s, marketable electricity. Hanford's location used the Columbia River as a cooling source for the reactors and consumed tremendous amounts of the hydroelectricity produced by the Grand Coulee and Bonneville dams. Thus, the Columbia River once again helped transform nature into a new energy form. Indeed, the plutonium produced at Hanford existed in only insignificant amounts in nature, so the engineering basically produced a new form of nature (White 1995, 81–84).

Harmful by-products produced by this process spread out of Hanford through the surrounding region. In 1949, the Atomic Energy Commission secretly and purposefully released large amounts of iodine-131, a radioisotope, into the atmosphere, an act known as Green Run. It released more than 500 times the amount discharged in the well-known Three-Mile Island disaster of 1979. Locally, levels were 1,000 times more than the acceptable level. Over a decade, half a million curies of this radioiodine were windswept over the inland Northwest. This poison bioaccumulated in food webs and concentrated in human and animal thyroids, causing cancer. Many have argued that Hanford managers deliberately poisoned northwesterners by their practices, as iodine-131 was a known carcinogen and cancer rates downwind seemed disproportionately high. Also dangerous to the surrounding ecology was the use of Hanford as a waste site. Buried, the waste eventually leached into groundwater and the river. Cleanup will cost many billions and will never eradicate the radioactive waste. But perhaps most ironically, the section of the Columbia that now flows past the Hanford Nuclear Reservation is the only section of free-flowing river left. And as a classified area, the reach became a wildlife refuge where endangered fish and wildlife could enjoy relative protection. In 2000, President Bill Clinton preserved it as part of Hanford Reach National Monument, demonstrating the sometimes paradoxical pathway to environmental protection (Gerber 2002; Hein 2003; White 1995, 83–88).

If nuclear power suggested the newest energy approach, then the old standby, oil, still provided much of the region's energy, as well as fodder for environmental protest. California's share of domestic oil production dropped sharply in the World War II era. Still, after the war, offshore oil development boomed when federal policies encouraged it. Santa Barbara residents had long worried about oil development offshore from their picturesque community. In January 1969, a Union oil well blew out, and nearly a quarter million gallons of petroleum spread into an oil slick 800 miles long and darkened 30 miles of beach. The Santa Barbara oil spill dramatically symbolized to a national audience that Americans' oil dependence rested on devastating ecological demands and potential problems. Moreover, the issue gained greater meaning, largely because it affected Santa Barbara's well-to-do residents. This attention intensified efforts to pass environmental legislation, like NEPA, which was signed less than a year after the spill, and in the 1980s residents resisted efforts to allow more offshore drilling. Once again, ecological calamity sponsored renewed reform efforts (Rothman 1998, 101–105; Williams 1997, 297–300).

As with agriculture and forestry, energy development transformed the environment in significant and sometimes harmful ways. Power development in the postwar era dramatically transformed nature from changing rivers into a series of reservoirs that radically disrupted existing ecologies to producing plutonium through engineered nuclear reactions. Ironically, increasing energy availability stimulated economic growth throughout the Far West that, in turn, allowed the beneficiaries of that growth to question energy consumption and the costs of energy production, as did environmentalists in the aftermath of the Santa Barbara oil spill or critics of the dams on the Columbia River. Energy historian James C. Williams explained in the Californian context: "Air pollution, offshore oil drilling, nuclear power, and other energy-related issues increasingly fell afoul of the rising environmental movement that was itself prompted by the good life engendered in part by abundant, cheap energy" (1997, 11). This condition largely prevailed in the modern era throughout the Far West. However, as more people questioned the costs of these energy developments, many searched for alternative energy sources. Smaller, local energy sources, as well as greater reliance on renewable sources such as solar or wind power promise to change the energy dynamics for the region. Indeed, California has led the nation in solar power use and wind-generated energy, as the massive hillsides of windmills at Altamont Pass and in the Tehachapi Mountains testify. Although hopeful, it is unclear how far-reaching those efforts will be (Williams 1997).

Volunteers clean up some of the 200,000 gallons of crude oil that leaked into the ocean from offshore oil drilling in the Santa Barbara Channel in 1969. The massive oil spill created an 800-square-mile slick that coated 35 miles of coastline with thick tar, killed great numbers of birds and other sea life, and ignited a public outcry that furthered the popular modern environmental movement. (Get Oil Out!)

CITIES IN THE MODERN FAR WEST

Some of the greatest energy (and other) demands, of course, came from the region's burgeoning cities. Urban concentrations had long shaped their ecosystems in powerful ways and influenced land use in their hinterlands. The last half of the twentieth century saw these trends continue with additional dynamics. Especially as cities grew spatially and encroached into rural areas, new environmental conflicts erupted. What also emerges is a picture of a highly polluted urban Far West and one where pollution affects populations differently. A few case studies of urban changes can highlight some of these various trends.

Southern California is the archetype of urban sprawl and attendant ecological problems. Because Angelenos had been so effective at acquiring water sources throughout the twentieth century, the metropolitan area grew, the population grew, and the geographic impact grew. Unfortunately, developers ignored ecological conditions that jeopardized this growth. Aridity, of course, was a primary deterrent to urban growth, but one that city leaders had solved by taking water from the Owens Valley, the Colorado River, and northern California. The fifty active faults underlying Southern California's most heavily urbanized counties similarly threatened urban growth. The 1994 Northridge earthquake, for instance, killed 72, made 25,000 homeless, and cost $42 billion. Government support and disaster amnesia, though, almost guaranteed rebuilding in areas shakily interwoven with fault lines. Devising ways to subvert or ignore ecological reality has been skillfully mastered in the Southland (Davis 1998, 25–39, 47–52).

Perhaps even more insidious has been the invasion of suburban development into the fire-prone chaparral environment of Southern California's foothills. The pyrophytic ground cover of Southern California has always burned regularly—with and without human influences. A century of fire suppression efforts substantially increased fuel loads in the foothills. Periodically throughout the twentieth century, fires burned in the foothill communities such as Malibu causing massive damages to the high-priced coastal homes abutting the mountains. According to historian Stephen J. Pyne, these "Malibu fires inaugurated a new fire cycle in which fires were media events, a suburban invention, started for reasons that had nothing to do with traditional fire practices" (1997, 423). Suburban development upended typical ecological regimes. Nonetheless, such a regime replacement became typical as urban development pushed into hinterlands. Indeed, this fire problem has become exemplary, named the intermix or urban-rural interface fire problem. The trends are worsening for property owners. Between 1920 and 1989, wildfire consumed 3,500 buildings in California, while the period between 1990 and 1993 saw 4,500 burn. And all of this took place as a disproportionate amount of the national fire control budget focused on Southern California. The combination of urban sprawl, climate, and the biological reality of fires spells a future in which burning buildings and brush will continue to plague not only Southern California but any place where conditions are ripe, such as the Northwest's inland mountain communities (Davis 1998, 99–112, 140–147; Pyne 1995, 224–231, 269–295; Pyne 1997, 404–423).

As cities ate into rural areas, consumed habitat, and grew new fuels for fires, they also generated pollution. Indeed, Los Angeles became particularly notorious for air pollution. This pollution came from the city's sprawling nature, which required hundreds of thousands of commuters to travel by automobile. In addition, its physical setting exacerbated air pollution because the surrounding mountains trapped foul and stagnant air full of hazardous particulates. The area's population mushroomed after World War II; in the 1950s, more than a thousand people a week relocated to Southern California. By the mid-1950s, there were approximately 2.5 million automobiles in Los Angeles Basin and twice that many people. Every day 3,000 tons of air pollutants spewed into the air from burning fossil fuels and incinerating garbage. Over the next two decades, the population doubled and automobile growth expanded at even greater rates. In the early 1960s, the air quality declined to a point where officials reported eye irritation about 200 days during a year. In 1967, automobiles pumped carbon monoxide into the air at a rate of 20 million pounds *daily*. Statistics also showed that heart attacks increased dramatically after days of high carbon monoxide levels, and studies found that drivers blamed for automobile accidents possessed higher levels of carbon monoxide in their blood, which suggested slowed reaction times. If human health did not persuade Angelenos of the problem, the air quality was so low that Los Angeles was threatened with declining growth, something anathema to the typical booster attitude of growth at all costs (Commoner 1998, 373; Davis 2002, 55–85).

Such economic and health problems demanded action, and California played a national role. California passed the first state pollution act in 1947, creating an Air Pollution Control District for every county; later, in 1955, the federal government passed a Federal Air Pollution Control Act. These laws suggested action, but they actually did little, in part because scientific understanding of the health effects of air pollution remained in its infancy. Those who wished to thwart pollution control—such as the automobile and petroleum companies—saw scientific uncertainty as a reason to not act. The political tide continued building to support environmental reform, though, and in 1970 Congress passed the Clean Air Act and President Richard Nixon created the Environmental Protection Agency (EPA). The Clean Air Act established National Ambient Air Quality Standards, which set acceptable pollution levels that would protect public health; the EPA enforced the standards. Significantly, these standards focused on the same pollutants that California had targeted earlier, demonstrating the state's leadership status. Further signifying its pacesetting ways, California in one striking measure required all vehicles sold in the state to maintain pollution standards far higher than any others, which the state would monitor. Because California constituted such a large market, as the most populous state in the country, the oil and automobile industries grudgingly responded and stopped blocking efforts at all reform, although the industries remained masters at delay tactics. Eventually in 1976, the EPA and Clean Air Act ended the manufacturing of automobiles engineered to run on leaded gasoline, thus ending one of the most deadly pollutant-delivery systems ever designed. Support for pollution reform remained high among the public, despite reluctance of some industries. That support achieved much. Devra Davis, an environmental epidemiologist, has reported a fifty-fold decrease in air pollution, a remarkable achievement by any standard. Nonetheless, reports suggest that air pollution continues to be a significant health threat to many (Davis 2002, 79–98, 156).

Pollution, of course, did not only affect large urban areas. Small urban enclaves experienced some of the worst poisoning. Kellogg, Idaho, home of the Bunker Hill Mining Company, provides such an example. The mining company in this region of northern Idaho spent the twentieth century extracting great wealth from the surrounding mountains and spewing pollution into the air and nearby rivers and lakes. Forced to respond to changing legal environment, the company hired pollution experts in the 1960s and worked to clean up hazardous conditions as mandated by the Clean Air Act, the Water Quality Improvement Act, the EPA, and numerous state regulations. Nevertheless, government studies in the early 1970s revealed high lead levels throughout the region that were increasing dramatically. Nearly a quarter of young children tested at levels qualifying for lead poisoning. Shockingly, a school located adjacent to a smelter was found to have lead levels 130 times the safe level, but the community voted to keep it open, reflecting their doubt that government officials understood the community. Bunker Hill continued to make some efforts to reduce pollution, but it focused more on battling and belittling the EPA. Various types of litigation followed for years, concentrating on health and pollutions problems, and then the mine and smelter closed in 1981. Two years later, the EPA named the Kellogg smelter a Superfund site. True to some Western traditions, many local residents resent the federal presence and its regulatory demands. Nevertheless, the de-vegetated hillsides, continued health problems, and elevated lead levels in rivers and lakes all suggest that cleanup remains necessary and illustrates that even small, isolated urban communities generated enormous and long-lasting pollution problems (Aiken 2005, 168–210).

Unfortunately, pollution affects different groups at different levels. When pollution disproportionately affects a marginalized group, it is called environmental racism, while the activism to reverse it is often called environmental justice. Examples of both phenomena can be found in the urban Far West. For instance, Portland has a history of environmental racism in the Columbia Slough of North Portland, the city's most polluted waterway. A number of toxics have leached into the slough for the better part of the twentieth century. World War II accelerated the slough's contamination as industrial uses there intensified. The North Portland area developed a reputation for industry, poor housing, and African Americans, a condition that abetted the process of treating the neighborhood as a throw-away zone. Meanwhile, the slough's conditions worsened as a 1948 flood removed the slough's outflow but did nothing to lower the level of toxics deposited in the waterway, concentrating pollution levels. In fact, the flood effectively cleared the way for more intensive industrialization. Soon, Portland built a sewage treatment plant in the neighborhood that did not contain sewage during periods of high rain—not an uncommon occurrence in western Oregon. When the city planned further industrial development, its Planning Commission considered population characteristics and housing conditions as factors for industrial zoning. Because of this, the commission associated poor and minority neighborhoods with areas designated for industry and its concomitant, industrial waste. By connecting a reputation for poor and non-white residents with a polluted neighborhood, Portland planners ensured that environmental racism would continue (Stroud 1999, 65–79).

Although by the 1970s Oregon and Portland had built a reputation for strong environmental sensibilities, the slough became a place sacrificed to pollution so other neighborhoods—those higher on the socioeconomic ladder and less diverse culturally—might enjoy greater environmental health. Urban development in the 1970s— based on Oregon's revolutionary land use reforms—protected upper and middle class neighborhoods, while industrial activities and their attendant pollution concentrated in North Portland (Robbins 2004, 281–313). In addition to the periodic overflow of sewage, the poor and often African-American or immigrant residents of that neighborhood often supplemented their income by fishing in the slough and thus ate from a contaminated food supply. As protests against the slough's filth increased in the 1970s, local activists recognized that the slough was being sacrificed by the city. Residents contributed essentially no voice in the planning process, as government experts worked to limit residents' testimony at hearings. Portland's environmentalists, explained historian Ellen Stroud, willingly sacrificed "the slough in order to secure gains elsewhere, and civic leaders' vision of North Portland as a 'natural' industrial site, a vision reinforced by ideas about who did and who should live in the area, profoundly shaped the peninsula. The result was a landscape of inequity" (1999, 70). Thus, the existence of environmental racism was a part of conscious decisions (Stroud 1999, 79–90).

Elsewhere, environmental justice movements responded vigorously to environmental racism. The Mothers of East Los Angeles (MELA) exemplified grassroots environmental justice developments. MELA consisted of several hundred middle-aged Mexican-American women, hardly the profile of typical militant environmental activists. Indeed, California officials targeted East Los Angeles for a toxic waste incinerator in part because its poor, working-class demographic was seen as unlikely to oppose it. Nevertheless, MELA used traditional social networks, such as church, toward political methods aimed at maintaining, preserving, or improving urban quality of life. The environmentally racist pattern of siting toxic waste dumps in neighborhoods populated by the poor and nonwhites was national; 60 percent of African Americans and Latinos lived near toxic areas. To counter this disturbing trend, MELA transformed their political and social identities in the mid-1980s and opposed the incinerator, citing potential health problems for MELA's families. Furthermore, many of the activists had been relocated in the 1950s as freeways disrupted their neighborhoods. Not wishing to be forcibly moved again, they struck back in the 1980s from a grassroots level. In this case, MELA succeeded in opposing the incinerator by becoming politically active and protesting when no one expected them to (Pardo 1998).

This example of an environmental justice movement demonstrated several aspects of modern environmental history. First, it illustrated the increasingly degraded urban environment many underrepresented groups experienced daily. Second, MELA exemplified an approach to environmental activism that defended community health and well-being rather than something more distant, such as wilderness, which was more common in mainstream environmental organizations. This focus on community health also tended to draw more women than men into environmental justice groups. Third, MELA represented participatory democracy at work, showing how environmental movements have often expanded democratic activism in the face of powerful agents of the capitalist economy, including the state. Many environmental justice movements have these attributes, and including them within the definition of environmentalism significantly broadens its scope (Gottlieb 1993; Pardo 1998, 254–255).

A related example of civic activism reflected urban problems for San Francisco. In 1960, three women sat over tea discussing a recent map that appeared in *The Oakland Tribune*. The map showed plans for filling in the edges of San Francisco Bay at a rate of 2,000 acres of wetlands per year. Over the course of a century, developers had reduced the bay from 780 to 550 square miles. Alarmed, Catherine Kerr, Sylvia McLaughlin, and Esther Gulick formed the Save the San Francisco Bay Association in 1961 to reverse the trend of filling in the bay. For several years, the organization grew and faced down numerous development plans, including one proposal to remove the top of San Bruno Mountain and use it to fill in a portion of the bay. The Save the Bay organization effectively lobbied local and then state legislative bodies. At first rebuffed, they eventually gained a state-sponsored study commission by 1964. The following year, responding to the numerous vocal critics these three women inspired, California legislators created the Bay Conservation and Development Commission in a remarkable turnaround that successfully reversed the trend of filling the bay to create more urban real estate. Although from the upper middle class, these local citizens were able to mobilize opinion and change political minds, much as MELA had done, illustrating another strategy for combating urban problems (Vileisis 1997, 212–215).

In the postwar era, cities increasingly dominated the Far West's culture and economy. At the same time, they faced significant environmental challenges as they grew spatially into ecological zones more vulnerable for problems like wildfire and as their populations produced more pollution than ever before. Southern California especially exemplified these trends, although places such as Portland, Oregon, and even Boise, Idaho, were not immune from problems associated with sprawl and pollution. Unfortunately, too, because of deliberate decisions, these ecological problems often affected classes and ethnic groups in disparate ways, making the least powerful the most likely to live near the most polluted and degraded locales. Environmental activists in the mainstream and at the margins fought against these problems and made some progress. As such, urban environments remain the place where the most unique environmental problems *and* solutions are found.

EPISODES OF CONFLICT: ENVIRONMENTAL POLITICS IN THE MODERN FAR WEST

The period since about 1960 is sometimes considered the environmental era. The last half-century experienced numerous and varied conflicts based on incompatible practices with evolving environmental values. Episodes of environmental politics and protest in the Far West in this modern era alone could fill a volume much larger than this, so only a few representative examples can be presented.

Wilderness Designations and Various Opponents

Wilderness preservation has animated much of the mainstream modern environmental movement. Preserving spaces where mining, logging, ranching, and urbanizing were impermissible focused a significant amount of activism and had roots in the 1950s. The first modern effort to preserve wilderness in the Northwest focused on the Three Sisters area of Oregon's central Cascades. In 1954, the Forest Service wished to reclassify the quarter-million-acre Three Sisters Primitive Area as a wilderness area, but in redrawing the boundaries the USFS excluded lower-elevation forests along the Horse Creek drainage from protection. This removal of 53,000 acres would allow timber interests access to

The Three Sisters volcanoes in Oregon (looking north). Left to right: South Sister, Middle Sister, and North Sister. The volcanoes are part of the Cascade Volcanic Belt and are included in the Three Sisters Wilderness Area. (Lyn Topinka/United States Geological Survey)

large stands of commercial timber, and the attendant road would increase access to higher-elevation recreational areas. To compensate recreation interests, the agency planned to add other areas for designated wilderness. USFS personnel touted the new management plan as particularly suited to the Forest Service's guiding multiple-use philosophy. Caught by surprise, foresters faced a barrage of criticism from scientists and hikers who organized themselves into Friends of the Three Sisters and worked to marshal scientific and aesthetic arguments against the removal. At public hearings, representatives from local and national environmental organizations, such as Howard Zahniser of the Wilderness Society, put the Forest Service plan into the national environmental spotlight. Friends of the Three Sisters also organized hiking trips into the backcountry. Although in this instance, wilderness proponents lost the battle at Three Sisters, the episode encouraged stronger strategizing and greater unity among regional wilderness activists who focused much of the next decade on creating the national wilderness preservation system that was finally embodied in the 1964 Wilderness Act. A significant motivator for this wilderness system was to stymie the Forest Service's seemingly capricious administrative decisions concerning wilderness management, such as at Three Sisters. All the strategies employed in the Three Sisters struggle—forming a focused organization, clamoring for a public hearing, demonstrating a national interest in local wilderness struggle, taking people into the wilderness to see it firsthand—represented a common pattern duplicated countless times throughout the Far West (Marsh 2002).

Other early struggles over wilderness targeted the National Park Service. In the 1950s, the NPS embarked on Mission 66, an effort to increase park visitation by constructing roads and hotels throughout the park system. On the far Northwest coast, Olympic National Park managed a relatively short stretch of ocean beach where park managers planned, with the support of local residents, a new road that would create a scenic byway along a primitive stretch of beach. In 1958, Polly Dyer, the Federation of Western Outdoor Clubs president, organized a hike of prominent conservationists to protest the road. Many of the leading lights of the postwar environmental movement participated, including Zahniser, wildlife biologist and Wilderness Society Executive Director Olaus Murie, Wilderness Society cofounder Harvey Broome, NPS Director Conrad Wirth, and Justice William O. Douglas, who was the nominal leader of the hike. These famous conservationists generated a media buzz surrounding the event. Indeed, much like the strategy in Oregon, this group needed and received national attention to preserve this place as roadless. Unlike in Three Sisters, wilderness advocates here succeeded, although a series of engineering problems were as significant as the protestors in staving off the blacktop (Sowards 2006, 25–32).

In Idaho, different threats emerged. For most of the twentieth century, various Idahoans desired a national park in the Sawtooth Mountains. This area of central Idaho is impressive with forty-two 10,000-foot peaks. Senator Frank Church took up the park cause in 1960, hoping to carve a national park out of part of the Sawtooth Primitive Area. Of course, the Forest Service never willingly gave away land, and grazers, hunters, and resource companies always feared greater restrictions that would ensue with national park status. After various studies, the park proposal faded behind an alternative Sawtooth National Recreation Area still to be administered by the USFS, solving the seemingly intractable bureaucratic political problem (Ewert 2000).

However, deeper conflict emerged in 1968 when the American Smelting and Refining Company (ASARCO) sought to develop an open-pit molybdenum mine in the White Clouds, a mountain range adjacent to the Sawtooths. The proposed mine would be 350 feet deep, 700 feet wide, and 7,000 feet long, and almost all of the waste would be deposited on site in alpine lakes to be used as tailings ponds. The project promised several hundred new jobs and millions of dollars in investment and tax revenues. The mine and associated roads also promised environmental damage. Moreover, ASARCO's atrocious environmental record and insensitivity to wildlife concerns made many Idahoans bristle. An ASARCO official addressed concerns about the effects of mine blasting on wildlife by reportedly stating: "Well, it sounds like thunder. And if it sounds like thunder, then it's a normal ecological feature, so why worry about it?" (quoted in Ewert 2000, 144). Idaho's governor, Don Samuelson, supported the mine and found environmentalist positions overstated. "They're not going to tear down the mountains; they're going to dig a hole," he protested (quoted in Andrus and Connelly, 1998, 82). In the meantime, *Life* magazine highlighted the proposal, once again furnishing valuable national attention. In the U.S. Senate, Frank Church and fellow Idahoan Len Jordan amended the Sawtooth National Recreation Bill to include the White Clouds. More locally, by the 1970 gubernatorial election, the fate of the White Clouds emerged as a leading issue, and Idahoans voted for preservation when they sent Cecil Andrus to the statehouse. The victory was a significant watershed, as Andrus recalled, "The battle over the White Clouds would help make me the first Western governor elected on an environmental platform" (Andrus and Connelly 1998, 19). Although Governor Andrus had no authority to grant or deny ASARCO's proposed mine, his election suggested that a new day was emerging in Idaho where environmental values other than exploitation could shape political outcomes. On 22 August 1972, Congress created the Sawtooth National Recreation, a total of 756,000 acres, of which 216,000 were designated wilderness. The legislation prevented mineral claims in the recreation area. Although the Sawtooths never achieved national park status, its defenders strengthened federal protection, defeated a powerful mining company, and created a powerful conservation symbol for the state (Andrus and Connelly 1998, 18–19, 82, 158; Ewert 2000; Neil 2005, ix-xiv).

In California, other quintessential conflicts emerged. For much of the twentieth century preservationists in California worked to preserve stands of coastal redwoods. The Save-the-Redwoods League and the Sierra Club led these campaigns. The Sierra Club especially viewed the remnant redwood stands as important for ecological reasons and consequently advocated to have large areas preserved. Indeed, their vision of preservation within a national park became so uncompromising that the club found itself at odds not only with politicians in Washington, DC, but also with the local league, who valued compromise. Throughout the 1960s, the Californian activists battled against timber companies who wanted to harvest the trees and some politicians who supported the companies or favored some concessions. While the Save-the-Redwoods League possessed strong fund-raising prowess and a willingness to work with industry, the Sierra Club mastered the publicity politics that are so necessary in environmental confrontations and maintained a stronger no-compromise approach. Yet to succeed in the political world required compromise, and by 1968, such a park bill was signed by President Lyndon Johnson. The middle ground forged in the compromise proved particularly flawed. For instance, parts of one valley were designated for wilderness protection and other parts for clear-cutting. Advocates of both complained about the situation, and battles continued. The evolution of this conflict helped fracture the Sierra Club and radicalized some environmentalists. Nevertheless, by 1978, President Jimmy Carter enlarged the national park through an extremely expensive land acquisition process and compensation package for workers who would lose their jobs in a time of budgetary crisis (Schrepfer 1983, 130–244). The long struggle and numerous concessions that required subsequent adjustments and often further compromises characterized this and many other environmental conflicts.

These wilderness case studies show the variations of wilderness politics. In the Far West, threats commonly have included Forest Service reclassifications and timber sales. Although national parks coped with overdevelopment, they often offered what environmentalists perceived to be the best protection option. Other backcountry areas faced mining exploration and development. As wilderness advocates confronted these challenges, they found that success came most readily when locals focused on a particular battle and enlisted national support and attention. From Three Sisters to the Olympic Coast, from the White Clouds to the redwoods, local activists and national coalitions faced down federal agencies and giant corporations with far superior financial resources. Success was never guaranteed and not always achieved. However, environmentalists increasingly managed to generate the necessary attention and assemble the political will to stop development plans. But wilderness preservation was not the only environmental conflict shaping modern Western politics.

The Reversal of Hydraulic Fortunes

Built largely by imperialistic methods of urban water acquisition, California's hydraulic society had faced setbacks since the 1960s. Numerous reasons explain this change, not least the rising power of environmentalists who challenged many of the prerogatives of the state's business as usual. Discussing new values, using the courts, and applying scientific insights put the state's water planners on the defensive (Hundley 2001, 303–364).

One of the first and more complicated setbacks came with the fight over the so-called Peripheral Canal. Saltwater continued intruding the delta and, in fact, worsened because of diversions from the Central Valley and State Water projects as well as the growth of the San Francisco Bay area. The saltwater adversely affected wildlife, and occasionally irrigators and towns, when the water moved so far inland to get into the southbound diversion canals. The Peripheral Canal promised help by ensuring there would be enough quality water heading south and enough stored water to release to counter any saltwater disturbance in the delta. Such potential to solve multiple problems attracted much support, but soon critics emerged citing the well-rehearsed arguments that it was the ever-imperialistic south taking more resources from the well-watered northern portion of the state. Other opponents faulted agribusiness and their never-satisfied thirst. Detractors suggested that it was time to break the cycle of water development and demand instead more efficient uses of water through existing but ignored irrigation technologies. Canal adversaries also argued that subsidized water policies were the root of the problem, and reforming those basic policies was necessary before any further hydraulic development. Delta residents feared that their water rights would be ignored, even though the Peripheral Canal ostensibly was to help them. Add environmentalists, who objected on a variety of grounds, to the mix of opponents and the Peripheral Canal faced a formidable challenge. Despite these arguments, the state passed a Peripheral Canal Bill in 1980 with a regional, bipartisan vote: all but one Southern California senator favored it; all but one northern California senator objected to it. But the passage of the act did not end the story (Hundley 2001, 313–325).

The next phase of the Peripheral Canal saga demonstrated a reversal of fortunes. Governor Jerry Brown, whose father Edmund G. "Pat" Brown played a critical role in developing the State Water Plan, attempted to neutralize environmentalists when his speech coinciding with the signing of the Peripheral Canal Bill touted environmental safeguards in the legislation. To offset potential opposition, Brown persuaded Secretary of the Interior Cecil Andrus to place more than 1,200 miles of northern rivers into Wild and Scenic River status, preventing subsequent dam-building in the streams. Despite this, environmentalists launched a successful campaign to put the canal law before the public in a referendum. Opponents marshaled a host of arguments against the canal as Norris Hundley summarized:

> the south did not really need the water and was subsidizing agribusiness, which continued to contribute market-glutting surpluses (and those surpluses included not only such commodities as forage crops and cotton but also sometimes even specialty crops like olives and almonds); the project contained no requirement for effective water conservation and groundwater management programs; it was too expensive; and, most important, the delta ecology—indeed, the ecology of the state—was too fragile to risk with such a complex and gargantuan undertaking (2001, 329).

Located in the Eastern Sierra region, Mono Lake is an alkaline and hypersaline terminal lake that provides a nesting habitat for several species of birds. Its decline due to hydraulic engineering caused notable protest that ultimately allowed the return of some water to the lake. (iStockPhoto.com)

Such a list illustrated the complexity of the opposition, resting on science, economics, and intraregional politics. As critics intensified their attacks, political support vanished. In 1982, the referendum to repeal the canal bill carried by an impressive 63 to 37 percent margin. It was the first time California voters had rejected a water development project since the 1920s. Although water development did not cease in California, the Peripheral Canal's defeat marked a serious watershed in the state's environmental history (Hundley 2001, 325–334).

Meanwhile, long accustomed to having its way with water, Los Angeles faced a setback when a variety of issues coalesced to restrict the city's access to Mono Lake water. The lake lay just north of Owens Valley and possessed no outlets, leaving it highly saline. Migratory birds—up to a million annually—depended on a type of brine shrimp and fly found only there on their flyway from the Arctic. When Los Angeles tapped the lake in 1941, the lake's level began dropping by about a foot a year; in 1970, the city completed a second aqueduct and the lake dropped another half a foot yearly. Declining water levels gave coyotes access to what had been islands, and the predators devastated the California gulls that bred on rookeries in the lake (Hundley 2001, 336–338).

Local university biologists began studying this transformation and found that this unique ecosystem was decimated. In 1978 these scientists organized the Mono Lake Committee to alert regional and national environmentalists to the devastation and to coordinate efforts to stop Los Angeles's assault on this unusual environment. The state responded with a task force that recommended an 85 percent reduction in the city's diversions, a recommendation Los Angeles effectively suppressed. The Mono Lake Committee and other environmental groups pursued a lawsuit that argued Los Angeles's diversion violated the public trust, the first time such an argument was deployed to protect ecological and scenic values. Environmentalists charged the state with a task of maintaining Mono Lake for the public's use as part of its public trust responsibilities. Predictably, the city disagreed, but the California Supreme Court determined that the public trust doctrine applied. After that, the state and courts had to devise a solution, an "interminably slow" process according to Hundley's apt characterization (Hundley 2001, 339–342).

Several scientific studies produced in the 1980s warned that the lake's ecosystem remained in peril. The lake was now an astounding and disturbing 37 feet lower than it had been before Los Angeles began its diversions, and the lake was at about 50 percent of its prediversion volume. Los Angeles did not dispute the studies; it merely pointed out the harm the city would face without the Mono Lake diversions. An impasse continued until 1993 when changing leadership at both the state and city levels led to discussions toward a settlement that was reached the following year. This settlement reduced the city's diversion by a third. After sixteen years of work, the Mono Lake Committee succeeded. An activist recognized the monumental nature of this achievement: "[F]or the first time ever in California, water was removed from the grasp of an anointed appropriative user and assigned, not to some rival diverter, but to an environmental purpose: the restoration of the lake and its feeder streams" (quoted in Hundley 2001, 346). Similarly, Los Angeles has had to return water to Owens Valley to prevent the toxic-laced alkaline soil from the dry lakebed from blowing and harming residents' health (Hundley 2001, 342–347, 353–360).

The Mono Lake episode reveals many aspects of the modern environmental movement. For instance, it showed the scientific basis that is at the root of many struggles; indeed, biologists, not simply activists, started the Mono Lake Committee and pursued the matter most consistently. More significantly, it showed that metropolitan power could be stopped through continuous campaigns, lawsuits, and new arguments about public trust. As the public recognized the unique attributes of the Mono Lake ecosystem and believed Los Angeles could both find alternative sources and improve its conservation efforts, it demanded the city respond creatively. Environmentalists recognized that persistence, publicity, science, and the law were essential tools for successful environmental campaigns. Combined with the shifting political tides that reversed the Peripheral Canal proposal, this episode marked a change in Far Western hydraulic developments.

The Klamath and Competing Priorities

The Klamath watershed of southern Oregon and northern California has long been a place of competing uses and schemes by various groups to reorganize its ecology to produce more—more timber, more crops, more salmon. The developments have involved American Indians, the federal government, local farmers, and small communities. Rival claims on resources and competing priorities have spelled intense conflict here and illustrate the profound challenges of multiple uses in a scarce landscape.

Native groups long used the various resources of the Klamath Basin, and newcomers also claimed access to the region's timber and water. Predictably, for decades the federal government has been an arbiter between groups, consistently aiding newcomers at the expense of the Klamath. In the 1950s, the Klamath Tribe faced termination, a federal policy that sought to end tribes' treaty rights and the federal government's obligations to them. Timber companies were deeply interested in termination, eyeing the Natives' timberland which amounted to nearly a million acres. Senator Richard Neuberger worked to place the Klamath lands into a new Winema National Forest. Conservation and civic groups backed this plan, believing federal ownership would lead to a sustained yield program and help protect the larger Klamath watershed. The hopes for better management with federal control fell flat. Lumber companies enjoyed essentially open access to the timber and quickly liquidated the available commercial timber. The tribe's desires or well-being did not enter much into the discussion; as one local recalled, it "wasn't a situation of a willing seller. It was a condemnation" (quoted in Robbins 2004, 107). Timber, however, has not been the most contentious issue (Robbins 2004, 227–232).

Even more troubling and problematic have been issues related to the water supply in the Klamath Basin. Before Europeans arrived, the Klamath watershed included rivers, lakes, and vast wetlands, perhaps 185,000 acres. The natural wetlands functioned as a stop on the Pacific flyway where 80 percent of migratory birds stop along their journeys to and from the Arctic. Further, the regional

Native peoples—the Klamath in Oregon, the Yurok in California—relied on the area's fish species, especially the coho salmon in the lower Klamath River and the short-nosed sucker in the upper Klamath Lake. The Klamath Reclamation Project changed all that. The project provided water to nearly a quarter-million acres, draining more than 100,000 acres of wetlands and redirecting water to irrigate the now-dry lake beds. This ecological rearrangement drastically reduced prime wetlands and jeopardized critical waterfowl and fish habitat. Compounding that, cutover timberland, overgrazed range, and pesticide pollution further degraded the area's watershed and rivers. Environmentalists demanded that the U.S. Fish and Wildlife Service manage the region's several national wildlife refuges better. They pointed specifically to the diminishing habitat that vanished under farmers' fields and pollution from toxic pesticides as key problems for fish, fowl, and wildlife populations (Robbins 2004, 104–112).

Such a scenario was not altogether uncommon in various parts of the West, but here they came to a boiling point when a recent drought revealed the real fragility of the area's ecology. On two occasions the Bureau of Reclamation stopped delivering water to farmers; this was accompanied by tremendous controversy, as it affected 1,400 farmers using 200,000 acres. In 2002, low water levels from drought and agricultural withdrawals led to an estimated 33,000 salmon deaths as they stacked up in shallow water. Earlier in the season local Indians had unsuccessfully asked for more water for salmon and afterward argued that lack of water caused the deaths, but federal officials speciously claimed that inadequate science existed to blame the salmon mortality on low water levels from irrigation withdrawals (Barcott 2003, 44–51; Robbins 2004, 104–112).

The combination of legal requirements of Indian water rights and the Endangered Species Act seemingly created this problem for the largely Euro-American irrigators. But the problem's true roots lay in the fact that the Bureau of Reclamation had for too long tried to do too much with too little water. Changing public environmental values, increasing political power of the Klamath, and weakening ecosystems altered the Klamath cultural and natural contexts so that the bureau could not continue its practices. Consequently, in recent years, non-Indian farmers who historically benefited from the reclamation project have found themselves in new and unpleasant circumstances. In a broader regional context of declining salmon runs, fewer forest jobs, and rural distrust of both cities and the federal government, the Klamath Basin controversy was explosive. Troubling social trends followed, including rising incidents of domestic violence and discrimination toward Klamath tribal members. Federal officials have had to resort to armed guards at diversion canals' headgates to keep them closed. Farmers have managed on occasion to force open

Dead fish on a Klamath River bank in California in 2002. Tens of thousands of fish died that year due to low water levels that resulted from drought and the diversion of water for irrigation. (Northcoast Environmental Center)

headgates and violate federal environmental regulations. Although some efforts are afoot to convert farms back to marshes and to ensure adequate water for the wetlands, much of the economic and historical precedents argue against that happening. However, without fundamental changes in watershed management in the Klamath Basin, one can expect continued problems—for farmers, or Indians, or the short-nosed sucker, or, most likely, all of them (Barcott 2003, 44–51; Robbins 2004, 104–112).

CONCLUSION

It is suitable to end with the Klamath controversy for it presents several trends that run through much of this book. The once sustainable and reciprocal practices of West Coast Native peoples have largely disappeared, as Natives' diminished political power has led to fractured ecosystems. Moreover, it shows the power of the federal government to aid the process of transforming the landscape, and its power to revise that strategy, even if incompletely, when laws and management priorities change. As in forestry and agriculture, reclamation efforts grew from a confidence that humans could engineer an economically productive landscape without reducing ecological sustainability. Many residents in the Far West found that confidence misplaced—with the Klamath Reclamation Project and the Peripheral Canal, at the Horse Creek drainage and Bodega Bay—for the faith promised too much to too many for too long. And long-term ecological sustainability became an ideal sacrificed for short-term profits. This spawned regret and reforms in forestry, agriculture, energy development, urban growth, and wilderness management, all of which had been inspired by the environmental movement at both local and national levels. Indeed, local challenges to typical management regimes depended more than ever on national recognition, mobilization, and reform. Ultimately, the Klamath episode symbolizes the long and challenging road to creating truly equitable and ecologically viable landscapes.

EPILOGUE

Environmental history can be a distressing discipline. After all, it examines the most basic human relationship—the one between human culture and the nonhuman environment—and reveals most often a severely unbalanced dynamic. Like many environmental histories, this book's basic narrative has shown a relative decline in sustainable human-nature interactions. It seems fitting, then, to end with two brief examples—one urban, one rural—of hope.

Although sitting within the beautiful San Francisco Bay area, Oakland has confronted significant environmental problems because of urban sprawl. Proliferating highways create these troubles, including air and noise pollution and lack of open space. In 1991, the Bay Area Rapid Transit authority (BART) proposed a gigantic 500-car parking lot in Oakland's Fruitvale district, a primarily Latino community. Hardly a project to revitalize the aesthetic or environmental qualities of the neighborhood, the proposed parking lot angered Fruitvale residents, especially the civil rights organization Unity Council. The Unity Council had long been interested in community environmental issues, such as cleaning up hazardous waste sites and restoring the few remaining riparian zones. Poor air quality because of the declining transportation infrastructure, however, dominated the ecological agenda and the issue would not be helped by a large parking lot. Nevertheless, the Unity Council and other Fruitvale residents recognized that a BART hub in their neighborhood might offer much-needed economic development. Consequently, the Unity Council proposed an alternative: the Fruitvale Transit Village (Shutkin 2000, 167–177).

This transit-oriented development would attract riders to public transportation not by building large parking lots—a common strategy that contributes to pollution by forcing drivers to drive to lots and encourages more driving within the neighborhood—but by improving retail and residential amenities near transit stations. The Fruitvale Transit Village would be an economic development plan that revitalized a poor community while promoting public transportation through partnership among the Unity Council, BART, and the city of Oakland. In the process, too, it would reverse air and noise pollution trends. Groundbreaking occurred in 1999, and the project actually improved various environmental conditions because developers cleaned up hazardous materials in the soil and reduced space devoted to parking. In addition, the civic environmentalism focused on the transit village inspired residents to work for increased open space along Oakland's waterfront, indicating the promise of environmental projects to engage the public in myriad ways. While hardly a typical environmental history episode, this Oakland experience demonstrates the urban environmental challenges of highly polluted neighborhoods that also house politically and economically marginalized groups. More interesting, though, the example suggests the potential for creative solutions originating within the community that would improve economic and environmental conditions, a key component to environmental justice movements. Still, whether the Fruitvale Transit Village can meet the promise of its vision over a long period remains to be seen (Shutkin 2000, 177–187).

In eastern Oregon, federal resource managers and local farmers and ranchers have tried for a century to control nature in the Malheur Basin. Massive engineering projects drained wetlands and irrigated dry lake beds, while predators have been actively reduced to increase the type of wildlife the public desires (e.g., antelope and ducks) or to protect the domestic animals on which ranchers depend. Frequent distrust colored the relationship between public agents and private landowners. In recent years, though, managers and Oregonians near the Malheur National Wildlife Refuge have tried to reverse that historic trend by instituting adaptive management, a management approach characterized by historian Nancy Langston as a "scheme that incorporates multiple human perspectives while responding to changing scientific understanding of dynamic ecosystems" (2003, 155). Ideally, such a strategy involves various actors from a particular location and consistently experiments with various tactics, discarding those that do not work. Adaptive management is far more flexible than traditional approaches in that it eschews ideological programs and forces those in a working landscape to periodically evaluate what is actually happening on the ground or in the water. Although this seems like a basic responsibility, a century of work in the Malheur Basin proceeded with surprisingly little inspection or introspection among federal agencies or local residents (Langston 2003, 151–159).

The collaborative, adaptive method taking place in the Malheur Basin has been rooted in fear. Secretary of the Interior Bruce Babbitt threatened to turn Steens Mountain into a national monument, a prospect local ranchers and farmers found unacceptable; environmentalists promised more lawsuits, ensuring inaction on the ground; and incoming Californians frightened both local agriculturalists and environmentalists who drew on personal experience to build their identities in the local landscape. These common threats essentially forced collaboration among regional environmentalists and ranchers. Although these groups shared little cooperative history, they listened to one another and recognized that much was at stake in this place both viewed as special. Of course, managing a complex landscape and social system like the Malheur is an ongoing and difficult process, but signs are promising. Creeks appear to be healing, and local residents are learning to live with more ecological variability than they once tolerated, a variability that allows riparian zones to support greater biodiversity than they have in decades (Langston 2003, 159–169).

At first glance, the Fruitvale Transit Village and the breakthrough in the Malheur Basin share few qualities. Nevertheless, collaboration and local involvement are two keystones that are evident and hold promise in resolving difficult environmental questions elsewhere. In both instances, it is too early to tell whether they will work over the long run. Indeed, it is always too early to tell if a human-nature relationship will work over the long run; our human vision is too brief and narrow. Still, we must act. And if marginalized groups in Oakland can envision and enact a development project that improves their lives and environment, and if ranchers and environmentalists can put aside their substantial differences and animosities to work to protect a fragile ecosystem, then hope remains for improving the relationship between people and their natural environment throughout the Far West and beyond.

REFERENCES

Aiken, Katherine G. *Idaho's Bunker Hill: The Rise and Fall of a Great Mining Company, 1885–1981.* Norman: University of Oklahoma Press, 2005.

Andrus, Cecil D., and Joel Connelly. *Cecil Andrus: Politics Western Style.* Seattle, WA: Sasquatch Books, 1998.

Barcott, Bruce. "What's a River For?" *Mother Jones* 28 (May/June 2003): 44–51.

Commoner, Barry. "Los Angeles Air." In *Green Versus Gold: Sources in California's Environmental History,* edited by Carolyn Merchant, 372–376. Washington, DC: Island Press, 1998.

Davis, Devra. *When Smoke Ran Like Water: Tales of Environmental Deception and the Battle Against Pollution.* New York: Basic Books, 2002.

Davis, Mike. *Ecology of Fear: Los Angeles and the Imagination of Disaster.* New York: Metropolitan Books, 1998.

Ewert, Sara E. Dant. "Peak Park Politics: The Struggle over the Sawtooths, from Borah to Church." *Pacific Northwest Quarterly* 91 (Summer 2000): 138–149.

Gerber, Michele Stenehjem. *On the Home Front: The Cold War Legacy of the Hanford Nuclear Site.* 2nd ed. Lincoln: University of Nebraska Press, 2002

Goble, Dale D. "Salmon in the Columbia Basin: From Abundance to Extinction." In *Northwest Lands, Northwest Peoples: Readings in Environmental History,* edited by Dale D. Goble and Paul W. Hirt, 229–263. Seattle: University of Washington Press, 1999.

Gottlieb, Robert. *Forcing the Spring: The Transformation of the American Environmental Movement.* Washington, DC: Island Press, 1993.

Guthman, Julie. *Agrarian Dreams: The Paradox of Organic Farming in California.* Berkeley: University of California Press, 2004.

Hein, Teri. *Atomic Farmgirl: Growing Up Right in the Wrong Place.* New York: Houghton Mifflin, 2003.

Hirt, Paul W. *A Conspiracy of Optimism: Management of National Forests Since World War Two.* Lincoln: University of Nebraska Press, 1994.

Hirt, Paul W. "Creating Wealth by Consuming Place: Timber Management on the Gifford Pinchot National Forest." In *Power and Place in the North American West,* edited by Richard White and John M. Findlay, 204–232. Seattle: University of Washington, 1999.

Hundley, Norris, Jr. *The Great Thirst: Californians and Water: A History.* Revised ed. Berkeley: University of California Press, 2001.

Langston, Nancy. *Forest Dreams, Forest Nightmares: The Paradox of Old Growth in the Inland West.* Seattle: University of Washington Press, 1995.

Langston, Nancy. *Where Land and Water Meet: A Western Landscape Transformed.* Seattle: University of Washington Press, 2003.

Marsh, Kevin R. "'This Is Just the First Round': Designating Wilderness in the Central Oregon Cascades, 1950–1964." *Oregon Historical Quarterly* 103 (Summer 2002): 210–233.

Neil, J. M. *To the White Clouds: Idaho's Conservation Saga, 1900–1970.* Pullman: Washington State University Press, 2005.

Neuberger, Richard L. *Our Promised Land.* New York: Macmillan, 1938.

Pardo, Mary. "Mexican American Women Grassroots Community Activists: 'Mothers of East Los Angeles.'" In *A Sense of the American West: An Environmental History Anthology,* edited by James E. Sherow, 243–260. Albuquerque: University of New Mexico Press, 1998.

Perkins, John. "California Scientists and the Pesticide Crisis." In *Green Versus Gold: Sources in California's Environmental History*, edited by Carolyn Merchant, 418–423. Washington, DC: Island Press, 1998.

Pisani, Donald J. *Water and American Government: The Reclamation Bureau, National Water Policy, and the West, 1902–1935.* Berkeley: University of California, 2002.

Pulido, Laura. *Environmentalism and Economic Justice: Two Chicano Struggles in the Southwest.* Tucson: University of Arizona Press, 1996.

Pyne, Stephen J. *Fire in America: A Cultural History of Wildland and Rural Fire.* Princeton, NJ: Princeton University Press, 1982. Reprint, Seattle: University of Washington Press, 1997.

Pyne, Stephen J. *World Fire: The Culture of Fire on Earth.* New York: Henry Holt, 1995.

Robbins, William G. *Landscapes of Conflict: The Oregon Story, 1940–2000.* Seattle: University of Washington Press, 2004.

Rothman, Hal K. *The Greening of a Nation? Environmentalism in the United States Since 1945.* Fort Worth, TX: Harcourt Brace, 1998.

Schrepfer, Susan R. *The Fight to Save the Redwoods: A History of Environmental Reform, 1917–1978.* Madison: University of Wisconsin Press, 1983.

Shutkin, William A. *The Land That Could Be: Environmentalism and Democracy in the Twenty-First Century.* Cambridge: Massachusetts Institute of Technology Press, 2000.

Sowards, Adam M. "William O. Douglas: The Environmental Justice." In *The Human Tradition in the American West,* edited by Benson Tong and Regan Lutz, 155–170. Wilmington, DE: Scholarly Resources, 2002.

Sowards, Adam M. "William O. Douglas's Wilderness Politics: Public Protest and Committees of Correspondence in the Pacific Northwest." *Western Historical Quarterly* 37 (Spring 2006): 21–42.

Steen, Harold K. *The U.S. Forest Service: A History.* Centennial ed. Durham, NC, and Seattle: Forest History Society in cooperation with University of Washington Press, 2004.

Stroud, Ellen. "Troubled Waters in Ecotopia: Environmental Racism in Portland, Oregon." *Radical History Review* 74 (Spring 1999): 65–95.

Sutter, Paul S. *Driven Wild: How the Fight Against Automobiles Launched the Wilderness Movement.* Seattle: University of Washington Press, 2002.

Taylor, Joseph E., III. *Making Salmon: An Environmental History of the Northwest Fisheries Crisis.* Seattle: University of Washington Press, 1999.

Udall v. FPC 387 U.S. 428 (1967).

Vileisis, Ann. *Discovering the Unknown Landscape: A History of America's Wetlands.* Washington, DC: Island Press, 1997.

Walker, Richard A. *The Conquest of Bread: 150 Years of Agribusiness in California.* New York: The New Press, 2004.

Wellock, Thomas. "The Battle for Bodega Bay." In *Green Versus Gold: Sources in California's Environmental History*, edited by Carolyn Merchant, 344–349. Washington, DC: Island Press, 1998.

White, Richard. *The Organic Machine: The Remaking of the Columbia River.* New York: Hill and Wang, 1995.

Wilkinson, Charles F. *Crossing the Next Meridian: Land, Water, and the Future of the West.* Washington, DC: Island Press, 1992.

Williams, James C. *Energy and the Making of Modern California.* Akron, OH: University of Akron Press, 1997.

Worster, Donald. *Rivers of Empire: Water, Aridity, and the Growth of the American West.* New York: Oxford University Press, 1985.

7

CASE STUDIES

CASE STUDY: ENVIRONMENT AND EMPIRE—
THE NORTHWEST FUR TRADE

In 1824, fur trader Peter Skene Ogden returned to the Snake River Plain for his second season of trapping for the Hudson's Bay Company (HBC). He reported, "this part of the Country tho' once abounding in Beaver is entirely ruined." Two years later, he likewise informed his company superiors, "We have onely [sic] one Beaver although upwards of fifty Traps—the Trappers certainly appear to have clean'd the river well" (quoted in Ott 2003, 179). For a trapper, such accounts would seem to indicate bad news, but to Ogden they constituted the opposite. This perspective represented the unique history of the Northwest fur trade, where economic and imperial motives combined in a deliberate strategy of wiping out the beaver population south of the Columbia River. Furthermore, the rapid success Ogden described demonstrated the potential rapidity of ecological change when Euro-Americans arrived in the Far West.

Indeed, perhaps no better example of the environmental consequences of cultural contact can be found than the fur trade. This economic exchange existed within several important contexts central to environmental and regional history: emerging global capitalism, changing marine and riverine ecosystems, competing economic and imperial relationships, and shifting demographics. The relationship between a dominant core and a subordinate periphery in the Northwest fur trade illustrates perfectly how distant factors—company headquarters in Montreal or New York, fashion trends in China or France, political agendas in Washington, DC, or London—impinged directly on local ecosystems. Even by the eighteenth century a process of globalization inexorably tied world markets and local environments together so tightly that it would be difficult to disentangle causes and effects. Placed properly in its multiple perspectives, the fur trade reveals much about the contact landscape, the dynamics of ecological change, and the ways in which environment has long been central to larger political, economic, and social forces.

Benjamin Franklin in his famous beaver-fur hat, a departure from the beaver-felt top hats of European fashion, in 1777. The demand for beaver-felt hats in Europe drove the trapping industry of North America well into the nineteenth century. (Library of Congress)

The logic of global capitalism governed the fur trade. Several historians have noted how capitalism shaped the initial exploitation of the Northwest's natural resources, beginning with fur-bearing animals. They have argued that Euro-American traders valued nature as a commodity—that is, as something to be traded. As one historian put it, "Nature that did not sell lost value" (Taylor 1999, 45). Certainly Native peoples traded goods extracted from their surrounding ecologies, but Euro-Americans arriving in the late eighteenth and early nineteenth centuries placed economic value on nature to a far higher degree and on far more items than indigenous groups. Moreover, Euro-Americans took bits of nature from their larger ecological context without recognizing that each component was part of a complex functioning system. Fortunately for Euro-American traders, and unfortunately for fur-bearing animals and the organisms dependent on them, the Northwest was initially abundant in furs. Markets in China for sea otter pelts and in Europe for beaver skins encouraged exploitation. These global market demands affected local ecosystems significantly. Before the global market influenced the Far West's economy and ecology, trade and consumption remained relatively localized, which minimized and confined environmental change. The fur trade, though, began the process that incorporated western ecologies into global economic systems, disrupting existing human-ecological regimes and paving the way for greater environmental, economic, cultural, and political changes (Bunting 1997; Langston 1995; Robbins 1997; Taylor 1999).

Ecology of Otters and Beavers

The two main animals coveted in the Northwest fur trade were the sea otter and the beaver. Although human actions depleted these mammals and caused significant ecological change, each species also functioned within ecosystems and modified their niche independent of humans.

Sea otters play a pivotal role in coastal marine ecosystems, and their removal resulted in cascading effects. Unfortunately for the otters, their social and reproductive characteristics made them easy prey for hunters. They typically produce only one offspring a year, allowing populations to deplete rapidly. Moreover, they are fairly docile, and mothers will carry their young, even when attacked. Such a combination of factors made killing them an easy job to those interested in their luxuriant pelts. Those pelts were the result of otters' adaptive ecology. Sea otters grow a thick pelt to keep them warm in cold coastal waters because they do not have the blubber other marine mammals do. Moreover, they eat prodigiously; on a daily basis, otters consume between one-fourth and one-third of their body weight. To reach that dietary intake, otters exploit abalone and other mollusks. Eating so many mollusks keeps their population down, allowing kelp to thrive, which otherwise would have been consumed by mollusks. The abundant kelp sheltered fish. So when sea otter populations plummeted, as they did because of the fur trade in the early nineteenth century, then mollusk populations subsequently skyrocketed, which caused a decline of kelp, which in turn reduced habitat for many fish species. Thus, otters were key components in a complex marine environment, and their fate was connected to the survival of other species (Calloway 2003, 412; Gibson 1992, 177; McEvoy 1986, 24, 76, 81–82).

Like otters, beavers also play a pivotal role in their ecosystems. In the best habitat, beavers can build between fifteen and twenty-five dams per mile. The attendant riparian ecological effects are significant. Environmental historian Nancy Langston explained their effects:

> Dams slowed the water flow, and by creating wetlands, they buffered floods and helped prolong the late summer flow of streams—both critical factors in allowing dry forests to persist. The dams retained tremendous amounts of sediment and organic matter in the stream channel—one dam alone could gather 229,500 cubic feet (6,500 cubic meters) of sediment behind it—and this too had critical effects on the streams as well as the surrounding forests. . . . Flooding the soil quadrupled the amount of nitrogen accessible to plants. Thus beaver activity enhanced nitrogen availability across the landscape, and because inland Northwest forests are nitrogen-limited, beavers probably indirectly increased forest productivity (Langston 1995, 57–58).

Besides changing the flow of streams, increasing sediment concentration, and enhancing nitrogen, beavers' activities also affected succession. It is possible that beavers could eliminate their favorite species in a particular reach of the stream and require relocation. For example, a twentieth-century study in Idaho found that a colony with a maximum of only five beavers cut more than 800 aspens in just three years. Their activities could also create openings for shade-intolerant shrubs to flourish. Other effects included reducing erosion along the edges and bottom of streambeds. Ecological studies have suggested that water quality increased in stream systems with beaver dams as nutrient levels stay higher. This process made streams better habitats for insects and other microorganisms. In addition, in arid environments, beaver dams and ponds created some of the only lush areas. This richness attracted other animals, such as migrating elk, deer, or antelope. The habitat also suited insectivorous birds and amphibians such as frogs or salamanders. The centrality of beaver populations in riparian ecologies is hard to overestimate (Langston 1995, 57–59; Ott 2003, 185–190).

Ecologists call animals that are critical in shaping their habitats, like otters and beavers, keystone species. Removing a keystone species invites havoc to the existing ecosystem. Given the large role these animals play in configuring an ecosystem's function, their removal often dramatically changed marine and riverine life in the Far West. Such changes mattered little to traders.

Initiating the Trade in Soft Gold

Reports of valuable furs, often called soft gold, came from the North Pacific beginning from the time Europeans first explored the region's waters. The fur trade had long been an important economic wedge in other parts of North America, usually preceding actual European settlement. European explorers noted the abundant furs and imagined the potential profits. Such attention attracted numerous traders—first to North Pacific waters and then throughout the region's rivers and mountains. The fur traders brought with them the values of capitalism, including a strong profit motive, and initiated ecological changes equal in importance only to the cultural changes the fur trade engendered in the Pacific West.

The earliest fur trade presence came from the Russians, who had moved into Alaska's Aleutian Islands by 1741. In subsequent decades they moved down the coast and had finally established Fort Ross north of San Francisco by 1812. The sea otter trade accompanied this imperial expansion. Although they mostly focused in Alaskan waters, Russian companies began hunting in California waters by the first decade of the nineteenth century, using Aleuts as the hunting crew. By the 1810s, they were annually taking several thousand otters. At the same time, Spain, and later Mexico, entered as a marginal player in the maritime fur trade. The Russian and Spanish presence in North Pacific and California waters may have initiated the trade but their efforts largely paled compared with the British and Americans who got involved in the maritime fur trade just slightly later. These groups not only affected fur-bearing animal populations, but they also played out imperial and economic rivalries with their associated environmental impacts (Gibson 1992, 12–21; Ogden 1998, 89–98).

British Captain James Cook's journey to the Northwest in 1778 fully initiated the British and American rush to obtain North Pacific furs. Cook's crew traded with Nuu-chah-nulth (Nootkans) on Vancouver Island, selling the pelts in Guangzhou (Canton), China, for $120 apiece, a fabulous profit of 1,800 percent. Cook's crew nearly mutinied to return for greater profits awaiting them in the North Pacific. Cook published accounts in 1784, publicizing the opportunities available for enterprising traders. For the next several decades

Europeans and Americans developed a commercial system based on exploiting the region's ecology, a system that devastated marine resources before moving to vastly rearrange inland riparian ecologies (Bunting 1997, 22–23; Gibson 1992, 22–23).

Competitive Environments and Declining Furs

Although the Northwest fur trade has been studied less rigorously and viewed less romantically than the New England, Great Plains, or Rocky Mountain trade in furs, the region's economic and political contexts make the Northwest's fur trade somewhat more intriguing. Imperial and commercial rivalries among Russian, Spanish, British, and American governments, companies, and individuals shaped the dynamics. Indeed, part of the trade in soft gold deliberately depleted a resource to advance particular political and economic goals—a cross somewhere between ecological warfare and environmental diplomacy. Two other factors played significant roles, as well, for traders required the labor of Native peoples, and they had to adapt to the ecology of fur-bearing animals. The history of the fur trade, viewed through the prism of environmental history, reveals the dependence markets always have on the natural world, the ways geopolitical concerns have long possessed ecological components, and the many and varied consequences of what seems at first glance to be simply an exchange of goods.

The maritime business decimated the sea otters rapidly, and it did not take much longer to deplete the beaver population throughout much of the Oregon country. Environmental historians have noted in passing the decline in sea otters and their ecological effects. However, they have not carefully detailed the process, while fur trade historians have largely been interested only in the business arrangements or relations between Natives and European traders (Fisher 1992, 1–23; Gibson 1992). Accordingly, much of our knowledge is incomplete; still, some of the environmental history of Northwest otters and beavers seems clear.

The enterprise functioned through an intercontinental trade triangle. European and New England merchants outfitted ships with manufactured goods and sailed around South America to the Northwest Coast. There, they participated in a vibrant trade with Native groups who readily adapted their own history of trading to new commercial circumstances. From the Northwest Coast, ships sailed with otter pelts to China, where they exchanged them for "silks, spices, and tea" (Calloway 2003, 402). Then they returned to Europe or New England. Merchants expected high profits based on exploiting fur-bearing marine mammals and connecting distant markets to Northwestern nature (Calloway 2003; Gibson 1992).

Over time, several developments altered this pattern. The first challenge to the fur trade came from the partners in the trade: the Northwest Coast tribes. From the beginning, Indian peoples were central to the trade. The Russian fur traders basically pressed Alaska Natives into service as forced laborers. Further south, Natives more willingly participated, incorporating new manufactured goods into their material culture. Some historians have maintained that Indians actually largely controlled the terms of the trade, demanding metals, cloths, and various tools before exchanging pelts (Fisher 1992, 1–23). The maritime trade with Europeans provided new avenues to wealth and prestige. With new possibilities for making wealth, more Indians participated in the fur trade either directly, with men hunting animals, or indirectly, with some women prostituting themselves to sailors. These new or intensified economic behaviors meant that some communities became short on foods that they had neglected to cache for winter. This situation often put them at a temporary disadvantage in the trade and could lead to intertribal violence. Indeed, European sea captains increasingly reported less cooperative interactions with Coastal tribes, especially as the trade slackened in the 1820s (Calloway 2003, 412; Gibson 1992, 173).

The second problem was a saturated market. The early voyages typically yielded high profits, in some circumstances approaching 300 percent. Such profitability predictably stimulated interest, leading to more ships trading in the waters off the Northwest Coast, driving up the price of furs there but driving down the price in China as early as the late 1790s in an increasingly saturated market. Eventually, prices in China increased but did so in response to the scarcity of furs (Gibson 1992, 176–178).

This third challenge, resource scarcity, was intimately related to the second. Because of the ease with which hunters took otters and the lure of profits, species declined with amazing haste. Begun in earnest only in the early 1780s, the maritime fur trade set the otter on a course toward extinction by the turn of the century. Localized extinctions in waters where economic competition was fiercest reached back to 1786, and by the 1820s, there were few otters left anywhere to harvest commercially. The Mexican government belatedly attempted to regulate the trading, but it did so weakly and after the damage was done (Calloway 2003, 412; Mackie 1997, 79; McEvoy 1986, 79).

The decline in the maritime fur trade did not spell the end of trading soft gold. Far from it. Instead, the land-based trade rose with even more interesting historical implications. In the shift, land-based traders solved some of the challenges faced by maritime traders. Although saturated markets continued to plague the trade, the shift to land-based trading bypassed the problems of otter

The sea otter's luxuriant pelt brought European traders to the Northwest Coast in the late eighteenth and early nineteenth centuries. Populations of otters were quickly decimated. (iStockPhoto.com)

depletion by shifting species. The difficulty caused by dependence on Native groups temporarily changed, at least for some companies, as Euro-Americans did some trapping themselves and established posts where they could maintain more consistent contact with Northwest Native peoples. Thus, the social and economic dynamics of the fur trade changed as it moved to land. Euro-American companies and individuals either trapped for themselves or were much closer to the trapping than they had been when engaged in maritime trading. As such, their records provide some of the best early information about the Northwest landscape.

Various fur traders provided ample evidence of the landscape's advantages. Hudson's Bay Company trader Duncan Finlayson, describing the west side of the Cascades near Fort Vancouver (present-day Vancouver, Washington), explained that "By its overpowering abundance all nature here demands attention" (quoted in Mackie 1997, 69). Fur trade narratives highlighted potential resource development besides simply beaver populations. For example, Ross Cox of the Pacific Fur Company described the Willamette Valley as possessing "a rich and luxuriant soil, which yields an abundance of fruits and roots" (quoted in Robbins 1997, 53). Cox and others noted climate, soil, timber, agricultural potential, and fish, focusing through it all on the potential profits for burgeoning global markets (Robbins 1997, 53–57).

The theme of abundance was common, even in the comparatively scarce Plateau country. Although some traders noted that the interior was "nothing but a sandy desert" or was "rude and hilly without woods for several miles, and destitute of deer, or the wild Sheep of the Mountains," others characterized some parts as "a Country well stocked with animals" (quoted in Vibert 1997, 84, 88). On the Snake River Plain, the site of intense fur trapping in the 1820s and 1830s, trapper Alexander Ross described the abundance of animals:

> Woods and valleys, rocks and plains, rivers and ravines, alternately met us; but altogether it is a delightful country. There animals of every class rove about undisturbed; wherever there was a little plain, the red deer [elk] were seen grazing in herds about the rivers; round every point were clusters of poplar and elder, and where there was a sapling, the ingenious and industrious beaver was at work. Otters sported in the eddies; the wolf and the fox were seen sauntering in quest of prey; now and then a few cypresses or stunted pines were met with on the rocky parts, and in their spreading tops the raccoon sat secure. In the woods, the martin and black fox were numerous; the badger sat quietly looking from his mount; and in the numberless ravines among bushes laden with fruits, the black, the brown, and the grizzly bear were seen. The mountain sheep, and goat white as snow, browsed on the rocks and ridges, and the bighorn species ran among the lofty cliffs. Eagles and vultures of uncommon size flew about the rivers (quoted in Arrington 1994, 94).

Although not lush like the descriptions of the west side of the Cascades, the landscape Ross described represented a landscape of potential, and one teeming with life that might be turned into profit.

Competing economic and political empires invaded these abundant and fur-laden landscapes. At various times and in various locales, the Spanish, Russian, British, and Americans had claimed parts of what became known as the Oregon Territory, which constituted most of present-day British Columbia, all of Washington, Oregon, and Idaho, and parts of western Montana. The United States was the latecomer to this imperial contest, establishing the basis of their claim with Robert Gray's "discovery" of the Columbia River in 1792 and the Lewis and Clark expedition's travels through the region in 1805–1806. In 1818, Great Britain and the United States agreed to jointly occupy the Oregon Territory without claiming sovereignty to the land between Russian America and Spanish (and later Mexican) California. The Adams-Onis Treaty (1819) between the United States and Spain saw the latter officially bowing out of the imperial contest in Oregon. During and even preceding this joint occupancy period, commercial interests basically functioned, in the absence of state power, as governments for Europeans in the area (Harris 1997, 31–67).

The Montreal-based North West Company (NWC) established the first European economic presence in the Northwest after Alexander Mackenzie traversed the continent in 1793, followed by his fellow Nor'westers Simon Fraser and David Thompson, who explored much of the region. The NWC faced American competition from the Pacific Fur Company, based out of Astoria, Oregon, although the Astorian presence in the Northwest was short-lived and largely unprofitable. The Americans and Canadian Nor'westers had pioneered a system whereby Euro-Americans trapped rather than traded for furs. The NWC soon faced a greater rival. At a time when the NWC extended itself too far geographically and financially, the global commercial empire of the Hudson's Bay Company soon dominated. Based out of London, the HBC enjoyed a significant competitive advantage over the NWC and in 1821 merged with the latter, effectively taking over the company, its employees, and all its buildings in the region (Mackie 1997; Schwantes 1996, 62–70).

The company initially did not involve itself much in the Columbia Department (the company's area west of the Rocky Mountains) and did so only after seriously reorganizing business practices. Although profits had eluded the NWC, the HBC traded nearly 10,000 beaver and otter pelts from the 1822–23 season, which attracted their interest and increased their commitment to the region (Meinig 1995, 68). HBC built on the series of trading posts throughout the region along major watercourses like the Columbia and Snake rivers that earlier trading enterprises had established. George Simpson, the HBC governor in charge of western British North America, devised a series of reforms to be pursued through his appointed manager, John McLoughlin. Relocating headquarters to Fort Vancouver, the HBC endeavored to diversify the economy at the trading posts, introducing European-style agriculture to the Northwest for the first time and eliminating much inefficient food waste. This would be the first step toward developing new commodities to be exchanged in the global market. Moreover, Simpson and McLaughlin worked to make Native people more dependent on the company with decidedly mixed results. This proved to be a difficult task, for Natives participated in the trade only on their own terms, which at times did not match HBC's goals. Perhaps the goals most important besides maintaining a profit were maintaining HBC's monopoly in the region and keeping Oregon Country British (Fisher 1992, 24–48; Harmon 1998, 13–42; Schwantes 1996, 68–75; White 1992, 30).

However, when Alexander Ross met American fur traders in southeastern Idaho in 1824, it was clear that the HBC's monopoly and Oregon's position as British were insecure. In the late 1820s through the 1830s, first Peter Skene Ogden and, then, John Work endeavored to create a fur desert in the Snake River country. Ogden and Work were carrying out the orders of Simpson and McLouglin. In Simpson's words:

> The greatest and best protection we can have from opposition is keeping the country closely hunted as the first step that the American Government will take toward Colonization is through their Indian Traders and if the country becomes exhausted in Fur bearing animals they can have no inducement to proceed thither. We therefore entreat that no exertions be spared to explore and Trap every part of the country (quoted in Bunting 1997, 32).

The purpose of decimating the resource on which the trade depended does not strike one as an organization attuned to conservation needs, as the HBC practiced with some success elsewhere (Ray 1992, 33–50).

Instead, political and economic motives combined to encourage the practice. The Anglo-American accord that declared Oregon a jointly occupied region was due to lapse in 1828, and the British anticipated that the international boundary would be the lower Columbia River. The fur desert strategy is comprehensible only from this geopolitical worldview, as historical geographer D. W. Meinig explained: "The purpose was not only to reap a maximum profit before relinquishing, but to exhaust the fur resources, thereby creating a buffer zone to shield the British operations on the Columbia" (1995, 74). Thus, imperial-economic rivalries gave rise to deliberate overhunting to keep Americans at bay.

Creating a fur desert in what is now southern Idaho succeeded with devastating ecological effects. As one scholar has observed, "To read the journals from the Snake Country expeditions is to read a story of creating scarcity" (Ott 2003, 178). And scarcity arrived quickly. During the years when the HBC practiced the fur desert policy, traders took out approximately 35,000 beaver. In the initial year alone, they harvested 4,500 pelts; within a decade, the yield declined to a paltry 665. Traders' accounts after the HBC adopted the policy made plain the rapidity with which ecological change could occur.

Hudson's Bay Company trappers used their knowledge of beaver ecology to hasten the destruction. By living in such obvious lodges, beaver populations were quite easy to locate. Trappers baited traps with castoreum, "a substance secreted by the glands of beaver with a scent that is individually unique," which piqued their territoriality and attracted the animals to their ultimate demise (Ott 2003, 181). Moreover, trappers exploited mating and reproduction behavior. By focusing their harvesting during winter and spring, when pelts were their thickest and most luxuriant, trappers exploited evolutionary adaptations beavers had developed as protective measures. Ponds covered by ice prevented predators from reaching kits, as did underwater lodge entrances. Since these defenses were so successful from an evolutionary perspective, beavers reproduced at low levels. Later, during spring and summer when young adult males would normally move on and join new lodges established by female beavers, there were too few animals left to repopulate lodges. That, coupled with the low reproduction rate, made them easier for trappers to eliminate. These ecological relationships were clearly known by Ogden who wrote in 1829:

> It is scarcely credible what a destruction of beaver by trapping this season, within the last few days upwards of fifty females have been taken and on an average each with four young ready to litter. Did we not hold this country by so slight a tenure it would be most to our interest to trap only in the fall, and by this mode it would take many years to ruin it (quoted in Ott 2003, 182).

Ogden's remarks revealed both the trader's ecological knowledge and awareness of the broader geopolitical motives of the fur desert policy. Undoubtedly, HBC trappers understood beavers and used that knowledge to their advantage in creating the fur desert (Ott 2003, 180–182).

Other environmental factors may have also contributed to the demise of the beaver population. Certainly they would not have approached near-extinction without the assistance of the HBC's policy; however, climatic conditions may have made beaver vulnerable. During the 1820s and 1830s, the Snake River country experienced warming and drying trends. Results of this may have included lower water levels behind dams, which possibly increased disease. In addition, higher temperatures and lower rainfall may have increased fires, destroying vital food and building sources. Trappers noted abandoned lodges, but they could not tell whether disease or fire were the causes. Such problems were likely not region-wide or there would have been greater records of them, but locally these conditions likely weakened certain beaver populations. Finally, fur-bearing animals experience natural cycles of population irruptions and crashes. These multiple factors suggest the challenge of basing an economy on a natural resource subject to logic outside the global marketplace (Mackie 1997, 247; Ott 2003, 182–183).

The Americans' presence encouraged the HBC to maintain their strategy and accelerated beaver decimation. Although the British and American governments extended their joint occupation twice, the boundary's ultimate insecurity did nothing to stave the competition. After all, Americans in the eastern section of the Snake River plain caused a faster extirpation of beaver there (Ott 2003, 184). Furthermore, American competition changed the economics of the trade. The HBC had once traded one blanket to Native traders for five beaver pelts. With Americans in the picture, it was reduced to a one to one ratio (Meinig 1995, 91). Still, in 1841 Simpson explained why there would be no conservation efforts in the Oregon country: "[I]n the present unsettled state of the boundary line it would be impolitic to make any attempt to preserve or recruit this once valuable country, as it would attract the attention of the American trappers" (quoted in Ott 2003, 180). Besides trappers, though, American missionaries began arriving in the 1830s, and their presence greatly increased American interest in the region. Moreover, they disconcerted the British traders who believed missionaries were agents of colonization as much as or more than agents of Christianization (Ott 2003, 180). British political calculations certainly hastened the devastation.

The End of the Fur Trade and Its Consequences

By the 1830s, the fur trade in the Pacific Northwest was nearing its end because of a combination of factors. Profits were falling, as silk hats became preferred to beaver hats in Europe (Mackie 1997, 245). More importantly, though, overhunting simply made trapping an impossible economy to sustain. The future of the trade was bleak, as the HBC governor George Simpson explained in 1843, "We cannot indeed contemplate either the present or future prospects of the fur trade on the west side of the [Rocky] Mountains without Anxiety . . . the trade has fallen off greatly for two successive years, marking, it is to be feared, a rapidly progressive diminution of the fur bearing animals" (quoted in Mackie 1997, 249). In two decades of intense trapping concentrated after the 1820s, local extinctions became common, and unrecoverable declines were common (Langston 1995, 57). Despite all of these challenges, the HBC remained productive and profitable in the region, primarily because they diversified their economy to include agriculture and timber, which set in motion a variety of other ecological changes that modified the region's landscapes (Mackie 1997, 249–256).

In its efforts to diversify its economic foundations, the HBC promoted two other industries that would ultimately become the agents of tremendous ecological change in the Northwest. Local mills provided lumber for construction and soon sawmills throughout the Northwest would provide timber for markets throughout the Far West and beyond to Hawaii and parts of Asia. Moreover, Fort Vancouver initiated Europeanized agriculture. At this site north of the Columbia River, they brought in herds of cattle and other animals, as well as orchard trees. They further developed their agricultural presence with the Puget Sound Agricultural Company, an HBC subsidiary that built sheep flocks greater than 4,000, cattle herds larger than 1,000, and abundant grains and legumes by 1841. From these places on the west side of the Cascades, these flora and fauna, the vanguard of Euro-American ecological imperialism, spread throughout the interior of the Northwest (Schwantes 1996, 72–73).

Cultural consequences could hardly be overstated. Native peoples benefited and lost because of the fur trade. One scholar has aptly characterized it as "a period of mixed blessings" (Frey 2001, 61). On the positive side, tribes and individuals acquired new wealth, at least temporarily. Some have argued that the increase in material wealth gave rise to an era of unprecedented cultural creativity (Fisher 1992, 48). In addition, marriage between Euro-American fur traders and Native women became widespread. Not only was this an economic arrangement that in large part determined the success of the land-based fur trade, but it also reflected mutual affection and helped enhance the multicultural nature of Northwest society (Harmon 1998, 37, 41; Van Kirk 2001, 184–194). Not least, new trade goods like glass beads, blankets, and metal tools introduced new aspects into Native material culture, and Indian peoples embraced many of the new goods on their own terms.

Although participating in the fur trade by their choice and with good reasons, Indian peoples were unable to control many aspects of the trade and faced a number of negative consequences. Most profoundly, the fur trade functioned as an effective vector for disease. Exchange between Native peoples and Euro-Americans spelled epidemiological disaster across the continent, and the fur trade was an important agent in spreading diseases such as smallpox (Fenn 2001, 224–258). Next in importance was the introduction of the market. Despite exchange being a long-standing tradition in the Native Northwest, global capitalism profoundly changed the nature of trade. As historians have shown, Native peoples understood trading as a personal exchange whereas Euro-Americans viewed it as an impersonal exchange. Such contrasting perspectives led inevitably to misunderstandings and frequently to conflicts (Harmon 1998, 27). Not all goods in the trade benefited Native peoples either, for alcohol and firearms arrived, often with disastrous effects. As trading in furs declined, Native peoples faced another new upheaval as Euro-Americans came to settle in the Northwest permanently, bringing with them an economy even more disruptive to Native human-ecological regimes and local environments. Moreover, being involved in an economic system over which one exercised only limited, if any, control contributed to unsettling consequences for Native peoples, as well as later residents who were subject to wildly fluctuating commodity prices or rapidly changing government policies and always dynamic ecosystems.

In addition to these important and significant cultural consequences, the varied ecological concerns cannot be overlooked. Sea otters lost genetic diversity because of the maritime trade, which may, to this day, be affecting the health of otter populations (Larson et al. 2002). Furthermore, the decline of otters appears to have disrupted marine ecosystems, replacing "one community dominated by otters, kelp, and abundant finfishes [with] another consisting of no otters, abundant mollusks, and fewer finfishes" (McEvoy 1986, 81). However, this may have increased the abalone industry, which spelled good economic fortune for Chinese harvesters in California (McEvoy 1986, 82). Without complete historical records and because of the complexity of historic ecosystems, it is impossible to know the full effect of the disappearance of sea otters off the North American coast.

A similar difficulty exists for precisely judging the impact of the near-extinction of beavers in the Northwest. Nancy Langston has explained that beavers had made "watersheds more resistant to disturbance," so their absence weakened that resilience. However, she warns against seeing succession as "straightforward: instead of moving in an orderly fashion from pond to marsh to meadow to forest, there were complex jumps and pauses. . . . Beaver ponds in various stages of creation and decay formed a shifting mosaic of diverse patterns across the landscape" and across time (Langston 1995, 58–59). But if we cannot determine with certainty the direction of change, we must recognize that the removal of 35,000 members of a riparian keystone species and their approximately 6,000 ponds had significant effects. Water flowed with greater velocity; erosion on the stream bottoms and stream banks worsened; insect habitat declined as surface area decreased; stream sediment loads increased; snow runoff accelerated; and the water table lowered (Ott 2003, 185–190). Clearly, nature changed because of the fur trade, and this accounting does not indicate the substantial concomitant ecological consequences of the beginning of timber and agriculture and the effects of disease.

CONCLUSION

The Northwest fur trade illustrates several important themes to the region's environmental history and the ecological past more generally, revealing patterns that repeated over and over in different places and distinct resources. First, it reiterates the theme of distant economics and geopolitics shaping local ecosystems. The fur trade in the Northwest initiated the capitalist transformation of nature into commodities designated for trade in a global market. Accelerating through the nineteenth century, and more so in the twentieth century, capitalist reconfiguring of nature disturbed and transformed ecologies in the Pacific West in far-reaching ways. And as shown here, it could be accomplished relatively easily, rapidly, and without industrialization.

Second, the fur trade episodes off the coast and in the interior Northwest demonstrate that ecological knowledge and values play a role in geopolitical and economic schemes. Rather than being simply an afterthought, the HBC recognized the environmental effects of creating a fur desert. It was a trade-off they were willing—indeed eager—to make. In the future, other companies and governments also made trade-offs that altered ecosystems for what they perceived to be a greater good. The political and economic context and decisions involved in the fur trade cannot be divorced from their environmental context.

Last, it set a pattern that would be repeated in various forms for the next two centuries. A population with capital and an interest discovered a resource, commodified it, traded or processed it for global exchange, rapidly depleted the resource, and belatedly or never began some sort of conservation practices. With varying details, this occurred in the timber, mining, fishery, and agricultural industries throughout the Far West. The Mexican government's weak and ineffective attempts to regulate sea otter harvesting is somewhat typical of the lackadaisical response governments gave to sustainability, while the economic and geopolitical calculus used by the HBC is all too typical of the way private enterprises respond to environmental decline (McEvoy 1986, 79). Thus, the Northwest maritime and terrestrial fur trade illustrates the interaction of ecology, economy, and culture. It decimated existing ecological systems and furnished an opening wedge to further environmental changes.

REFERENCES

Arrington, Leonard J. *History of Idaho: Volume I.* Moscow and Boise: University of Idaho Press and Idaho State Historical Society, 1994.

Bunting, Robert. *The Pacific Raincoast: Environment and Culture in an American Eden, 1778–1900.* Lawrence: University Press of Kansas, 1997.

Calloway, Colin G. *One Vast Winter Count: The Native American West before Lewis and Clark.* Lincoln: University of Nebraska Press, 2003.

Fenn, Elizabeth A. *Pox Americana: The Great Smallpox Epidemic of 1775–82.* New York: Hill and Wang, 2001.

Fisher, Robin. *Contact and Conflict: Indian-European Relations in British Columbia, 1774–1890.* 2nd edition. Vancouver: University of British Columbia Press, 1992.

Frey, Rodney, in collaboration with the Schitsu'umsh. *Landscape Traveled by Coyote and Crane: The Worlds of the Schitsu'umsh (Coeur d'Alene Indians).* Seattle: University of Washington Press, 2001.

Gibson, James R. *Otter Skins, Boston Ships, and China Goods: The Maritime Fur Trade of the Northwest Coast, 1785–1841.* Seattle: University of Washington Press, 1992.

Harmon, Alexandra. *Indians in the Making: Ethnic Relations and Indian Identities around Puget Sound.* Berkeley: University of California Press, 1998.

Harris, Cole. *The Resettlement of British Columbia: Essays on Colonialism and Geographical Change.* Vancouver: University of British Columbia Press, 1997.

Langston, Nancy. *Forest Dreams, Forest Nightmares: The Paradox of Old Growth in the Inland West.* Seattle: University of Washington, 1995.

Larson, Shawn, Ronald Jameson, Michael Etnier, Melissa Fleming, and Paul Bentzen. "Loss of Genetic Diversity in Sea Otters (*Enhydra lutris*) Associated with the Fur Trade of the 18th and 19th Centuries." *Molecular Ecology* 11 (October 2002): 1899–1903.

Mackie, Richard Somerset. *Trading Beyond the Mountains: The British Fur Trade on the Pacific, 1793–1843.* Vancouver: University of British Columbia Press, 1997.

McEvoy, Arthur F. *The Fisherman's Problem: Ecology and Law in the California Fisheries, 1850–1980.* New York: Cambridge University Press, 1986.

Meinig, D. W. *The Great Columbia Plain: A Historical Geography, 1805–1910.* 1968. Reprint, Seattle: University of Washington Press, 1995.

Ogden, Adele. "Sea Otters Encounter Russians." In *Green Versus Gold: Sources in California's Environmental History,* edited by Carolyn Merchant, 89–98. Washington, DC: Island Press, 1998.

Ott, Jennifer. "'Ruining' the Rivers in the Snake Country: The Hudson's Bay Company's Fur Desert Policy." *Oregon Historical Quarterly* 104 (Summer 2003): 166–195.

Ray, Arthur J. "Some Conservation Schemes of the Hudson's Bay Company, 1821–1850: An Examination of Resource Management in the Fur Trade." In *The American Environment,* edited by Lary M. Dilsaver and Craig E. Colten, 33–50. Lanham, MD: Rowman and Littlefield, 1992.

Robbins, William G. *Landscapes of Promise: The Oregon Story, 1800–1940.* Seattle: University of Washington Press, 1997.

Schwantes, Carlos Arnaldo. *The Pacific Northwest: An Interpretive History.* Revised and expanded ed. Lincoln: University of Nebraska Press, 1996.

Taylor, Joseph E., III. *Making Salmon: An Environmental History of the Northwest Fisheries Crisis.* Seattle: University of Washington Press, 1999.

Van Kirk, Sylvia. "The Role of Native Women in the Creation of Fur Trade Society in Western Canada, 1670–1830." In *Women in Pacific Northwest History,* revised ed., edited by Karen J. Blair, 184–194. Seattle: University of Washington Press, 2001.

Vibert, Elizabeth. *Traders' Tales: Narratives of Cultural Encounters in the Columbia Plateau, 1807–1846.* Norman: University of Oklahoma Press, 1997.

White, Richard. *Land Use, Environment, and Social Change: The Shaping of Island County, Washington.* Seattle: University of Washington Press, 1980. Reprint with a new foreword, Seattle: University of Washington, 1992.

CASE STUDY: LOS ANGELES AND THE PROBLEM OF WATER

"If you don't get the water, you won't need it." Those were the prescient words of a Los Angeles water official who recognized the necessary commitment to water that the growing metropolis required (Hundley 2001, 139). In the coming decades, Los Angeles waged a relentless battle to acquire ever more water to supply an ever greater population. This quest revealed extensive urban power, as Los Angeles drastically affected landscapes and water supplies hundreds of miles outside the city limits—a practice several scholars have characterized as imperialistic (Brechin 1999; Hundley 2001; Hurley 2005; Starr 1990). In addition, it is a story scholars and writers have retold countless times, including a fictionalized version presented in the 1974 feature film, *Chinatown* (Hoffman 1981; Kahrl 1982; Mulholland 2000; Reisner 1993; Sauder 1994; Walton 1992).

Despite its ubiquity in Western history annals, the history of Los Angeles and the problem of water merits a retelling here, in part because the dramatic story of acquiring water has obscured other important environmental themes. For example, historians are beginning to pay serious attention to water within the city, and the history of urban water reveals significant ecological changes, as well as alternatives not pursued. This case study also emphasizes the growing power of cities to shape nature in two ways. Much like William Cronon illustrated with nineteenth-century Chicago, this story demonstrates the reach of a metropolis far into its hinterland (Cronon 1991). As an unparalleled economic engine, the metropolis drew resources from distant ecosystems transforming them and itself simultaneously. Ironically, too, the history of Los Angeles and water is not just a story of overcoming scarcity but also of dealing with abundance in the form of frequent floods. Besides these chief factors, histories of Los Angeles and water in the decades around 1900 reveal changing relationships between government—federal and local—and nature, especially as embodied in the new Progressive conservation movement. Finally, neither the story about acquiring water nor the story about water in the city is about following nature's cues before constructing a metropolis.

The Nature of Los Angeles River Development

Like all rivers, the Los Angeles River has a history. As such, providing simple figures about how long it runs and how much water it contains depends on *when* one discusses the Los Angeles River. It has long been a meandering river.

The flow sometimes entered into one bay, sometimes another. Often, much of it never reached any bay at all. The river's course was neither deep nor permanent. And storms could radically disrupt what was once there. For example, a flood in 1825 shifted the channel significantly, reorienting the bulk of the river's outflow from Santa Monica Bay to San Pedro Bay, 20 miles to the south. Its origins are in the San Gabriel Mountains, which can receive up to forty inches of annual precipitation, even though the Los Angeles Basin averages about one-third of that. Such averages, however, are misleading because the Southern California Mediterranean climate is one of extremes. Floods in the American East can sometimes double a river's flow, but in Southern California, floods can increase a river's typical flow 3,000 times (Davis 1998, 17). In February 1914, a three-day storm dumped a whopping twenty inches of rain in the San Gabriel Mountains (Deverell 2004, 111–113). These are just some examples of nature's regional unpredictability.

Compared with other Western rivers, such as the Columbia, the Snake, or even the San Joaquin, the Los Angeles is puny but important. Nevertheless because of its semiarid surroundings, the watercourse always constituted an important part of settlement in Southern California as it was the only stream in the area that typically flowed year-round. As such, it attracted settlers. The river made Los Angeles a productive home for the Gabrielino people, then Spanish and Mexican immigrants, followed by Anglo Americans. Living in the Los Angeles Basin before the twentieth century depended almost entirely on the Los Angeles River. The river's tendency to flood, alter channels, or occasionally run dry caused minimal problems while the basin's population remained relatively small and mobile. However, when the Southern Pacific Railroad arrived in 1876, urban boosters accelerated economic development and initiated dramatic changes to Los Angeles that ultimately rendered the Los Angeles River an entirely different natural system (Gumprecht 1999; Gumprecht 2005).

Ignoring the lack of water, many civic leaders promoted instead the abundant beaches, sunshine, and mountains, successfully attracting hundreds of thousands of immigrants. Los Angeles had about 128,000 residents in 1902, but hundreds more arrived every week. With so many incoming people, the river could no longer meet the population's needs. Although the river fed the agricultural settlements and created a charming bucolic landscape that attracted midwesterners and others, the residential demands of the new migrants quickly displaced agricultural uses for the river despite building more reservoirs and increasing the size of existing ones. The river was set firmly on an urban course (Gumprecht 2005, 118–123; Hundley 2001, 124).

A bird's eye view of Los Angeles taken from a balloon, ca. 1902. As the city expanded, it required more water, which it acquired from vast distances. (Library of Congress)

Metropolitan growth contained not only increased consumption but also an industrial component that polluted the river. Concerns about pollution became common by the 1890s and centered on the river because railroad development and concomitant industrial development concentrated there. Of course, nineteenth-century industrialization had polluted many urban waterways throughout the world. However, with so little water in the Los Angeles River, refuse did not flush as easily from the river as in larger streams. Accordingly, rubbish and waste accumulated in the riverbed. So precarious and polluted was the river's ecosystem that it was not even fit for water. Geographer Blake Gumprecht explained, "the stated goal of the city in managing the river was to keep any water from flowing in its channel. The city water department sought instead to intercept any water destined for the river while it was still underground and clean, pumping it from the groundwater basins that were its natural source" (2005, 123). In 1910, the city council finally passed an ordinance to stop garbage dumping in the riverbed. However, industries widely ignored it. A survey in 1912 indicated that 27 truckloads of garbage were dumped in the river every day. Violating its own ordinance, the city used the riverbed as a dump until 1925. Furthermore, construction companies used the riverbed as a quarry, taking between 1,000 and 1,200 truckloads of sand and gravel from the riverbed daily as early as 1907. In short, the river was a hard-used place with accumulating industrial problems, increasing public health concerns, and diminishing resemblance to a river with water running through it (Deverell 2004, 108–110; Gumprecht 2005, 123–129). Despite its importance to early settlements, the river could not sustain the dreams of subsequent Angelenos. Just as the river's banks could not always contain the water, the river itself could not support the boosters' dreams.

Solving the Problem of Too Little Water

Los Angeles had long been expansionistic in using its water. According to Spanish custom, the town council managed water, including all necessary taxes and regulations. Initially, management focused on irrigating agricultural tracts; however, by the 1870s, Los Angeles recognized water demands beyond irrigation. By the 1880s, commerce became the dominant use of water and agriculture followed. Subsequent real estate booms stressed the scarce water supplies and initiated what became a central process to twentieth-century California life: the transformation of agricultural tracts to suburban tracts. To ensure the city's future water supply, officials worked hard to assert and secure water rights (Hundley 2001, 123–124).

The city began its crusade in 1874. As was common at the time, the city turned to the courts, which were where water law developed, not in legislatures. The city sought an injunction to end upstream uses of the Los Angeles River, claiming it possessed a pueblo water right that gave Pueblo de Los Angeles ownership of all the water in the river. Historian Norris Hundley has called this an "audacious claim," because Spanish water law never provided outright ownership of water, only usufructuary rights (2001, 130). Although the state legislature, bowing to metropolitan pressure, approved a municipal charter that asserted this pueblo right, the city's specious evidence and claims initially failed to persuade the courts. However, in a series of actions and subsequent court decisions over the next two decades, Los Angeles's claim to the river's water strengthened. Finally, in the 1895 decision, *Vernon Irrigation Co. v. Los Angeles*, a court confirmed the pueblo right to all of the Los Angeles River. In addition, because of an earlier decision, the city could not sell water to customers outside the city limits, a regulation that spurred annexation (Hundley 2001, 127–139). With the river's water secure, Los Angeles appeared poised for unrestrained growth.

As the twentieth century dawned, several factors coalesced to make the city's water demands not only increase but also seem increasingly urgent. Los Angeles experienced a water shortage. Records indicate that the river's capacity declined from 100 cubic feet per second in the 1880s to only 45 cubic feet by 1902. Moreover, dramatic population increases multiplied water demands. Los Angeles doubled in size from 100,000 to 200,000 residents in just four years, from 1900 to 1904. A drought beginning around 1900 exacerbated the problem. Thus, demographic factors spurred by economic prosperity and a cultural predilection for Southern California's sunny climate and ocean breezes collided with nature's limits and produced anxiety. Nevertheless, city boosters, who dreamed of even larger populations, and city engineers, who dreamed of engineering their way out of problems, sought another water source that the city could control. Bringing more water, after all, meant attracting more people, and more people meant greater power for city leaders. And so the cycle began (Reisner 1993, 60–62).

Civic leaders inside and outside municipal government developed a vision that satisfied their needs. Undoubtedly, the most important character in the evolving story was William Mulholland. This self-taught Irish immigrant worked as a ditch-tender for Los Angeles's private water supplier. Soon, Mulholland's ambition and natural ability facilitated a rise through the ranks so that when Progressive Era reforms municipalized the water supply he remained on the job and became superintendent. With the entire distribution system committed to memory, he enjoyed immense power presiding over a "virtual fiefdom" while becoming the highest salaried public official in California (Hundley 2001, 144). Unlike some, Mulholland did not simply crave power for its own sake; he genuinely believed Angelenos required more water to maintain their livelihoods and to feed beneficial economic expansion. Along with other civic leaders, Mulholland identified a reliable water source in the distant Owens Valley, and he and other Los Angeles boosters maneuvered to achieve their civic dreams (Hundley 2001, 141–151; Mulholland 2000; Reisner 1993, 57–58).

The Owens Valley lay east of the Sierra Nevada Mountains and west of the White Mountains about 250 miles northeast of Los Angeles. With Owens Lake and the Owens River, it was surprisingly well-watered in an otherwise desert environment. Resident Paiutes first irrigated the valley, but Euro-Americans arriving in the 1860s overtook those irrigation works, eventually irrigating more than 40,000 acres by 1900 while growing alfalfa, various grains, and fruit trees. Desiring stability and economic prosperity, local society searched for ways to expand this agricultural base. At the same time, the newly formed U.S. Reclamation Service sought locales for federally sponsored reclamation projects. Local farmers anticipated a bright future with federal assistance extending irrigation to an additional 100,000 acres. Thus, the city of Los Angeles was not the only group interested in using Owens Valley water (Reisner 1993, 58–60, 63–64; Sauder 1994; Walton 1992).

Fortunately for Los Angeles's interests, the Reclamation Service moved slowly. Besides Mulholland, two individuals loom large in the story of how Los Angeles secured Owens Valley water. The first was Fred Eaton, a native Angeleno, an erstwhile hydrological engineer, a consummate booster, and a former Los Angeles mayor. The second was Joseph B. Lippincott, a federal reclamation engineer who set aside professional scruples by operating a private consulting business and retaining Los Angeles as a client. Lippincott gave Eaton access to valley land and water right records that would normally have been unavailable. Eaton hurriedly secured options for the valley's premier land and water rights, including the only viable reservoir site in the valley. He planned to sell the rights to the city at cost; but if the city would not buy the land, he would sell to others. It constituted blackmail, facilitated by a renegade federal employee. Although at first hesitant, the city eventually bought Eaton's land as the secrecy of their plan was breaking down and Owens Valley residents were growing suspicious (Reisner 1993, 60–71; Hundley 2001, 144–151).

In the hot summer of 1905, then, the Los Angeles Board of Water Commissioners announced its plan, and the *Los Angeles Times* declared: "Titanic Project to Give City a River" (Reisner 1993, 70; Hundley 2001, 151). Before the city could get the river, however, it faced difficult obstacles. First, it needed the public to pay for constructing a 250-mile aqueduct. Fortunately, Mulholland's and Eaton's constant dire predictions about imminent water shortages ensured widespread support. In two votes—one to purchase rights and the other to fund the construction—Los Angeles voters overwhelmingly approved the measures by respective margins of fourteen to one and ten to one. This wide gap symbolized the general popularity of Mullholland's plans. The vote illustrated the ways in which access to municipal water had become a popular public policy (Reisner 1993, 78; Hundley 2001, 151).

A second obstacle could have derailed the entire project. The federal government had to approve the right-of-way, as it owned most of the land along the aqueduct route. There was also the Reclamation Service project, which was still being considered for Owens Valley and that depended on the Owens River. The entire reclamation movement during the Progressive Era promised government support for agricultural development, particularly assistance for small farmers in the arid West. Owens Valley epitomized a perfect case for federal development, and abandoning it would symbolize a defeat of the reclamation ideal. Mullholland would not allow this criticism to stand, though, traveling to Washington, DC, and personally lobbying President Theodore Roosevelt and Gifford Pinchot, the nation's chief forester and most prominent conservationist. Mullholland argued that it was more important for Los Angeles to grow, especially on America's weak western shore, than creating farms in the Owens Valley. Pinchot did not object, and Roosevelt approved the move, exclaiming: "It is a hundred or thousand-fold more important to the State and more valuable to the people as a whole if [this water is] used by the city than if used by the people of the Owens Valley" (quoted in Hundley 2001, 155; Reisner 1993, 82). Subsequently, much of the land once reserved for a reclamation project found itself protected as a national forest, through which federal officials gladly furnished a right-of-way for Los Angeles's aqueduct. The federal government capitulated to virtually all of Los Angeles's needs, and Owens Valley residents felt understandably betrayed (Hundley 2001, 151–155; Reisner 1993, 79–84).

Myth holds that the city illegally stole Owens Valley water. This is technically incorrect, but as writer Marc Reisner aptly put it, "Los Angeles employed chicanery, subterfuge, spies, bribery, a campaign of divide-and-conquer, and a strategy of lies to get the water it needed" (1993, 62). Clearly, it was not an open process. Still, if all this had been done simply to get necessary water to Los Angeles, Owens Valley residents might still have been angry, but it is unlikely the vitriol that subsequently developed would have reached the level it did. But an additional event ensured greater outrage.

At the same time Eaton moved secretively around Owens Valley purchasing options, a group of the most influential individuals in Los Angeles organized a land syndicate that bought 16,000 acres for a mere $35 per acre in the neighboring San Fernando Valley the same day Eaton secured the key site in Owens Valley. The planned aqueduct would travel through San Fernando Valley, provide the valley with water, and land values would skyrocket. Meanwhile, these well-placed and wealthy individuals would become much wealthier. Because of the enormous profits realized, many then and now suspected corruption and a conspiracy between the city and these investors. The most careful research suggests that there were no bribes. Nevertheless, some people clearly used inside information to profit, as the syndicate included some members of the Board of Water Commissioners who were well aware of the aqueduct plan. Although Mulholland was not above suspicion, he did not personally profit from the San Fernando Valley scheme. To Owens Valley residents and sympathizers, the entire episode left a bitter taste, especially because Owens Valley water irrigated farms in the San Fernando Valley instead of in the watershed in which the water originated (Hundley 2001, 156–162; Reisner 1993, 71–74).

So with federal blessings and Los Angeles voter approval, along with a little clandestine planning, the city constructed the Los Angeles aqueduct and further fed urban growth. Construction lasted from 1908 until 5 November 1913, when Mulholland opened the aqueduct, stating simply, "There it is. Take it!" (quoted in Hundley 2001, 156; Reisner 1993, 86). Take it, they did. The aqueduct could deliver 258 million gallons of water daily (Hurley 2005, 16). Meanwhile, the city grew to 500,000, and Mulholland instituted some important water conservation measures. One effect of those measures was that the aqueduct was not necessary to meet the city's needs, despite Mulholland's earlier dire predictions of an imminent water shortage. However, it was essential for the type of growth metropolitan boosters envisioned. Laws restricted Los Angeles to selling water only within the municipality, and so a kind of annexation frenzy ensued to enlarge the city's borders. By 1915, Los Angeles tripled its space and would

The "Deadman Syphon" section of the Los Angeles Aqueduct, ca. 1923. The aqueduct, a steel pipe eleven feet in diameter, carried water from the Owens River to Los Angeles. (Philip Brigandi/Library of Congress)

continue to grow both geographically and demographically in the coming years. By 1925, the city counted 1.2 million residents. The success of boosters' growth mentality could hardly be questioned, except nature dealt a cruel blow: drought (Hundley 2001, 151–156; Reisner 1993, 84–89).

This change in ecological conditions brought social tensions to a boiling point. The urban population boom corresponded to a period of unusually high precipitation. The subsequent drought caused particular harm in Owens Valley. Even with the exponential increase in Los Angeles's population, Owens Valley farmers had been able to continue their work and valley towns grew, in part, ironically, because of the larger market in Los Angeles. In the 1920s, the growth, the drought, and the city's continued practice of annexing more land finally meant that water no longer flowed to Owens Valley farmers. In retaliation, valley residents first diverted the water from the aqueduct and then repeatedly dynamited the aqueduct, participating in the region's long-standing tradition of extralegal protest (Walton 1992). Mulholland and officials used armed detectives, court orders, and even issued shoot to kill orders. Perhaps surprisingly, the long-suffering valley residents received fairly sympathetic press coverage throughout Southern California, suggesting a turn against the imperialistic policies of Los Angeles that had long been hailed popularly. It was a last gasp effort, though. In 1927, all the valley's banks closed. Owens Valley became a water colony of Los Angeles, an occasional tourist destination, and not much else (Hundley 2001, 164–167; Reisner 1986, 92–96; Walton 1992, 131–197).

What does this episode teach us about environmental history? Like the previous case study, this one demonstrates the centrality of nature to political and economic power struggles. Furthermore, it shows the muscle cities could muster to shape distant landscapes—in this case to eventually destroy the economic system in place in Owens Valley. Indeed, although Los Angeles's growth was exceptionally fast, similar examples exist of urban power voraciously coveting and consuming the hinterland's resources (Hundley 2001, 171–194). Climate played a significant role in this case study, too. Without the region's aridity, the city would not have required such an expansionistic approach. Also, without the drought, Owens Valley residents may not have resorted to dynamiting aqueducts. In the end, Los Angeles's insistent drive to acquire ever more water from distant places indicates a community unwilling to exist within the limits its environment imposed. This basic narrative became a recurring one. Even William Mulholland recognized it; by the late 1920s, he eyed the Colorado River to fuel more urban growth. Like so many others of the time and since, Mulholland believed one could engineer one's way beyond ecological imperatives.

Solving the Problem of Too Much Water

Although the Owens Valley saga is the best known and most dramatic, acquiring water is not the complete story of Los Angeles and the problem of water in the early twentieth century. Indeed, some of the most fascinating and unexpected developments with water in Los Angeles focused on what to do about too much water, not too little. Mulholland and his peers focused so much on obtaining enough water to fuel their urban expansion that they seldom considered flooding and associated problems. Throughout the twentieth century, too much water in this semiarid city has proven far more costly in lives and property than too little.

Despites residents' most fervent desires, Southern California's Mediterranean climate is erratic. Massive downpours and week-long storms visit the Los Angeles Basin periodically. Such precipitation has often wreaked havoc in the metropolis. Recall the 1825 flood that carved a new channel 20 miles to the south. Floods in the 1880s shifted the channel again, albeit less dramatically. Heavy rains in 1913 and 1914 inundated the city. In February 1914, when twenty inches of rain fell in the San Gabriel Mountains over three days, an estimated $10 million damages occurred. After the rain fell, as it always did, rainwater flowed into the city. But urban development changed the Los Angeles Basin, as historian William Deverell noted: "Roofs and roads had replaced open land. Water collected only long enough to gather more strength and more vertical slope" (Deverell 2004, 113). In the 1914 storm, the river discharged water comparable to the normal flow of the comparatively mighty Colorado River. Four million cubic yards of silt went into the harbors. Rail and telegraph lines were no match for the power of floodwaters. Just as drought had helped convince Los Angeles voters to authorize the Owens River aqueduct, the 1914 flood produced political changes (Deverell 2004, 111–114).

In the aftermath, the city created the Los Angeles County Flood Control District (LACFCD), an entity that represented a typical Progressive Era response to conservation issues. The Progressive conservation movement placed faith in experts, such as engineers, and Progressives enjoyed broad support in California government (Pincetl 1999, 25–73). Not surprisingly, then, the LACFCD promoted a strictly engineering solution to the problem of too much water by designing massive flood-control projects. The city's flood control plans from their inception were "technocratic" (Orsi 2005, 136). Technology and technical concerns guided the efforts, and floods were seen almost exclusively as an engineering problem, extracting the floods from their ecological, political, and cultural contexts. Such Progressive approaches to resource questions led to organized experts managing perceived problems in systematic and arguably efficient ways often within a centralized political authority (Hays 1975).

So confident was the LACFCD that it predicted an end to floods within five years. Such forecasts exemplified engineers' convictions, as well as the short-term memory of Angelenos characteristic of the recent arrival of the population. After all, by 1910, almost half a million residents had not witnessed a flood in the basin (Deverell 2004, 114). Such self-assurance was vastly exaggerated, and technical approaches proved insufficient to address the myriad facets to flooding. Environmental historian Jared Orsi argued,

"Flood control engineering turned out to be less an exercise in replacing a disorderly natural hydrology with an orderly engineered one than in substituting the disorder of nature for the disorder of artifice" (2005, 139). In other words, the LACFCD produced a no more controlled or efficient hydrological system than nature had.

To be certain, the flood-control plan was ambitious, even if it was insufficient. Municipal bonds passed in 1917, in the aftermath of another tremendous flood, and again in 1924 helped to pay for the projects. Federal money supplemented this investment and vastly increased with federal New Deal spending programs by the mid-1930s (Hurley 2005, 17). With these resources, engineers redesigned the river, straightening its course, increasing channel capacity, and paving it. Between 1935 and 1959, engineers constituting a "concrete cult" used 3 million barrels of concrete to create a paved river; indeed, urban designers saw concrete rivers as beautiful, as an achievement of mastery over nature (Deverell 2004, 111; Gumprecht 2005, 116). But engineering solutions did not work, at least not entirely. Big floods came 1 January 1934, killing poor migrants who had unfortunately made their temporary homes in the dry riverbed. Four years later the biggest flood on record killed, depending on the source one consults, between fifty-nine and eighty-one people and caused between $62 million and $78 million in property damage (Deverell 2004, 122–123; Hurley 2005, 17; Orsi 2005, 143). As investment in flood control increased, damage paradoxically mounted.

All these floods were not simply natural disasters; human activity caused some of the worst floods and exacerbated others. Perhaps the most tragic example came when the St. Francis Dam failed on 12 March 1928. William Mulholland had built this dam in San Francisquito Canyon to regulate river flow. This measure was necessary in part because displacing the irrigators in Owens Valley removed one of the most effective regulators of water from the ecosystem, making the river more prone to flooding. Hours before the failure Mulholland inspected and declared the dam sound. A subsequent investigation blamed the political structure for giving Mulholland too much unchecked power (Hundley 2001, 167–169). The failure of the dam resulted in the loss of as many as 450 lives, the majority of them, according to environmental historian Andrew Hurley, "poor Mexican citrus workers who were permitted to live in harm's way despite suspicions of the dam's structural weakness" (2005, 17). The experience crushed Mulholland personally and professionally. Four years later, a series of small dams failed, delivering 600,000 cubic yards of sludge loosened by a combination of fire and rain. Again, a number of people died and many homes were destroyed. Such dam failures clearly represented human failures (Hurley 2005, 17).

Other less obvious cultural factors increased the frequency and damage of floods. Perhaps most significant were cultural attitudes toward water consumption. Much of the water Los Angeles acquired was used to make the semiarid environment conform to suburban ideas developed in more humid environments; these preferences included water-intensive lawns and eventually swimming pools. Moreover, subdivisions and pavement meant more storm runoff went to the river, as did water used in the irrigation in the San Fernando Valley. As a result, the San Fernando aquifer recharged, and the Los Angeles River no longer ran dry. In a roundabout way, then, the Owens River flowed in the Los Angeles River bed instead of its own (Gumprecht 2005, 123). The massive hydraulic engineering projects—the vaunted solutions to Los Angeles's problem of water—often exacerbated the problems, and thus aqueducts produced the need for drains (Hundley 2001, 200–201; Hurley 2005, 12). Flood-control projects that manipulated water was the favored approach, but alternatives existed that received no hearing or only incomplete applications.

Despite the overwhelming focus on technical solutions, some reformers proposed alternatives. Some conservationists advocated measures to be taken upstream, such as building small check-dams, preventing fires, replanting trees, and conserving soil and water. Others called for more municipal parks in riparian areas that would include flood-control measures through ecological rather than engineering means. Perhaps the most radical in conservative, growth-oriented Southern California was the suggestion to end building in floodplains. Collectively, these ideas gained little traction because the political structure hid them beneath the commitment to urban development (Davis 1998, 66; Orsi 2005, 139–143).

Other proposals for addressing broader issues that also connected to flooding included the combination of comprehensive zoning and city planning. In 1927, the Los Angeles Chamber of Commerce hired two prominent firms: the Olmsted Brothers, a leading landscape architecture company, and Harland Bartholomew and Associates, a top urban planning business. After three years of careful study, Olmsted and Bartholomew produced *Parks, Playgrounds and Beaches for the Los Angeles Region,* a landmark document in Southern California history (reprinted in Hise and Deverell, 2000). The report expressed dire warnings about the future of Los Angeles given its rapacious growth. This growth resulted in what urban theorist Mike Davis has called "the underproduction of public space" (Davis 1998, 61–67). The Olmsted-Bartholomew report proposed a remedy that would address several perceived shortcomings: develop open spaces as park and recreational spaces for the public. Rapid urban development, which consumed so much water, also left too little space for leisure. The Olmsted-Bartholomew report urged the city to reclaim riparian areas, prevent further development there, and create parks and parkways that could double as spillover zones to minimize potential property damage when the inevitable floods came (Hise and Deverell 2000, 96–99).

To be sure, the Olmsted-Bartholomew report was more concerned with the public's moral well-being than with the overall environmental issues. For example, they wrote: "[I]n great urban areas bitter experience proves that, *without adequate parks, the bulk of the people are progressively cut off from many kinds of recreation of the utmost importance to their health, happiness, and moral welfare.* Public agencies, therefore, must progressively fill the more and more numerous gaps left by commercial and other private agencies" (Hise and Deverell 2000, 85, original emphasis). This representative statement indicates the connection between health and recreation, as well as highlighting the public and civic duty for providing such places for the burgeoning urban population. These themes were central to the era's broader relationship to nature and Progressive Era conservation.

It would be a mistake, however, to think the Olmsted-Bartholomew report did not address anything about ecology. Key parts of the study advised reclaiming and reserving riparian areas for park development. By keeping riparian zones and floodplains relatively undeveloped, the city could reduce costly flood damage. "The combination of parks with flood-control necessities is frequently possible," the report explained, "and wherever practiced it not only will yield a double return on the investment in land but also may lead to *an ampler and better solution of both problems at a much lower cost of construction than either would separately pay*" (Hise and Deverell 2000, 98, original emphasis). Olmsted and Bartholomew recognized the long-term ecological and property damage floods would inevitably cause and argued that their alternative would be the economical and ecological long-term solution. Proper planning could keep real-estate development out of vulnerable floodplains and simultaneously improve recreational opportunities (Davis 1998, 66). Despite such apparent promise, the plan received a cold reception by city leaders.

Although hazard zoning and park development would have saved money and prevented much damage, Los Angeles instead pursued the technical solution of flood control by paving the Los Angeles River. Wetlands identified by the Olmsted-Bartholomew plan were paved over, exacerbating floods. Conservative growth ideologies, relying on real-estate speculation, expected higher profits from private development than from public parks. From the longer-term perspective of flood damage, though, the path of flood control through engineering cost the public much more (Davis 1998, 69–70). And ironically, as Orsi explained, "engineering works to control floods often created a sense of security that induced more development, which in turn necessitated more flood control works" (2005, 140). Simply put, faith in technology overrode the logic of broader planning efforts. And in that respect, this approach largely characterized Progressive (and later) approaches to environmental management.

Postscript

Predictably, the problems continued. Los Angeles pursued more distant water sources. Los Angeles managed to tap the Colorado River with a 242-mile aqueduct as part of federal reclamation development on the river, which by the 1940s gave the city the capacity to use Colorado River water. Later, by the 1970s, the unimaginatively named State Water Project delivered northern California water to burgeoning populations in Southern California. In both cases, the main beneficiaries—Angelenos and other Southern Californians—gained valuable water through taxpayer revenues concentrated elsewhere. Just as in the Owens Valley episode, others paid for the costs of Los Angeles's growth (Hundley 2001, 203–302).

Meanwhile, floods continued. In 1938, the biggest flood on record killed 59 and damaged $62 million in property. Continuing its practice, the Army Corps of Engineers built more and more flood-control projects, paving more and more riverbeds. Orsi calculated that "Before it could harm the fortified metropolis, a flood striking at the end of the 1960s would have had to fill the 106 mountain debris basins, spill over the five valley flood control dams, or make a breach somewhere along the 350 miles of concrete river channels" (2005, 144). This massive engineering, of course, reorganized the city's ecology with a willing political culture and generous federal subsidies. Still, floods in 1978 and 1980 surprised technocrats with much more damage than they had imagined possible. The technical approach did not solve the flooding problem. One reflective engineer captured the likely truth, explaining that engineering was "necessary but not sufficient" (quoted in Orsi 2005, 151). In recent decades, environmentalists and local planners have pursued strategies that would reduce flood damage by restoring some wetlands along the Los Angeles River. These efforts represent a small but important shift in thinking about Los Angeles and the problem of water (Hurley 2005, 20–21; Orsi 2005, 143–151).

CONCLUSION

What themes do Los Angeles and the problem of water illuminate? Fundamentally, this experience demonstrates how environmentally dependent even cities are. In the semiarid environments of the Far West, water was and is the most obvious deficiency. Without adequate water, cities cannot grow—a fate most communities at this time desperately sought to avoid. Gaining water from distant areas invariably took water out of the hands of others, exemplifying the political and ecological power of rising metropolises like Los Angeles. Ironically, such resource imperialism fueled ever greater growth, which in turn created ever greater desires, a cycle of ecological demands that seemingly knew no bounds. And the even greater irony for Los Angeles—that driest of American cities—became the perennial flooding problem. Once more, nature exerted its power to thwart and even wreck human plans. However, those empowered to decide the city's fate chose repeatedly to proceed as if nature mattered little. Engineering, the politicians (and voters) maintained, would transcend ecological limits. They promised to bring more water or to divert excess water. What Angelenos refused to do, and in this they reflected broad tendencies, was rethink how many residents a semiarid city could support or significantly restrict building to conform to likely flood disasters. In the early twentieth century, relying on technical experts while ignoring larger ecological parameters represented a common *modus operandi.*

REFERENCES

Brechin, Gray. *Imperial San Francisco: Urban Power, Earthly Ruin.* Berkeley: University of California Press, 1999.

Cronon, William. *Nature's Metropolis: Chicago and the Great West.* New York: W. W. Norton, 1991.

Davis, Mike. *Ecology of Fear: Los Angeles and the Imagination of Disaster.* New York: Metropolitan Books, 1998.

Deverell, William. *Whitewashed Adobe: The Rise of Los Angeles and the Remaking of Its Mexican Past.* Berkeley: University of California, 2004.

Gumprecht, Blake. *The Los Angeles River: Its Life, Death, and Possible Rebirth.* Baltimore, MD: The Johns Hopkins University Press, 1999.

Gumprecht, Blake. "Who Killed the Los Angeles River?" In *Land of Sunshine: An Environmental History of Metropolitan Los Angeles,* edited by William Deverell and Greg Hise, 115–134. Pittsburgh, PA: University of Pittsburgh Press, 2005.

Hays, Samuel P. *Conservation and the Gospel of Efficiency: The Progressive Conservation Movement, 1890–1920.* 1959. Reprint, New York: Atheneum, 1975.

Hise, Greg, and William Deverell. *Eden by Design: The 1930 Olmsted-Bartholomew Plan for the Los Angeles Region.* Berkeley: University of California Press, 2000.

Hoffman, Abraham. *Vision or Villainy: Origins of the Owens Valley-Los Angeles Water Controversy.* College Station: Texas A&M University Press, 1981.

Hundley, Norris, Jr. *The Great Thirst: Californians and Water: A History.* Revised ed. Berkeley: University of California Press, 2001.

Hurley, Andrew. "Aqueducts and Drains: A Comparison of Water Imperialism and Urban Environmental Change in Mexico City and Los Angeles." *Journal of the West* 44 (Summer 2005): 12–21.

Kahrl, William L. *Water and Power: The Conflict over Los Angeles' Water Supply in the Owens Valley.* Berkeley: University of California Press, 1982.

Mulholland, Catherine. *William Mulholland and the Rise of Los Angeles.* Berkeley: University of California, 2000.

Orsi, Jared. "Flood Control Engineering in the Urban Ecosystem." In *Land of Sunshine: An Environmental History of Metropolitan Los Angeles,* edited by William Deverell and Greg Hise, 135–151. Pittsburgh, PA: University of Pittsburgh Press, 2005.

Pincetl, Stephanie S. *Transforming California: A Political History of Land Use and Development.* Baltimore, MD: The Johns Hopkins University Press, 1999.

Reisner, Marc. *Cadillac Desert: The American West and Its Disappearing Water.* Revised and updated ed. New York: Penguin Books, 1993.

Sauder, Robert A. *The Lost Frontier: Water Diversion in the Growth and Destruction of Owens Valley Agriculture.* Tucson: University of Arizona Press, 1994.

Starr, Kevin. *Material Dreams: Southern California through the 1920s.* New York: Oxford University Press, 1990.

Walton, John. *Western Times and Water Wars: State, Culture, and Rebellion in California.* Berkeley: University of California Press, 1992.

CASE STUDY: THE FORESTS, THE OWLS, AND THE RADICALS

A bombing. Acts of civil disobedience. Endless court challenges. Dissension among federal employees. FBI investigations. Scientific hesitation. These have not been the typical scenarios in environmental controversies. However, by the late 1980s and early 1990s, these situations and more animated the West Coast's ancient forests as what became the quintessential modern environmental battle developed. Its story became well known.

A cynical version went something like this: Once upon a time, hardworking rural communities depended on access to nearly inexhaustible forests for employment that fed their families. These families worked in the huge West

Coast forests for several generations and considered themselves true environmentalists who knew the woods as their home. Their jobs were unexpectedly pitted against spotted owls, as radical, urban environmentalists who were afraid of work invaded the forests. These environmentalists were un-American and misanthropic, caring more for the life of a tree or an ugly owl than they cared for the lives of logging families. Furthermore, these outsider activists dangerously broke laws by pounding spikes in trees, which could kill loggers and mill workers. They also staged ridiculous antics such as chaining themselves to logging machinery, blocking roads by cementing their feet in roadways, and sitting high up in trees. Worse still, after President Bill Clinton's election, the environmentalists persuaded the federal government and its agencies of their point of view, and the government capitulated and shut down logging in West Coast forests to placate the radicals, save a few birds, and doom rural communities.

Another cynical version resembled this: Once upon a time, rapacious and greedy timber companies came to the West Coast searching for a source of timber wealth they could exploit as quickly as possible. By clear-cutting, the companies attacked the land as if it were a mortal enemy. They devastated the landscape, leaving it with eroded hillsides and ugly stumps. Knowingly, they ruined habitat for endangered species like the northern spotted owl and Pacific salmon, because they did not believe these animals had the continued right to exist. These callous companies paid off politicians to ensure that their ravenous cutting could continue apace. Into this scene came altruistic environmentalists who put their own lives on the line to save the last remnants of old-growth forests and endangered species. Like civil rights activists before them, these environmentalists cared not for their own safety, just their cause. The companies' hired thugs and official law enforcement routinely harassed these rebels who selflessly put themselves on the side of justice and right. In the end, the federal government capitulated to timber companies and doomed the ancient forests to continued, albeit reduced, cutting.

Not surprisingly, the real stories were more complex.

The Setting

The Forest

West Coast forests are particularly rich biological entities, valuable economic resources, and deep cultural symbols. Old-growth forests generate complex ecologies with dense understories and high canopies, creating various levels of biodiversity and unique ecologies. The forests formed in part from the wet marine climate. Along the coast, temperate rain forests can receive 200 inches of rain a year, while that precipitation can be reduced 90 percent just a few miles inland. "The mix produces some of the most varied ecology and biggest trees in the world," explained journalist William Dietrich about the Olympic Peninsula. "Down in the Queets River drainage is the world's largest Douglas fir. Up in the Quinault is the world's largest western hemlock and yellow cedar. On the remote Bailey Range of the Olympics is the world's largest subalpine fir" (1992, 16). The coastal forests of Oregon and California exhibit similar diversity and size, with California redwoods boasting trees several thousand years old (Dietrich 1992, 16–19). Furthermore, nearly 500 species of plants and animals may find their only habitat in such forests (Hirt 1994, 289). Besides being valuable as habitat, a single tree in the early 1990s might be worth $15,000 for local timber merchants. By the late 1980s, estimates indicated that only 2.5 million acres still stood, more than 10 percent of their original domain (Wilkinson 1992, 157). Perhaps it was their size, perhaps their value, perhaps their age, or perhaps their rarity. Whatever the reason or combination of reasons, they became a battleground.

The Owl

Among the animal species that lived in the forests was the northern spotted owl *(Strix occidentalis caurina),* and its presence has caused the uproar. Eric Forsman, a young Oregon State University student, discovered the owl while working summers in the late 1960s for the Forest Service. He was not the first—a couple dozen other accounts of the owl existed—but his became the important discovery because he pursued the encounter, and as a graduate student he studied the small bird. Almost nothing was known about the bird then; by 1990, it was one of the most studied species in the world. This shift occurred because the owl had apparently no fear of humans, yet they lived in dense forests often far from people, which explained the paucity of previous sightings. Two of Forsman's first discoveries were that spotted owl populations were declining, and they required vast ranges, preferably in old-growth forests. Soon, a number of researchers formed the Oregon Endangered Species Task Force and recommended that 300 acres of old-growth forest should surround each known nest, a requirement bitterly contested but eventually acceded to in 1977. The more owls found, the more forest biologists claimed as necessary for habitat preservation, and the fewer trees sent to local mills. This small owl certainly seemed to force change (Chase 1995, 131–135; Dietrich 1992, 72–77; Wilkinson 1992, 159).

A Spotted Owl on the cover of the June 25, 1990 editon of TIME magazine. The battle in the Northwest woods had become national news in the early 1990s. (Time & Life Pictures/Getty Images)

The Laws

At the time Forsman conducted his research, changes emanating from Washington, DC, transformed the legal landscape for West Coast forests. In 1973, a time when environmental laws passed Congress regularly, the Endangered Species Act became law. Idealistically, the Endangered Species Act valued biological diversity and species survival above economic considerations—a fairly revolutionary stance. The U.S. Fish and Wildlife Service administered the act, deciding which species were endangered or threatened. Unfortunately, conducting the necessary studies and devising the required recovery plans was time-consuming and expensive. Then, once a species was listed, the law demanded action to keep the species population viable. Consequently, progress was slow, protection often difficult, and opposition fierce. The spotted owl made an initial list of potentially endangered species (Dietrich 1992, 76–77). Meanwhile, the 1976 National Forest Management Act directed the Forest Service to "provide for diversity of plant and animal communities" (quoted in Wilkinson 1992, 160). Because it would be prohibitively expensive and impracticable to survey all species, the USFS used indicator species. Such species were essentially proxies for entire ecosystems; if the indicator species had a healthy population, the ecosystem was likely healthy. For West Coast old-growth forests, the northern spotted owl served that function, as owls depended on small mammals that depended on fungi that depended on old growth. The owls' position in the food web allow them to be good markers of overall ecosystem vigor (Wilkinson 1992, 160; Zakin 1993, 242). Long before environmentalists began active protests, then, these laws were in place—and unambiguous—in requiring owl and habitat protection.

And because the law was clear, judges decided in favor of owls. Among the many important suits concerning these matters, one stood out for its significance. *Seattle Audubon v. Evans*, decided in 1991 by Federal District Judge William Dwyer, temporarily halted 135 planned timber sales, or 80 percent, in the west-side forests of the Pacific Northwest. Dwyer found that the U.S. Fish and Wildlife Service and the USFS had failed to finish an owl management plan as required by law. He wrote that this effort, or lack thereof, "exemplifies a deliberate and systematic refusal by the Forest Service and Fish and Wildlife Service to comply with the laws protecting wildlife" (quoted in Dietrich 1992, 213). Furthermore, he explained, "The problem here has not been any shortcoming in the laws, but simply a refusal of administrative agencies to comply with them. . . . This invokes a public interest of the highest order: the interest in having government officials act in accordance with law" (quoted in Wilkinson 1992, 335). Dwyer's reasoning laid bare the crux: the law was clear and federal agencies were breaking it. The ruling shocked the region. Some Northwest politicians were so desperate by the turn of events that they passed a law that forbade environmentalist lawsuits for a year—an action not surprisingly declared unconstitutional. Thus, because of earlier Congressional mandates, the rule of law favored those who would protect old-growth owl habitat (Dietrich 1992, 216; Wilkinson 1992, 162).

Forestry Reform: New Forestry and Dissent in the Forest Service

At a basic level, this controversy stemmed from the legacy left by decades of intensive forestry practices in the region. Clear-cuts had long been the preferred harvest method in the Northwest's Douglas-fir forests, making the woods look, in the colorful words of one observer, "like big, harmless victims of a psychotic barber" (Zakin 1993, 242). While environmentalists had long urged reforming such practices, calls for change also came from within the forestry community.

What became known as New Forestry largely came from the innovative forester Jerry Franklin. He spent most of his career as a scientist for the Forest Service in the Northwest before leaving to work at the University of Washington. He devised an approach that sought to integrate ecology with timber production. Older forestry theories speciously argued that clear-cuts mimicked how Douglas-fir forests naturally grew, but Franklin's New Forestry called for more selective logging that left trees in random patterns that furnished habitat for insects and wildlife, as well as providing seed sources for the next forest. This practice more accurately represented natural disturbance than clear-cuts did. Franklin realized the need for reform when he recognized that normal forestry practices contradicted the biological evolution of forests. Thus, embarking on extensive research in old-growth forests and synthesizing available research, Franklin worked to organize New Forestry on ecological principles that valued biological diversity and complexity, an approach that countered forestry's traditional simplifying tendencies. Challenging the commodity-driven forest practices did not win Franklin widespread admiration among those in the timber industry. And because he advocated logging in some environments under certain circumstances, many environmentalists also mistrusted him. Still others have argued that his management tactic is based on social values only masquerading as science. Despite these controversies, though, Franklin's New Forestry became an important management strategy the USFS adopted in part when it moved toward ecosystem management in 1992 in a program known as New Perspectives (Chase 1995, 165–175; Dietrich 1992, 97–115).

A surprising development occurred within the Forest Service when employees broke openly with the agency's traditional management priorities. Political

and economic pressures for high production blunted much of the nominal reform like the National Forest Management Act. One employee explained this
problem: "There is a tremendous pressure on timber planners, wildlife biologists, hydrologists and others to suppress their professional expertise and modify their documentation to ensure that short-term political expediencies . . . are
met" (quoted in Hirt 1994, 271). Especially as the USFS employed more biologists and ecologists to supplement the corps of traditionally trained foresters
and engineers, the agencies' understanding of the ecological problems increased
and its willingness to maximize commodity production declined. Still, in most
cases, the agency continued business as usual. Indeed, throughout the 1980s,
politicians demanded that the Forest Service increase harvests in the Northwest
against an array of contrary evidence. This practice rankled many of the rank
and file. So much so, they formed the Association of Forest Service Employees
for Environmental Ethics. This association's existence and numerous polls and
studies at the time revealed that the agency was changing to reflect the different disciplinary perspectives of newly educated foresters and the evolving public values that regarded forests as something far greater than board feet.
According to one study, the fifteen years after the National Forest Management
Act passed saw wildlife biologists in the USFS more than double, the number of
fisheries biologists more than triple, and ecologists increase twelve times (Dietrich 1992, 99). Reflecting these changes, President Clinton's appointee to head
the USFS, Jack Ward Thomas, was trained in wildlife biology, the first chief
forester who was neither a forester nor an engineer (Hirt 1994, 271–272,
281–286, 291).

That is, more or less, the institutional setting. A more dramatic tale occurred closer to the trees and reveals more about environmental activism and
the shape of authority in such modern environmental contests.

Activism in the Woods

Grassroots environmental activism in the West Coast woods illustrates arguably the most prominent protest movement on behalf of the environment in
U.S. history. Activists attempted truly radical reform through direct actions
that embraced a biocentric philosophy. Their approach caused controversy and
produced backlash. Nevertheless, activists created publicity that nationalized
the fight against intensive harvesting on old-growth forests and slowed what
had been an almost unstoppable economic force in the region. To that extent,
old-growth activists must be counted as among the most important groups in
the Far West's recent environmental history.

Beliefs

Environmentalism had evolved since its roots in the conservation movement at the turn of the twentieth century. Although some within the movement still largely embraced the wise-use mantra of Gifford Pinchot, most challenged that anthropocentric vision that nature's worth must be measured by human values. Instead, a biocentric alternative shaped activists' perspectives. The leading edge of this movement centered on the West Coast forests.

As much as anything, biocentrism was a philosophy that certain environmentalists sought to put into action. The leading American proponent was Bill Devall, a sociology professor at Humboldt State University in the heart of California's redwood country. He adapted the Norwegian philosopher Arne Naess's concept of deep ecology, which posited a new relationship between humans and the rest of nature that was harmonious and, even more, egalitarian. This notion helped push environmentalism into a stronger ethical position. In some forms, biocentrism affected national policy. After all, the Endangered Species Act indicated that species enjoyed the continued right to exist. Rather than focusing on policy, deep ecologists and other biocentrists advocated a new way of thinking that might radically reorient fundamental relationships. Besides philosophical changes, some argued the necessity for dismantling capitalist economies and replacing them with a more ecologically based economy. For example, Judi Bari, a prominent activist and organizer for Earth First! in northern California, explained, "I don't believe that it's possible to save the Earth under capitalism because I think that capitalism is based on the exploitation of the Earth, just like it's based on the exploitation of workers. . . . We need a new way to live on the Earth without destroying the Earth or exploiting lower classes. It needs to be socially just and it needs to be biocentric" (Bevington 1998, 270). Proceeding from these assumptions, radical environmental activists challenged basic institutions in addition to advocating new intellectual directions (Chase 1995, 119–130; Zakin 1993, 244–245).

Actions

Biocentrism animated individuals and grassroots organizations like Earth First!, perhaps the best-known radical environmental group in the nation. The rise of Earth First! symbolized a fracturing of mainstream environmentalism that had been effective in lobbying for national legislation like the ESA or the National Environmental Policy Act. The fracturing had been occurring for some time. David Brower's ouster from the Sierra Club in 1969 for being too uncompromising constituted the opening round. The new radicals dispensed with political compromise and moved more toward direct action. Such actions could run the gamut from guerrilla theater antics to tree-spiking.

In the Pacific West, perhaps the first action by Earth First! occurred at Bald Mountain. This short peak stood in the Siskiyou National Forest, part of an area Bob Marshall had recommended for wilderness in the 1930s. A small portion of it—77,000 acres—was official wilderness, the Kalmiopsis, a far cry from Marshall's hope for more than a million acres. Much of the region remained undeveloped, but logging roads began piercing the forest in the late 1960s, and by the 1980s the situation had worsened. The Sierra Club largely abandoned the project, leaving it in the hands of more confrontational activists. In the spring of 1983, as a road crew built a road on Bald Mountain, four protestors blocked a bulldozer. Despite threats, the protestors returned the next day, and this time the dozer operator pushed a pile of dirt to the protestors' feet and reportedly called them a "bunch of communist bastards" (quoted in Zakin 1993, 251). The press arrived and then local law enforcement arrested the activists. For the next several months, Earth First! staged seven blockades, and forty-four activists were arrested, including Dave Foreman, the organization's cofounder, in a violent skirmish that ended with Foreman beneath a truck. Collectively, these actions amounted to the opening round in the forest wars (Chase 1995, 179–191; Zakin 1993, 228–259).

Such civil disobedience inspired more. The most notorious was tree-spiking. If a chainsaw or a saw in a mill struck a metal spike, it could shatter and dangerously launch shrapnel through the immediate area. Earth First!'s official policy concerning tree-spiking was to notify timber companies (and the media) when it had occurred. According to the activists, spiking slowed down timber production by forcing companies to methodically search through stands with metal detectors. As Mike Roselle, an Earth First! leader, explained, "At Hardesty Mountain, we flagged the trees with survey tape. In the Middle Santiam, we spray-painted them. All they had to do was locate and remove the nails. The big deal is that it's a very marginal business, harvesting on public land, and spiking threatens its economic viability" (quoted in Zakin 1993, 260). Despite popular misunderstanding, most environmentalists did not spike trees to hurt timber workers. Still, it happened, such as when a young millworker, George Alexander, was hit by a shattering bandsaw that broke his jaw multiple times, knocked out teeth, and cut his jugular vein. He survived, and in the aftermath, some of the more radical Earth First!ers betrayed a serious lack of compassion and a decided indifference to Alexander's injuries. One Earth First! activist incensed by this violence was Judi Bari, who became one of the more radical activists for different reasons (Chase 1995, 229–230; Zakin 1993, 259–260).

Bari brought to Earth First! a history of labor organizing and tried to combine environmental and labor activism in the redwood region of northern

California. Such a coalition did not come easily, despite what Bari saw as a common culture of domination, over the worker and over nature. Seeking to ally the Industrial Workers of the World, or Wobblies, with Earth First!, Bari tapped into worker anger against local timber companies that had been taken over with insider trading by corporate raider Charles Hurwitz, who promptly doubled timber harvests and ransacked West Coast redwood forests to pay off other debts (Chase 1995, 203–214; Zakin 1993, 345–349). Although local workers were offended by such corporate actions, they mostly failed to find common cultural ground with Earth First! activists, many of whom were ex-hippies or young college radicals. Bari insisted the local interest lay in expelling outside corporations, which threatened the long-term ecology and economy of northern California communities. Nevertheless, her advocacy led to only a small number of workers joining the activist coalitions (Bevington 1998, 252–258; Zakin 1993, 354–364). Still, some have suggested that Bari's efforts to bridge social justice and ecological issues constituted the real radicalism of Earth First!'s efforts in California, not the biocentrism or the direct action (London 1998). That social and environmental linking, though, remained incomplete.

Predictably, Bari's actions prompted backlash. On 24 May 1990, a car bomb exploded in Bari's car. The blast shattered her pelvis, other bones, and paralyzed her for a short time. The Federal Bureau of Investigation became involved immediately in the investigation and labeled Bari and her passenger, fellow activist Darryl Cherney, terrorists. Later the same afternoon, police charged Bari with possession of the bomb. However improbably, authorities asserted that Bari had built the bomb herself and became its victim. A letter claiming responsibility and full of details about the bomb cast doubt on the charges against Bari, as did the simple illogic of knowingly driving around with an armed bomb. The aggressive FBI investigation backfired, as even critics recognized that Bari was innocent. Charges were eventually dropped. The entire episode demonstrated the level of frustration in the region, and it dampened West Coast activism (Chase 1995, 322–333; Zakin 1993, 389–394).

Nevertheless, Earth First! continued with its largest campaign to date, Redwood Summer. This event did nothing to dispel the chasm between timber workers and environmentalists. And in fact, it exacerbated the backlash. With Bari as the initial organizer, Earth First! brought a mass movement to northern California during the summer of 1990 with a promised series of nonviolent direct actions. The event attracted activists of all ages from across the nation, who hoped to bring attention to the old-growth forest issues through civil disobedience. In that respect, Redwood Summer was a success—more than 100 were arrested, and the media faithfully reported the theatrics,

Earth First! environmental activist Judi Bari was seriously injured by a bomb that detonated in her car in May of 1990. (David J. Cross/Time & Life Pictures/Getty Images)

including blockades and tree-sitting. Nothing much substantive happened to preserve redwood stands during the summer, but several legacies did emerge. First, Earth First! fractured into competing ideologies with those who focused primarily on biocentrism on one side and those who embraced a broader social justice component on the other. Second, it tutored thousands of students in countercultural environmental politics. Last, after the bombing and media attention of Redwood Summer, mainstream environmental organizations, such as the Sierra Club, finally took greater interest in old-growth campaigns.

The radicals had sparked more conservative activism, a development that ultimately helped reach more of the American public (Chase 1995, 334–350; Zakin 1993, 378–80, 396).

Direct action continued. Lawsuits proliferated. Timber companies cut trees, closed mills, and fired workers. Then they blamed environmentalists and the spotted owl, a strategy that served well to muddy the waters and misdirect local anger away from the company policies and technological changes that were responsible for most of the lost jobs. Meanwhile discourse in the woods soured. As the battled devolved, it became clear that all sides in the woods saw issues in irreconcilable ways.

Competing Authorities

All participants in the controversy presented their positions as resting on superior claims. They did so by appealing to different values. The distinct perspectives generally made compromise unlikely, if not impossible. Furthermore, the standoff hardened as the sides retreated to increasingly ideological positions. By appealing to different authorities, participants hoped to persuade others and the public that their position, at least, was not ideological. This strategy generally only convinced themselves.

Not surprisingly, timber communities found the changing circumstances exceedingly frustrating. Residents countered environmentalists' claims that loggers were destroying forest in two ways. First, having lived and worked in the forests for decades, loggers claimed deeper knowledge of them. By pointing out that many environmentalists were outsiders, timber workers suggested a type of exploitive carpetbagging generally associated with the corporate targets of environmentalists, not the activists themselves. Moreover, it made radicals' claims to be working for local communities seem specious. Second, timber workers argued that logging actually helped wildlife. At a public hearing in Eugene, Oregon, about whether to list the northern spotted owl as a threatened species under the Endangered Species Act, one timber worker's comments represented these perspectives well:

> I was born and raised in Oregon from pioneer stock. For generations my family has been involved in the timber industry in one aspect or another. We have always depended on timber for our livelihood, and because of this dependence, we have probably gained a respect for the forest and the land that few people will ever know. . . . As long as I can remember, loggers have been accused of ruining wildlife habitat. From past experience, I would disagree. Unless a person has actually sat quietly at a logging site and watched and listened, they cannot

appreciate the amount of wildlife that is around. . . . Ask any logger who daily shares his lunch with a raccoon, a chipmunk, a raven, or even a doe and her fawn if he is destroying habitat or enhancing it. The timber industry has done more to perpetuate our natural resources than any other group I can think of (quoted in Proctor 1995, 270–271).

Such a perspective asserted a moral authority that timber workers found unimpeachable. It appealed to history and fairness and located timber workers as part of the woods—nearly organically so. This particular worker continued his testimony by indicting environmentalists: "I find it hard to believe anyone would ever support throwing all of this away for the sake of saving a small bird who is not even in trouble and doesn't give a damn how big the tree he is living in or how old it is. I would urge you to do your best to protect the people that are doing their best to protect the forest" (quoted in Proctor 1995, 271). The worker challenged environmentalists' knowledge about the true state of spotted owl habitat and habits. Moreover, he assumed that environmentalists did not care about the forest or its inhabitants, much less the human communities who depended on them. In such a characterization, what motivated environmentalists remained a complete mystery.

Environmentalists appealed not to the moral authority of their length of residence or their intimate knowledge of the forest. Instead, they often appealed to a broad biocentrism, the idea that all species enjoyed rights to exist with none having a superior claim. At the same hearing in Eugene, one resident stated simply, "I don't feel that any one species has the right to condemn any other species to extinction" (quoted in Proctor 1995, 270). Such a position placed the struggle on ethical terms about rights. From this rather simple foundation, many activists radicalized. Indeed, Earth First!'s motto was "No compromise in defense of Mother Earth." It is always difficult, after all, to compromise fundamental and universal rights. For activists who adopted deep ecology's biocentrism, saving the northern spotted owls in viable populations or even saving a single tree became a moral duty. This ethical imperative could also justify ecotage. As Earth First! cofounder Dave Foreman explained, "When we fully identify with a wild place, then, monkeywrenching becomes self-defense, which is a fundamental right" (quoted in Chase 1995, 189). For those who accepted deep ecology's principles, such justifications made sense, intuitively and logically. However, not everyone accepted biocentrism. For some critics, deep ecology masked under philosophical drivel an elitist, misanthropic, primitivist agenda that had little to do with species' inherent rights and much to do with individual activists' egos (Chase 1995). Needless to say, biocentrism did not persuade loggers or much of the public; its claim to authority therefore was limited.

To persuade more people, less radical environmentalists, sympathetic politicians, some land managers, and others appealed to science. This strategy caused more problems than may have been expected. Because of the legal requirements of the Endangered Species Act and the National Environmental Policy Act, science had to answer questions about forests and owls, questions that were newly posed. The science of old-growth ecosystems was young and shifted constantly from the 1970s to the 1990s as ecologists increasingly asked new questions, developed new theories, and proposed new management. Many scientists studied owls and old-growth forests trying to determine their interrelationship. It was a difficult task fraught with not only the typical challenges of scientific research in the field but also mired in political significance from the onset. Using ecosystem models, scientists seemed to discover that the northern spotted owl survived best in old-growth conditions and required vast acreages to breed. However, many scientists recognized that with this research just beginning in earnest, any conclusions would have to be tentative, a situation not conducive to the rapid pace of politics and generally unsatisfactory to loggers and activists. Thus, scientists felt pressured and responded cautiously by urging timber cuts to be slowed so that scientists could discover more about habitat requirements. Some have argued that this caution amounted to a "rising tyranny of ecology," although the physician's directive to first do no harm seems a more apt characterization (Chase 1995). Such criticism, though, shows how the contest in the woods rested on intense demands on shifting, imperfect science (Chase 1995, especially 148–164, 244–260; Dietrich 1992, 47–59). But that is a truism for most environmental controversies, as science has always been an evolving process, responsive to myriad social factors. Still, even if flawed and incomplete, scientific authority met legal requirements in ways the moral authorities of loggers and activists could not. Thus, science enjoyed tremendous, and perhaps ultimate, prestige in the story of competing authorities, despite its obvious imperfections.

Resolution: Timber Summit and the Northwest Forest Plan

The standstill in the woods became a prominent political problem by the presidential election in 1992. When Bill Clinton was elected, he promised action and organized the Northwest Forest Conference in Portland, an event soon commonly referred to as the timber summit. As was his style, Clinton wanted to please everyone. In his opening remarks at the summit, he said, "This is not about choosing between jobs and the environment but about recognizing the importance of both" (quoted in Hirt 1994, 290–291). A consummate politician,

Jack Ward Thomas, chief forester of the United States Forest Service during the Clinton administration, brought deep experience with the spotted-owl issue to his leadership position. (Joel W. Rogers/CORBIS)

Clinton wanted everyone's votes and that meant persuading them that a policy could be devised to keep cutting economically and ecologically viable, a desire that continued the "conspiracy of optimism" historian Paul W. Hirt argued had guided the Forest Service throughout the postwar era. Central to the Clinton approach was ecosystem management, the new policy inspired by Franklin's ideas and reputed to be ecologically sound. Clinton's appointee, Jack Ward Thomas, was already deeply involved in multiple scientific studies concerning old-growth and owls. Many expected dramatic change (Hirt 1994, 288–291).

There were several outcomes; none pleasing to everyone. One was the Forest Ecosystem Management Assessment Team, which was charged with finding an ecologically sound management plan. Its report presented ten options. President Clinton and his advisers selected Option 9 and through an official Record of Decision in February 1994 it became the Northwest Forest Plan (Haynes and Perez 2001, 4–7). The choice angered all sides. This approach opened 20 percent of old-growth forests to full-scale logging, and the rest would be opened for salvage and thinning practices. Fisheries protection received only half the protection scientists recommended. Inland forests would be harvested at an accelerated pace to compensate for the lower cutting in coastal forests. The plan also required something called adaptive management, a strategy that, in the words of a Forest Service report, serves "as a built-in self-evaluation mechanism for changing the standards and guidelines—and potentially the strategy itself, over time" (Haynes and Perez 2001, 7). Such self-reflection has been rare in the agency and could constitute a fundamentally changed approach to natural resource management. Predictably, most environmentalists believed the plan was fatally flawed. But so did timber interests who found the reduced availability of forests economically devastating and believed the science was manipulated (Chase 1995, 383–394; Hirt 1994, 291–292). To the extent that ecosystem management and not some other scientific construct guided managers, the science was manipulated. But this manipulation was no different from the methods by which high production goals had shaped scientific inquiry for the entire history of the Forest Service. The Northwest Forest Plan did not satisfy many of the principals close to the ground. It did, however, meet Judge William Dwyer's requirements and thus became the legal plan in December 1994.

CONCLUSION

By the end of the 1990s, nothing in the forests seemed the same. Environmental activism radicalized at a grassroots level and pioneered new ways to protest, including dangerous tree-spiking and publicity-seeking tree-sitting. Federal legislation did not promote full-scale economic development but encouraged preservation for the sake of species that did not financially profit anyone. The old-growth and spotted owl controversy paralyzed the timber industry for many years. This paralysis represented a shift in the Far West's environmental history. Never before had environmentalists been able to so completely slow down business as usual, especially in such a traditionally powerful industry. Now, moving beyond the standstill while protecting environmental and economic values remains the main challenge of the day. If it can be achieved in the Pacific West, it will mark a new era and herald the leadership of the region in environmental issues.

REFERENCES

Bevington, Douglas. "Earth First! in Northern California: An Interview with Judi Bari." In *The Struggle for Ecological Democracy: Environmental Justice Movements in the United States*, edited by Daniel Faber, 248–271. New York: Guilford Press, 1998.

Chase, Alston. *In a Dark Wood: The Fight Over Forests and the Rising Tyranny of Ecology.* New York: Houghton Mifflin, 1995.

Dietrich, William. *The Final Forest: The Battle for the Last Great Trees of the Pacific Northwest.* New York: Penguin, 1992.

Haynes, Richard W., and Gloria E. Perez, technical eds. *Northwest Forest Plan Research Synthesis.* Portland, OR: U.S. Department of Agriculture, Forest Service, Pacific Northwest Research Station, 2001.

Hirt, Paul W. *A Conspiracy of Optimism: Management of National Forests since World War Two.* Lincoln: University of Nebraska Press, 1994.

London, Jonathan K. "Common Roots and Entangled Limbs: Earth First! and the Growth of Post-Wilderness Environmentalism on California's North Coast." *Antipode* 30 (April 1998): 155–176.

Proctor, James D. "Whose Nature?: The Contested Moral Terrain of Ancient Forests." In *Uncommon Ground: Toward Reinventing Nature,* edited by William Cronon, 269–297. New York: W. W. Norton, 1995.

Wilkinson, Charles F. *Crossing the Next Meridian: Land, Water, and the Future of the West.* Washington, DC: Island Press, 1992.

Zakin, Susan. *Coyotes and Town Dogs: Earth First! and the Environmental Movement.* Tucson: University of Arizona Press, 1993.

DOCUMENTS

EXCERPTS FROM "THE CHANNELED SCABLAND OF EASTERN WASHINGTON"

The interpretation of landscapes, like all sciences, undergoes various transformations in understanding. The unusual landforms in the Columbia Plateau caused by massive floods thousands of years ago were misunderstood for years. J. Harlen Bretz offered this interpretation in the 1920s and was largely ridiculed by his peers. Belatedly, he received his due for uncovering the mystery of these channeled scablands.

No one with an eye for land forms can cross eastern Washington in daylight without encountering and being impressed by the "scabland." Like great scars marring the otherwise fair face of the plateau are these elongated tracts of bare, or nearly bare, black rock carved into mazes of buttes and canyons. Everybody on the plateau knows scabland. It interrupts the wheat lands, parceling them out into hill tracts less than 40 acres to more than 40 square miles in extent. One can neither reach them nor depart from them without crossing some part of the ramifying scabland. Aside from affording a scanty pasturage, scabland is almost without value. The popular name is an expressive metaphor. The scablands are wounds only partially healed—great wounds in the epidermis of soil with which Nature protects the underlying rock.

With eyes only a few feet above the ground the observer today must travel back and forth repeatedly and must record his observations mentally, photographically, by sketch and by map before he can form anything approaching a complete picture. Yet long before the paper bearing these words has yellowed, the average observer, looking down from the air as he crosses the region, will see almost at a glance the picture here drawn by piecing together the ground-level observations of months of work. The region is unique: let the observer take the wings of the morning to the uttermost parts of the earth: he will nowhere find its likeness.

The Thesis

Conceive of a roughly rectangular area of about 12,000 square miles, which has been tilted up along its northern side and eastern end to produce a regional slope approximately 20 feet to the mile. Consider this slope as the warped surface of a thick, resistant formation, over which lies a cover of unconsolidated materials a few feet to 250 feet thick. A slightly irregular dendritic drainage pattern in maturity has been developed in the weaker materials, but only the major steamways have been eroded into the resistant underlying bed rock. Deep canyons bound the rectangle on the north, west, and south, the two master streams which occupy them converging and joining near the southwestern corner where the downwarping of the region is greatest.

Conceive now that this drainage system of the gently tilted region is entered by glacial waters along more than a hundred miles of its northern high border. The volume of the invading water much exceeds the capacity of the existing streamways. The valleys entered become river channels, they brim over into neighboring ones, and minor divides within the system are crossed in hundreds of places. Many of these divides are trenched to the level of the preexisting valley floors, others have the weaker superjacent formations entirely swept off for many miles. All told, 2800 square miles of the region are scoured clean onto the basalt bed rock, and 900 square miles are buried in the debris deposited by these great rivers. The topographic features produced during this episode are wholly river-bottom forms or are compounded of river-bottom modifications of the invaded and overswept drainage network of hills and valleys. Hundreds of cataract ledges, of basins and canyons eroded into bed rock, of isolated buttes of the bed rock, of gravel bars piled high above valley floors, and of island hills of the weaker overlying formations are left at the cessation of this episode. No fluviatile plains are formed, no lacustrine flats are deposited, almost no debris is brought into the region with the invading waters. Everywhere the record is of extraordinary vigorous subfluvial action. The physiographic expression of the region is without parallel; it is unique, this channeled scabland of the Columbia Plateau.

The Causal Episode

It is difficult to convey by words and pictures a visualization of the observable relationships of channeled scabland. Such terms as "canyon," "cataract," "scarp," "gravel deposit," calling up images of the usual topographic forms, suggest the usual explanation, which involves far less water and far more time than the writer's hypothesis demands. Put a glacial Columbia first in one channel, then in another, then in a third, instead of running all of them at the same time; use more than one glacial episode if necessary; upwarp the transected divides to give the present high altitudes above valley floors. Procedures such as described here are unheard-of. Other ice sheets have come and gone and left no such record. How could such quantities of water be yielded from so small a front and with so little retreat?

It may be that there are other significant facts yet to be discovered. But the writer is convinced that the relations outlined in this paper do exist and that no alternatives yet proposed by others or devised by himself can explain them.

That unique assemblage of remarkable physiographic forms on the Columbia Plateau of Washington, described here as the channeled scabland system or complex, records a unique episode in the Pleistocene history. Special causes seem clearly indicated. But what these causes were is yet an unsolved problem.

Source: J. Harlen Bretz, "The Channeled Scabland of Eastern Washington," *Geographical Review* 18 (July 1928): 446–447, 476–477.

EXCERPT FROM "NATIVE PLANT STORIES"

This oral tradition from the Salish of the Columbia Plateau is one of many that explains the origins of foods for Native peoples. Oral history provides one of our best sources for understanding indigenous conceptions of the natural world. Much of the Native meaning, however, gets lost in the translation from oral performance to written word. Nevertheless, this story about the creation of the bitterroot demonstrates the belief that foods are for particular purposes and are gifts to Native peoples.

It was the time just after winter in the valley in the mountains. There was no food and the people were starving. The fish had not yet returned to the streams and the game animals had moved far away into the mountains. The men had gone out to seek game and they had been gone a long time. It was not yet time for berries to ripen, and the women had gathered what plants they could find that could be eaten, but the ones that were left from the winter were tough and stringy.

In one of their lodges, an old woman was grieving because there was no food for her grandchildren. She could no longer bear to look at their thin, sad faces, and she went out before sunrise to sing her death song beside the little stream which ran through the valley.

"I am old," she sang, "but my grandchildren are young. It is a hard time that has come, when children must die with their grandmothers."

As she knelt by the stream, singing and weeping, the Sun came over the mountains. It heard her death song and it spoke to that old woman's spirit helper.

"My daughter is crying for her children who are starving," Sun said. "Go now and help her and her people. Give them food."

Then the spirit helper took the form of a redbird and flew down into the valley. It perched on a limb above the old woman's head and began to sing. When she lifted her eyes to look at it, the bird spoke to her.

"My friend," the redbird said, "your tears have gone into Earth. They have formed a new plant there, one which will help you and your people to live. See it come now from Earth, its leaves close to the ground. When its blossoms form, they will have the red color of my wings and the white of your hair."

The old woman looked and it was as the bird said. All around her, in the moist soil, the leaves of a new plant had lifted from Earth. As the sun touched it, a red blossom began to open.

"How can we use this plant?" said the old woman.

"You will dig this plant up by the roots with a digging stick," the redbird said. "Its taste will be bitter, like your tears, but it will be a food to help the people live. Each year it will always come at this time when no other food can be found."

And so it has been to this day. That stream where the old woman wept is called Little Bitterroot and the valley is also named Bitterroot after that plant, which still comes each year after the snows have left the land. Its flowers, which come only when touched by the sun, are as red as the wings of a red spirit bird and as silver as the hair of an old woman. And its taste is still as bitter as the tears of that old woman whose death song turned into a song of survival.

Source: Joseph Bruchac, *Native Plant Stories* (Golden, CO: Fulcrum Publishing, 1995), 39–42.

HOW THE CAYUSE GOT FIRE

This story explains how the Cayuse of the interior Northwest obtained fire from the Creator and Mount Hood. It is clear that one of central characteristics that make the Cayuse human is their need for and ability to manipulate fire. Such tropes are common throughout the world. This story also suggests some of the broader ecological functions and relationships with fire.

The Cayuse call themselves Te-taw-ken (we, the people), and they say at one time all the fire in the world was inside of Mt. Hood. From the top of the mountain fire and smoke used to come as if from a chimney, and all inside of the mountain was a great lake of fire. Hon-ea-woat (the Creator) had given this fire into the care of an old demon, and he had under him many demons—more than can be counted or even thought of. He and his devil army kept the fire from every one; from all animals and from all men. The animals and birds did not need fire, and the [people] did not need fire while they were animals; and they did not need fire while they could change from animals to men and back to animals.

But one time in the autumn the people had a great feast, and they put off all their skins as animals and came away to dance naked as men, leaving their warm animal hides folded up on the prairie. While they danced as men and women, Hon-ea-woat sent a great eagle to carry away the skins which the people had put off. He blackened the sky with his wings and made the air to shake. The people were afraid and ran for their skins, as a gopher runs for his hole; but when the sky cleared they found themselves naked men and women forever, and they were very cold. It was then that they first began to kill their brethren and take their hides to keep themselves warm; but nevertheless they were cold, and they sickened of raw food.

Takhstspul was the most knowing of all the tribe, and he was homonick (a brave). Ipskayt was the most cunning; he could steal the hair from your head. These two decided to steal fire for the people from the demon of the fire mountain. At that time all the country was bare of trees, covered only with grass, as the Snake River country is now. Takhstspul and Ipskayt journeyed together toward the mountain top. Takhstspul wished to travel right along, but Ipskayt would travel only in the night-time; so in the day they lay in a ditch or on the prairie like two stones.

When they had finally reached the top of the fire mountain, Ipskayt said to Takhstspul: "You wait here, and I will creep in and steal the fire; then you must carry it away." Ipskayt disappeared just as the fog does in the morning: it is just gone, you know not how. He crept into a crevice in the mountain; and, making himself flat, he slipped in little by little. It was a long time; sometimes he would move only an inch; sometimes he would not move for two hundred heartbeats.

At last he came to where the fire was, and the great Demon was walking about watching it. There was a pile of wood from which other devils fed the fire. Ipskayt crept behind this wood and covered himself all over with the gray bark and moss. When he was covered he rolled himself toward the fire—Oh, ever so little at a time, watching when the demons' backs were turned. Just as he was getting near to the fire, so that he could almost take some, one of the fire devils said, "Here is a log for the fire!" But as he took hold of it, Ipskayt jumped up, leaving the bark in the claws of the fire devil; and snatching a piece of the fire, he flew down the dark crevice out to the light.

The Fire Demon followed him, calling all his servants to come, and they crowded into this crevice so it was like a swollen stream under drift logs; no more could come until the first had gone through. Ipskayt burst into the light and gave the fire to Takhstspul, who fled down the mountainside. After him followed the Fire Demon, still calling his devils; and they flocked from everywhere, thicker than blackbirds. They darkened the sun! Tatkhstspul fled so fast he melted the snow as he ran. The John Day River marks his path. But very fast also came the Fire Demon. Takhstspul was out of breath; he turned and shot his arrows at the Demon, but they only made him stop for a very little while, until he could cure himself. Then the arrows were all gone, and Takhstspul stood on the banks of the Columbia. The river his footsteps had made was flowing into it. He lifted up his hands to Hon-ea-woat and asked for help; then he jumped into the water.

The Great Spirit changed all the demons instantly into pine trees, and that is why the trees are so thick high up on the mountain; those which scatter out toward the river are the demons that were ahead of the others. The Great Spirit was willing his children should have the fire they had stolen; so he changed Takhstspul into a beaver, as he is to this day, the most knowing of animals. The beaver swam across the Columbia and spat the fire into a willow log lying on the bank; that is why the willow is chosen as one of the woods from which to rub fire. Ipskayt was changed by Hon-ea-woat into the little gray woodpecker; and today he hugs the tree, so much like the bark that if he is quiet you cannot see him. He goes creeping up and down, and stealing around the tree, tapping with his bill to show that fire is in the wood.

Source: Jarold Ramsey, *Coyote Was Going There: Indian Literature of the Oregon Country* (Seattle: University of Washington Press, 1977), 15–17.

COYOTE INVENTS THE FISHING RITUALS

This Chinook story suggests the importance of salmon for subsistence and culture, and it clearly indicates the strict traditions that governed salmon harvesting and use and how gender organizes these taboos. Negative consequences were sure to accompany any violation of these strictures. Finally, note the salmon biology present in this story.

Coyote was coming. He came to Got'a't [Clatsop]. There he met a heavy surf. He was afraid that he might be drifted away and went up to the spruce trees. He stayed there a long time. Then he took some sand and threw it upon the surf, saying: "This shall be a prairie and no surf. The future generations shall walk on this prairie." Thus Clatsop became a prairie.

At Nia'xaqsi [Neacoxie], a creek originated. He went and built a house at its mouth. Then he speared a silver-side salmon, a steelhead, and a fall salmon. Then he threw away the steelhead and the fall salmon, saying: "This creek is too small. I do not like to see steelhead and fall salmon here. It shall be a bad omen when a fall salmon is killed here; somebody shall die. When a female salmon or fall salmon is killed a woman shall die; when a male is killed a man shall die." Now he carried only the silver-side salmon to his house. When he arrived there he cut it at once, steamed it, and ate it.

On the next day he took his harpoon and went again to the mouth of [the creek]. He did not see anything, and the flood tide set in. He went home. On the next day he went again and did not see anything. Then he became angry and went home. He defecated and said to his excrements, "Why have these silver-side salmon disappeared?"—"Oh, you with your bandy legs, you have no sense. When the first silver-side salmon is killed it must not be cut. It must be split along its back and roasted. It must not be steamed. Only when they go up river can they be steamed." Coyote went home.

On the next day he went again, and speared three. He went home and made three spits. He roasted each salmon on a spit. On the next day he went again and stood at the mouth of the creek. He did not see anything until the flood tide set in. Then he became angry and went home. He defecated. He spoke and asked his excrements, "Why have these silver-side salmon disappeared?" His excrements said to him, "We told you, you with your bandy legs, when the first silver-side salmon are filled, spits must be made, one for the head, one for the back, one for the roe, one for the body. The gills must be burnt."—"Yes," said Coyote.

On the next day he went again. He again killed three silver-side salmon. When he arrived at home he cut them all up and made many spits. He roasted them all separately. The spits of the breast, body, head, back, and roe were at separate places. Coyote roasted them. On the next morning he went back again. He speared ten silver-sides. Coyote was very glad. He came home and split part of the fish. The other part he left and went to sleep. On the next morning he roasted the rest. Then he went fishing again. He did not see anything before the flood tide set in. He went home. On the next morning he went again, but again he did not see anything. He went home angry.

He defecated and asked his excrements: "Why have these silver-sided salmon disappeared?" His excrements scolded him, "When the first silver-side salmon are killed, they are not left raw. All must be roasted. When many are caught, they must all be roasted before you go to sleep." On the next morning Coyote went and stood at the mouth of the river. He speared ten. Then he made many double spits, and remained awake until all were roasted that he had caught. Now he had learned all that is forbidden in regard to silver-side salmon when they first arrive at [Neacoxie Creek]. He remained there and said, "The Indians shall always do as I had to do. If a man who prepares corpses eats a silver-side salmon, they shall disappear at once. If a murderer eats a silver-side salmon, they shall at once disappear. They shall also disappear when a girl who has just reached puberty or a menstruating woman eats them. Even I got tired."

Now he came this way. At some distance he met a number of women who were digging roots. He asked them, "What are you doing?" "We are digging camas." "How can you dig camas at Clatsop? You shall dig some roots here, but no camas." Now they gathered only thistle roots and wild onions. He left those women and spoiled that land. He changed the camas into wild onions.

Then he came to Clatsop. It was the spring of the year. Then he met his younger brother the snake. Coyote said to him, "Let us make nets." The snake replied, "As you wish." Now they bought material for twine, and paid the frog and the newt to spin it. . . . It got day. Then they [Coyote and Snake] went to catch jack salmon in their net. They laid the net and caught two in it. Coyote jumped over the net. Now they intended to catch more salmon, but the flood-tide set in. So they went home. Coyote said that he was hungry, and split the salmon at once. They roasted them. When they were cooked they ate. The frog and the newt were their cousins. The next morning they went fishing again with their net. Newt looked after the rope, the snake stood at the upper end of the net, Coyote at the lower end. They intended to catch salmon, but they did not get anything before the flood-tide set in. They went home.

Coyote was angry. He defecated and spoke to his excrements, "You are a liar!" They said to him, "You with your bandy legs. When people kill a salmon they do not jump over the net. You must not step over your net. And when the first salmon are caught, they are not cut until the afternoon."—"Oh," said Coyote, "you told me enough." On the next morning they went fishing again. When they killed a [jack] salmon they did not jump over the net. They laid their net twice. Enough salmon were in the net. Coyote ordered the newt, "Bail out the canoe, it is full of water." She bailed it out. Then they intended to fish again, but the flood-tide set in. They went home and put down what they had caught in the house. In the afternoon Coyote split the salmon. He split them in the same way as the silver-side salmon. He placed the head, the back, the body, and the roe in separate places and on separate double spits. They were done.

The next morning they went fishing. They did not catch anything. Coyote became angry and defecated. He said to his excrements, "Tell me, why have these [jack] salmon disappeared?" His excrements scolded him, "Do you think their taboos are the same as those of the silver-side salmon? They are different. When you go fishing for [jack] salmon and they go into your net, you may lay it three times. No more salmon will swim into it. It is enough then. Never bail out your canoe. When you come home and cut the salmon, you must split it at the sides and roast belly and back on separate spits. Then put four sticks vertically into the ground [so that they form a square] and lay two horizontal sticks across them. On top of this frame place the back with the head and the tail attached to it." Coyote said to his excrements, "You told me enough. . . ." [Poor Coyote blunders his way through seven more taboos, being excrementally lectured each time on how the salmon should be killed, how they should be distributed among the people—a crow must be allowed to carry one off beforehand—how at a certain place on the Washington shore the first salmon to be caught should be offered salmonberries, and so on. At last the lessons seem to run their course, and Coyote says:]

"Even I got tired. The Indians shall always do it in the same manner. Murderers, those who prepare corpses, girls who have just matured, menstruating women, widows and widowers shall not eat salmon. Thus shall be the taboos for all generations of people."

Source: Jarold Ramsey, *Coyote Was Going There: Indian Literature of the Oregon Country* (Seattle: University of Washington Press, 1977), 135–138.

EXCERPTS FROM JUAN CRESPÍ'S LETTERS AND DIARY

Juan Crespí was one of the most significant missionaries in Spanish California. He surveyed California on a maritime voyage in 1769–1770 and helped found San Diego, California, in 1769. His accounts of mission life, mission Indians, and mission landscapes are some the best sources for glimpsing the colonial era. In the excerpts below, taken from his letters and diaries of the Portola Expedition, it is apparent the importance Fray Crespí placed on finding a mission site with plenty of water, as well as ample pasturage. In addition, he observes evidence of Native burning, the effects of introduced diseases, and the wonder of earthquakes.

Letter to Fray Francisco Palou (9 June 1769)

. . . When we reached this port, since there was no fresh water near, we went back about a league, still in sight of it, where the fathers who arrived the first of the month had already investigated. We found there a good sized river which empties into the sea through an estuary which the ships use as a watering station. This river has a very large, broad plain on its banks, which seems to be of very good soil, with many willows, some poplars, and some alders, although so far it has not been possible to examine it properly. If the river is permanent it may prove in time to be the best of those discovered in all California. On the banks of this river, which are thickly covered with willows, there are many Castilian rose bushes with very fragrant roses, which I have held in my hand and smelled. All the plain is dotted over with wild grapevines, which look as if they had been planted, and at present their many branches are in bloom. We have been in this port since the second day of the Feast of Espíritu Santo, but we have not yet been able to begin the mission, and we are much troubled because the river, which flows through the plain and which has very good, clear water, as we have observed every day, is drying up to such a degree that although two weeks ago when we arrived we saw it flowing with an abundant stream, it has now diminished so that it hardly runs at all, and they say now that they can cross it dry shod. If this continues it will be necessary to look for another place to establish the mission and obtain irrigation.

This port is a large, level place in the midst of great meadows and plains, with very good pasturage for all kind of cattle, and not a stone is found for variety. All the port is well populated with a large number of villages of Indians, too clever, wide awake, and business-like for any Spaniard to get ahead of them.

Letter to Fray Francisco Palou (6 February 1770)

. . . Where the Point of Pines begins there is a little cove; it may be that it cuts into the land a quarter of a league, and from here the Point of Pines extends. The pines are very dilapidated, and not as the accounts describe them, as I can assure your Reverence that I did not see a single one on the whole point that would do for masts or spars for these ships. This point ends where it merely touches the port of Monterey, and from its terminal there extends a very large cove, of some twenty leagues at least, to the Point of Año Nuevo. Along this cove runs a range of very high sand dunes; they are mountains in height and cover five leagues in extent. Near this point where they begin we found lakes, all of poor, brackish water, so that from only one of them could one drink in case of need. Where the lakes are there is a plain which extends as far as the dunes run, and turns perhaps two leagues toward a mountain. Some large and small oaks and same scraggly pines are found near some dunes where the lakes are. This, according to the opinion which we all formed at this place, is the spot where Monterey ought to be, but we found it closed with high sand dunes.

About four or five leagues from where these dunes end, on the same beach, there is a large estuary, and near it much moist level ground, and many large ponds, but without any trees or any signs of a harbor, although we explored it twice. Near this spot we not only saw tracks of the mountain animals, but the captain succeeded in seeing on a trail about twenty eight as large as cows, with their young, but it seems to me, as I heard them described, that the antlers or horns were like those of deer, although the points were like blunt sticks. Not far from this spot we saw manure like that of horses and we saw bears' dung everywhere. I did not succeed in seeing any of these animals, but the soldiers saw several, and testified that the animals with the manure like horses were like mules. They said that they cannot have been, according to the skins and the pictures which I have seen.

Letter to Joseph De Galvez Describing an Area Near San Francisco Bay (8 February 1770)

. . . We moved near the estuary or arm of the sea, which must penetrate the land at least ten leagues, all surrounded by high mountains, and must be three leagues wide in the narrowest place. We pitched camp in a plain some six leagues long, grown with good oaks and live oaks, and with much other timber in the neighborhood. This plain has two good arroyos with a good flow of water, and at the southern end of the estuary there is a good river, with plenty of water, which passes through the plain mentioned, well wooded on its banks. We all hold the opinion, without a doubt, that this is a very great and magnificent port, with shelter from all winds. At one side, about six leagues before reaching it, I took the latitude, and it came out 37 degrees and 49 minutes.

Diary (25 March 1769)

. . . This arroyo has many willows, poplars, and alders in its bed, and some pools of water. The soldiers told me that lower down it had much level land on both banks; and the California Indians, who went further up the arroyo than the soldiers, told me that lower down it runs with a good stream of water. This being the case it may be suitable for a mission. When we reached this arroyo we found a village of heathen, who fled as soon as they saw us. Our California Indians ran after them and caught a youth whom they brought to the camp, naked and all painted. He was regaled, in order to dispel the fear felt by him and the rest. We have had with us nearly all the way trees called cirios and cocobas. The California Indians are getting sick on our hands. As soon as I arrived I confessed one who is very ill.

26 March, Easter

I said Mass, which was attended by everybody, and we stopped until afternoon. I buried the Indian from Santa Gertrudis whom I confessed and gave extreme unction last night, and a cross was planted over his grave. This day I took latitude, and it came out for me thirty degrees and forty-six minutes.

Thursday, 20 July

We set out about seven in the morning, which dawned cloudy, and taking the road straight to the north, we traveled by a valley about one league long, with good land, grassy and full of alders. This passed, we ascended a little hill and entered upon some mesas covered with dry grass, in parts burned by the heathen for the purpose of hunting hares and rabbits, which live there in abundance. In some places there are clumps of wild prickly pear and some rosemary. A league and a half from the camping place we saw another beautiful green valley, well grown with alders and other smaller trees. On going down to it we saw a lagoon which the explorers said was salt water. We pitched camp in this valley near a pool of fresh water; the reason for stopping, although the march has only covered a league and a half, is because, since the departure from San Diego, we have had on the right a very high mountain range, and we are now apparently going to meet it, and it is necessary to explore it before crossing it, for it seems as though it is going to end on the beach. The pool of water, which I just saw, is more than a hundred varas in length, and its water is very clear and good. Besides this one the explorers say that lower down the arroyo from the north, there are some more pools, and that a good stream of water runs from them, and they have good lands on which crops might be raised by irrigation. According to this, the place is better suited for a town than the preceding. Because we arrived at this place on the day of Santa Margarita, we christened it with the name of this holy virgin and martyr. As soon as we arrived, the heathen of the village, and counting men, women and children, they made not less than sixty, who have their own town on the same plain, came to the camp. We gave them presents of beads and sent them off.

Tuesday, 1 August

This day was one of rest, for the purpose of exploring, and especially to celebrate the jubilee of Our Lady of Los Angeles de Porciúncula. We both said Mass and the men took communion, performing the obligations to gain the great indulgence. At ten in the morning the earth trembled. The shock was repeated with violence at one in the afternoon, and one hour afterwards we experienced another. The soldiers went out this afternoon to hunt, and brought an antelope, with which animals this country abounds; they are like wild goats. I tasted the roasted meat, and it was not bad. To-day I observed the latitude and it came out of us thirty-four degrees and ten minutes north latitude.

Wednesday, 2 August

We set out from the valley in the morning and followed the same plain in a westerly direction. After traveling about a league and a half through a pass between low hills, we entered a very spacious valley, well grown with cottonwoods and alders, among which ran a beautiful river from the north-northwest, and then, doubling the point of a steep hill, it went on afterwards to the south. Toward the north-northeast there is another river bed which forms a spacious water course, but we found it dry. This bed unites with that of the river, giving a clear indication of great floods in the rainy season, for we saw that it had many trunks of trees on the banks. We halted not very far from the river, which we named Porciúncula [now the Los Angeles River]. Here we felt three consecutive earthquakes in the afternoon and night. We must have traveled about three leagues to-day. This plain where the river runs is very extensive. It has good land for planting all kinds of grain and seeds, and is the most suitable site of all that we have seen for a mission, for it has all the requisites for a large settlement. As soon as we arrived about eight heathen from a good village came to visit us; they live in this delightful place among the trees on the river.

They presented us with some baskets of pinole made from seeds of sage and other grasses. Their chief brought some strings of beads made of shells, and they threw us three handfuls of them. Some of the old men were smoking pipes well made of baked clay and they puffed at us three mouthfuls of smoke. We gave them a little tobacco and glass beads, and they went away well pleased.

Thursday, 3 August

At half-past six we left the camp and forded the Porciúncula River, which runs down from the valley, flowing through it from the mountains into the plain. After crossing the river we entered a large vineyard of wild grapes and an infinity of rosebushes in full bloom. All the soil is black and loamy, and is capable of producing every kind of grain and fruit which may be planted. We went west, continually over good land well covered with grass. After traveling about a half a league we came to the village of this region, the people of which, on seeing us, came out into the road. As they drew near us they began to howl like wolves; they greeted us and wished to give us seeds, but as we had nothing in hand in which to carry them we did not accept them. Seeing this, they threw some handfuls on the ground and the rest in the air. We traveled over another plain for three hours, during which we must have gone as many leagues. In the same plain we came across a grove of very large alders, high and thick, from which flows a stream of water about a buey in depth. The banks were grassy and covered with fragrant herbs and watercress. The water flowed afterward in the deep channel towards the southwest. All the land that we saw this morning seemed admirable to us. We pitched camp near the river. This afternoon we felt new earthquakes, the continuation of which astonishes us. We judge that the mountains that run to the west in front of us there are some volcanoes, for there are many signs on the road which stretches between the Porciúncula River and the Spring of Alders, for the explorers saw some large marshes of a certain substance like pitch; they were boiling and bubbling, and the pitch came out mixed with an abundance of water. They noticed that the water runs to one side and the pitch to the other, and that there is such an abundance of it that it would serve to caulk many ships. This place where we stopped is called the Spring of Alders of San Estévan.

Source: Herbert Eugene Bolton, *Fray Juan Crespi: Missionary Explorer on the Pacific Coast, 1769–1774* (New York: AMS Press, 1971), 3–5, 26–27, 44, 64, 132–133, 146–149.

EXCERPT FROM "A VOYAGE OF DISCOVERY TO THE NORTH PACIFIC OCEAN AND ROUND THE WORLD"

After first traveling with Captain James Cook to the Pacific Northwest, George Vancouver led his own expedition to the region for the British Navy. From 1792–1794, he carefully surveyed the Northwest Coast, including the Puget Sound area. He observed the landscape and inhabitants with a keen eye for future settlement. In this excerpt from May 1792 describing the area of Hood Canal in Puget Sound, he notes that the region likely benefited from human influence to a greater degree in the past. Vancouver speculated, correctly, that the Native peoples had already encountered European diseases, which depopulated this place.

From these circumstances alone, it may be somewhat premature to conclude that this delightful country has always been thus thinly inhabited; on the contrary, there are reasons to believe it has been infinitely more populous. Each of the deserted villages was nearly, if not quite, equal to contain all the scattered inhabitants we saw, according to the custom of the Nootka people; to whom these have great affinity in their persons, fashions, wants, comforts, construction of these their fixed habitations, and in their general character. It is also possible that most of the clear spaces may have been indebted, for the removal of their timber and underwood, to manual labour. Their general appearance furnished this opinion, and their situation on the most pleasant and commanding eminences, protected by the forest on every side, except that which would have precluded a view of the sea, seemed to encourage the idea. Not many years since, each of these vacant places might have been allotted to the habitations of different societies, and the variation observed in their extent might have been conformable to the size of each village; on the site of which, since their abdication, or extermination, nothing but the smaller shrubs and plants had yet been able to rear their heads.

In our different excursions, particularly those in the neighborhood of port Discovery, the scull, limbs, ribs, and back bones, or some other vestiges of the human body, were found in many places promiscuously scattered about the beach, in great numbers. Similar relics were also frequently met with during our survey in the boats; and I was informed by the officers, that in their several perambulations, the like appearances had presented themselves so repeatedly, and in such abundance, as to produce an idea that the environs of port Discovery were a general cemetery for the whole of the surrounding country. Notwithstanding these circumstances do not amount to direct proof of the extensive population they indicate, yet, when combined with other appearances, they warranted an opinion, that at no very remote period this country had been far more populous than at present. Some of the human bodies were found disposed of in a very singular manner. Canoes were suspended between two or more trees about twelve feet from the ground, in which were the skeletons of two or three persons; others of a larger size were hauled up into the outskirts of the woods, which contained from four to seven skeletons covered over with a broad plank. In some of these broken bows and arrows were found, which at first gave rise to conjecture, that these might have been warriors, who after being mortally wounded, had, whilst their strength remained, hauled up their canoes for the purpose of expiring quietly in them. But on further examination this became improbable, as it would hardly have been possible to have preserved the regularity of position in the agonies of death, or to have defended their sepulchres with the broad plank with which each was covered.

The few skeletons we saw so carefully deposited in the canoes, were probably the chiefs, priests, or leaders of particular tribes, whose followers most likely continue to possess the highest respect for their memory and remains: and the general knowledge I had obtained from experience of the regard which all savage nations pay to their funeral solemnities, made me particularly solicitous to prevent any indignity from being wantonly offered to their departed friends. Baskets were also found suspended on high trees, each containing the skeletons of a young child; in some of which were also small square boxes filled with a kind of white paste, resembling such as I had seen the natives eat, supposed to be made of the saranne root; some of these boxes were quite full, others were nearly empty, eaten probably by the mice, squirrels, or birds. On the next low point, south of our encampment, where the gunners were airing their powder, they met with several holes in which human bodies were interred slightly covered over, and in different states of decay, some appearing to have been recently deposited. About a half a mile to the northward of our tents, where the land is nearly level with high water mark, a few paces within the skirting of the wood, a canoe was found suspended between two trees, in which were three human skeletons; and a few paces to the right was a cleared place of nearly forty yards round; where, from the fresh appearance of burnt stumps, most of its vegetable productions had lately been consumed by fire. Amongst the ashes we found the sculls, and other bones, of nearly twenty persons in different stages of calcinations; the fire, however, had not reached the suspended canoe, nor did it appear to have been intended that it should. The Skeletons found thus disposed, in canoes, or in baskets, bore a very small proportion to the number of sculls and other human bones indiscriminately scattered about the shores. Such are the effects; but of the cause or causes that have operated to produce them, we remained totally unacquainted; whether occasioned by epidemic disease, or recent wars. The character and general deportment of the few inhabitants we occasionally saw, by no means countenanced the latter opinion; they were uniformly civil and friendly, without manifesting the least sign of fear or suspicion at our approach; nor did their appearance indicate their having been much inured to hostilities. Several of their stoutest men had been seen perfectly naked, and contrary to what might have been expected of rude nations habituated to the warfare their skins were mostly unblemished by scars, excepting such as the small pox seemed to have occasioned; a disease which there is great reason to believe is very fatal amongst them. It is not, however, very easy to draw any just conclusions on the true cause from which this havoc of the human race proceeded: this must remain for the investigation of others who may have more leisure, and better opportunity, to direct such an inquiry: yet it may not be unreasonable to conjecture, that the present apparent depopulation may have arisen in some measure from the inhabitants of this interior part having been induced to quit their former abode, and to have moved nearer the exterior coast for the convenience of obtaining in the immediate mart, with more ease and at a cheaper rate, those valuable articles of commerce, that within these late years have been brought to the sea coasts of this continent by Europeans and the citizens of America, and which are in great estimation amongst these people, being possessed by all in a greater or less degree.

Source: W. Kaye Lamb, ed. *A Voyage of Discovery to the North Pacific Ocean and Round the World* (London: The Hakluyt Society, 1984), 2: 538–540.

EXCERPTS FROM "FUR HUNTERS OF THE FAR WEST: A NARRATIVE OF ADVENTURES IN THE OREGON AND ROCKY MOUNTAINS"

Alexander Ross worked as a North West Company fur trader in the interior Northwest for several years. His published accounts of the trade, after several revisions, presented to the English-speaking world a number of observations about the people and place. The excerpts below from his 1855 book, Fur Hunters of the Far West, *demonstrate his attention to the commodity (i.e., beavers) of most interest to him. However, they also celebrate the landscape and the region's wildlife. Like some earlier explorers and later promoters, Ross identifies the potential of nature in the region.*

While following this little stream we passed several beaver lodges, and observed many marks of the ravages of that animal. In many places great trees had been cut down, and the course of the water stopped and formed into small lakes and ponds by the sagacious and provident exertions of the beaver. In one place we counted forty-two trees cut down at the height of about eighteen inches from the root, within the compass of half an acre. We now began to think we had found the goose that lays golden eggs; this, however, was a delusion. Some low points were covered with poplar and other soft wood and wherever that timber and water were plentiful, there were beaver, but not in great numbers.

• • •

The place selected was commanding. On the west is a spacious view of our noble stream in all its grandeur, resembling a lake rather than a river, and confined on the opposite shore by verdant hills of moderate height. On the north and east the sight is fatigued by the uniformity and wide expanse of boundless plains. On the south the prospect is romantic, being abruptly checked by a striking contrast of wild hills and rugged bluffs on either side of the water, and rendered more picturesque by two singular towering rocks, similar in color, shape, and height, called by the natives "The Twins," situated on the east side; these are skirted in the distance by a chain of the Blue Mountains, lying in the direction of east and west. To effect the intended footing on this sterile and precarious spot was certainly a task replete with excessive labour and anxiety.

In the charming serenity of a temperate atmosphere, Nature here displays her manifold beauties, and, at this season, the crowds of moving bodies diversify and enliven the scene. Groups of Indian huts, with their little spiral columns of smoke, and herds of animals, give animation and beauty to the landscape. The natives, in social crowds, vied with each other in coursing their gallant steeds, in racing, swimming, and other feats of activity. Wild horses in droves sported and grazed along the boundless plains; the wild fowl, in flocks, filled the air; and the salmon and sturgeon, incessantly leaping, ruffled the smoothness of the waters. The appearance of the country on a summer's evening was delightful beyond description.

• • •

The general features of the Snake country present a scene incomparably grateful to a mind that delights in varied beauties of landscape and in the manifold works of Nature. Lofty mountains, whose summits are in the clouds, rise above wide-extending plains, while majestic waters in endless sinuosities fertilize with their tributary streams a spacious land of green meadows, relieved by towering hills and deep valleys, broken by endless creeks with smiling banks. The union of grandeur and richness, of vastness and fertility in the scenery, fills the mind with emotions that baffle description.

The Rocky Mountains, skirting this country on the east, dwindle from stupendous heights into sloping ridges which divide the country into a thousand luxurious vales, watered by streams which abound with fish. The most remarkable heights in any part of the great backbone of America are three elevated insular mountains, or peaks, which are seen at a distance of 150 miles. The hunters very aptly designate them the Pilot Knobs.

Source: Alexander Ross, *Fur Hunters of the Far West: A Narrative of Adventures in the Oregon and Rocky Mountains* (London: Smith, Elder and Co., 1855), 1: 146, 175-176, 266-167.

EXCERPT FROM "THE RESOURCES OF CALIFORNIA, COMPRISING AGRICULTURE, MINING, GEOGRAPHY, CLIMATE, COMMERCE, ETC. ETC. AND THE PAST AND FUTURE DEVELOPMENT OF THE STATE"

In the nineteenth century, boosters from the Far West routinely published books describing the various attributes of the particular region they wished to promote. While always painting a positive picture, these sources reveal important ideas about regional development and the environment that fueled that economic expansion. This source celebrates the California landscape and bemoans the fact that too few people are taking advantage of the diverse opportunities. The belief that pervades this excerpt is the abundance and inexhaustibility of California's resources and potential for profit.

Chapter XIV: Present and Future Development of the State

Section 306. General Summary

Twelve chapters of this book have been filled with a detailed statement of the nature and characteristics of the resources, industry and society of CALIFORNIA. In this chapter, I shall present a summary of their main features.

We have, then, before us a state, lying in the midst of the temperate zone, on the western coast of North America; bounded on one side by the Pacific Ocean, and on the other by a high range of mountains; reaching through nine degrees of longitude and three of latitude; with a coast-line eight hundred miles long, and a total area of about one hundred and sixty thousand miles. The heart of the state is drained by two large rivers, which run from north and south, unite midway, and in their course to the sea form three large and deep bays, with secure and spacious harbors. On these bays and their tributaries, there are nearly one thousand miles of navigable streams now used by steamboats and sailing-vessels.

The climate near the ocean is the most equable in the world. At San Francisco, there is a difference of only seven degrees between the mean temperatures of summer and winter—the average of the latter season being 50° and the former 57° Fahrenheit. Ice and snow are never seen in the winter; and in summer the weather is so cool, that heavy woolen clothing is worn every day. There are not more than a dozen days in the year too warm for comfort at mid-day, and the oldest inhabitant cannot remember a night when blankets were not necessary for a comfortable sleep. The climate is just of the character most favorable to the constant mental and physical activity of men, and to the unvarying health and continuous growth of animals and plants. In the interior, the summers are much warmer than near the ocean; while in the mountains the winters are much colder. By traveling a few hundred miles, the Californian can find almost any temperature that he may desire—great warmth in winter, and icy coldness in summer.

The rocks of the state are chiefly granite and tertiary sandstone; the former occupying the high mountains, the latter the valleys. In former eras there were several, or perhaps many, volcanoes in the range of the Sierra Nevada. Mount Shasta was one of them, and it now has hot springs on its summit, and sends up sulphurous vapors. On the western slope of the Sierra Nevada, about half way between the summit and the foot, are numerous beds of slate and veins of quartz. The same formations are found in the Klamath basin and in other parts of the state; and in nearly every case they are auriferous. There is scarcely a county which does not contain gold. The districts which contain enough gold to support a mining population, have an area of about ten thousand square miles. The gold-yield of the state is about forty-three million dollars annually—more than that of any other country, save the colony of Victoria, in Australia.

The number of men engaged in mining may be estimated at eighty thousand. Our placers and auriferous quartz veins are almost inexhaustible; there are great mountains of gold-bearing gravel which cannot be washed away for a century to come; and the quartz-lodes will last still longer.

The gold-mining of California is conducted in the most thorough and enterprising manner. Although the main principles of the sluice and hydraulic washing were known and used, on a small scale, long before the discovery of gold in California, it was here that those modes of working were first perfected, applied on an extensive scale, and brought into universal use. Large rivers are turned out of their beds; mountains are pierced by tunnels; hills are washed away; and the rivers roll thick with mud to the sea through summer and winter.

The silver-mines of the state were discovered only a short time ago, and their value is not yet fully known; but that some of the ore is wonderfully rich, is established beyond a doubt. The silver districts are in the basin of Utah, at an elevation of five thousand feet or more above the level of the sea, in the midst of a desert country.

In quicksilver, California is the richest country in the world. There are extensive beds of sulphur, asphaltum, and plumbago, and large lakes and springs impregnated with borax.

The natural scenery of California is varied and grand. The Yosemite valley is a chasm ten miles long, two miles wide, and three thousand feet deep, in the heart of the Sierra Nevada, without its equal in the world for sublime and picturesque scenery. It has a dozen great cascades, the highest of which has a fall of thirteen hundred feet. The Mammoth Trees are the largest known growths of the vegetable kingdom. There are likewise in the state mud-volcanoes, natural bridges, many caves, and numerous hot springs, some of which throw out great columns of steam.

The animals and plants of California are peculiar to this coast. The finest group of coniferous trees in the world is that of this state. The mammoth tree, the redwood, the sugar-pine, the red fir, the yellow fir and the *Thuja gigantea*, all reach the wonderful height of three hundred feet; the mammoth tree grows to be thirty feet in diameter, the redwood twenty, and the others from eight to twelve.

The grizzly bear is the largest and strongest indigenous animal of the continent; and the Californian vulture is, next to the condor, the largest bird that flies. The sea near our coast teems with halibut, turbot, mackerel, herring, sardines, anchovies and smelts; while sturgeon and salmon are abundant in our rivers.

There are 40,000,000 acres of tillable land in the state, but not more than 1,000,000 acres are now cultivated. In 1860, the aggregate product of grains and roots of annual growth amounted to 14,470,000 bushels, being an average of twenty-four bushels to the acre cultivated, and of thirty-eight bushels to each inhabitant of the state. The crop of barley was the largest, measuring 5,700,000 bushels; that of wheat 5,000,000 bushels; oats and potatoes, each, 1,500,000 bushels; and maize 500,000 bushels. The barley forms thirty-nine per cent of the 14,470,000 bushels; wheat, thirty-four per cent; and beans, peas, sweet potatoes, buckwheat, and rye, one-half of one per cent each.

Farmers in California have many advantages over men of the same occupation in other parts of the United States. The winter is never cold as to interrupt their work, and there are no storms of rain and hail to destroy their grain and hay. They need no barns. Barley thrives better than in any other part of the world. The soil and climate are also particularly favorable to the growth of wheat, which unites the valuable qualities of whiteness, dryness, and glutinousness, to a greater degree than any other wheat in the world. Our average crops are also larger than any other place where manure is not used extensively. The yield of hops is large, and the facilities for drying them, so as to preserve their strength, are better than in any other land where they are cultivated. Our kitchen vegetables grow to an unparalleled size. Nowhere else have pumpkins been seen to reach two hundred and fifty pounds in weight each, beets one hundred and twenty pounds, white turnips twenty-six pounds, solid-headed cabbages seventy-five pounds, carrots ten pounds, water-melons sixty-five pounds, onions forty-seven ounces, Irish potatoes seven pounds, sweet potatoes fifteen pounds, and so forth. Some cabbages and beets have spontaneously become perennials here, continuing to grow from year to year and remaining green throughout winter and summer; and many of our kitchen vegetables might be converted into perennials by preventing them from going to seed.

The abundance, excellence, and variety of our fruit astonish the stranger, though he may have come from the markets of London or New York, which draw tribute from whole hemispheres. No market on the globe surpasses ours in variety, and yet it is not ten years since we began to import fruit-trees direct from the Eastern states and Europe. Our mild winters permit the trees to grow during nine or ten months in the year, and they grow more rapidly, and reach maturity more speedily, than any other country where they are so healthy, and bear so abundantly. The pear and apple trees which were planted by the missionaries thirty or forty years ago, are still in perfect health, and some of them produce as much as a ton of fruit to the tree every year. The apple and pear seem to have found here their most congenial clime. There are no worms in our apples; no curculios in our plums or cherries; no Hessian fly or weevil in our wheat. The olive and fig grow luxuriantly beside the apple and pear. We can produce olives better than any of the olive-producing regions of the Mediterranean, because we have none of the storms of thunder and hail and rain, which frequently destroy the crops in southern Europe and Asia Minor. The vine produces more abundantly than in any other part of Europe, and the crop has never failed or been destroyed here, as often happens there. A yield of one thousand gallons of wine to the acre is as frequent, proportionately, in California, as of four hundred in France or Germany. Our gardens are, in time, to be the most beautiful in the world, resplendent with conifers and deciduous trees, with the flowers of the temperate zone, and the luxuriant plants of the tropics. The shrubs which in New York remain small, and live only under shelter, as delicate exotics, are naturalized in San Francisco, grow almost to tree-like size, remain green throughout the year, and bloom during most of the months. The rosebush is covered with flowers from January to December.

Domestic herbivorous animals live and increase without shelter, and without cultivated food. They reach their full growth a year earlier than in Eastern states. The absence of extreme cold gives them a more rapid growth, and exemption from many disease. Sheep produce more wool, are healthier, increase more rapidly, and are kept at far less cost in California than in any American state east of the Rocky Mountains. Bees increase more rapidly, and make more honey than there is any record of their doing elsewhere. Thunder and rain storms kill a large proportion of the silkworms in Italy, France, Turkey, and China every year; in the valleys of California we never have any lightning, and no rain during the season when the silk-worms feed.

The wages of labor in California are higher than in any other part of the world. Mechanics' wages are generally from two dollars and fifty cents to four dollars per day; common laborers, from one dollar and seventy-five cents to two dollars and fifty cents per day; farm laborers, and men and maid servants, from twenty dollars to thirty dollars per month. Our imports and exports of treasure are larger in proportion to our population than those of any other state. Our chief city is favorably situated for commerce, and its harbor always contains vessels of the largest size from every sea. It has an undoubted supremacy in the commerce of the north Pacific. We have no paper money, and no current coin less than a dime.

The inhabitants of the state, numbering nearly four hundred thousand, represent in their nativities every American state, and every continent, and every country of Europe, and many of the countries of Asia and Africa. Our population is unsurpassed in intelligence, experience in travelling, and skill in the arts. Our society is liberal in tone, and free in intercourse.

With many drawbacks, which have been set forth clearly and unreservedly, California is still the richest part of the civilized world. It possesses most of the luxuries of Europe, and many of the advantages which the valley of the Ohio had forty years ago. It offers an open career to talents. In the few years of its history it has astonished the world, and its chief glories are still to come. The arts, the sciences, the refinements of life, are to find a favored home in California.

Why is it then that the permanent population of the state has not increased more rapidly? Why have so many of the early immigrants left her shores, never to return, by their departure depriving her of the greatest element of wealth? The great cause is the mismanagement of land-titles by the federal government, and the consequence is, that the people have been unable to obtain secure homes, and therefore have gone to the Eastern states, where they could find permanent residences. This mismanagement has prevailed both in the mineral and agricultural districts, and has produced incalculable evils.

Source: John S. Hittell, *The Resources of California, Comprising Agriculture, Mining, Geography, Climate, Commerce, Etc. Etc. and the Past and Future Development of the State* (San Francisco: A. Roman & Company, 1863), 431–437.

EXCERPTS FROM "WOODRUFF V. NORTH BLOOMFIELD GRAVEL MINING CO. AND OTHERS"

During California's Gold Rush, hydraulic mining caused enormous ecological changes and frequently harmed downstream residents. In 1884, sufficient power mobilized against the hydraulickers to successfully pursue a lawsuit. The judge, Lorenzo Sawyer, investigated claims and issued a perpetual injunction against the mining company. The excerpt from his opinion issued in Woodruff v. North Bloomfield Gravel and Mining Co. *(1884) in the Ninth U.S. Circuit Court in San Francisco reveals the extensive power of hydraulic mining to wreak havoc. The injunction resulting from this decision proved a powerful change from the courts that had theretofore largely supported corporate mining operations.*

This is a bill in equity to restrain the defendants, being several mining companies, engaged in hydraulic mining on the western slope of the Sierra Nevada mountains, from discharging their mining *debris* into the affluents of the Yuba River, and into the river itself, whence it is carried down by the current into the Feather and Sacramento Rivers, filling up their channels and injuring their navigation; and sometimes by overflowing and covering the neighboring lands with debris, injuring and threatening to injure and destroy, the lands and property of the complainant, and of other property owners, situate on and adjacent to the banks of these water-courses. . . .

Hydraulic mining, as used in this opinion, is the process by which a bank of gold-bearing earth and rock is excavated by a jet of water, discharged through the converging nozzle of a pipe, under great pressure, the earth and *debris* being carried away by the same water, through sluices, and discharged on lower levels into the natural streams and water-courses below. Where the gravel or other material of the bank is cemented, or where the bank is composed of masses of pipe-clay, it is shattered by blasting with powder, sometimes from 15 to 20 tons of powder being used at one blast to break up a bank. . . .For example, an eight-inch nozzle, at the North Bloomfield mine, discharges 185,000 cubic feet of water in an hour, with a velocity of 150 feet per second. The excavating power of such a body of water, discharged with such velocity, is enormous; and unless the gravel is very heavy or firmly cemented, it is much in excess of its transporting power. At some of the mines, as at the North Bloomfield, several of these Monitors are worked, much of the time night and day, the several levels upon which they are at work being brilliantly illuminated by electric lights, the electricity being generated by water power. A night scene of the kind, at the North Bloomfield mine, is in the highest degree weird and startling, and cannot fail to strike the stranger with wonder and admiration.

• • •

The portion of the valley here referred to as covered with sand is that portion of the Yuba River extending across the Sacramento valley from the foot-hills to its junction with Feather River at Marysville—a distance of about 12 miles. Formerly, before hydraulic mining operations commenced, the Yuba River ran through this part of its course in a deep channel, with gravely bottom from 300 to 400 feet wide, on average, with steep banks from 15 to 20 feet high, at low water, on either side. From the top of the banks, on each side, extended a strip of bottom lands of rich, black, alluvial soil, on average a mile and a half wide, upon which were situated some of the finest farms, orchards and vineyards in the state. Beyond this first bottom, was a second, constituting a basin between the higher lands on either side of from a mile and a half to three miles wide. Not only has the channel of the river through these bottoms been filled up to a depth of 25 feet and upwards, but this entire strip of bottom land has been buried with sand and debris many feet deep, from ridge to ridge of high land, and utterly ruined for farming and other purposes to which it was before devoted, and it has consequently been abandoned for such uses.

• • •

The waters of the Yuba are so charged with debris that they are wholly unfit for watering stock, for any of the uses, domestic or otherwise, to which water is usually applied, without being first taken out of the stream and allowed to stand in some undisturbed place and settle. As it comes down to Marysville it is so heavily charged with sand as to render it unfit even for surface irrigation.

• • •

The North Bloomfield Mining Company, defendant, has constructed a dam to impound *debris*, 50 feet high, near the junction of Humbug canyon with the south Yuba. The dam, not having been carried higher as it filled up, is now full, and the *debris* that has passed over the dam has filled the canyon and the south Yuba below the dam to a level with the debris above, so that now the debris passes along down the canyon over the dam without obstruction, as though no dam at all existed at that point.

• • •

About 1868 the people of Marysville found it necessary to build levees around the city and along the north bank of the Yuba River to protect it from the rapid encroachment of the *debris* coming down the Yuba; and levees were built. It has been found necessary to increase these levees in height and thickness from year to year ever since. In 1875 the levee on the north side of the Yuba broke, some three or four miles above the city, and the city and other lands were not only flooded, but a large amount of *debris* was deposited.

• • •

The defendants have attempted to show that much of the danger from the overflows results from the acts of the people themselves, inconsequence of the improper system of levying adopted, and the cutting off by such means of some outlets of water, available at high water. There is, as might be expected, some conflict in the testimony of experts and others on these points; but it is probable that they have not in all instances adopted the wisest plan possible in their efforts to protect life and property. These works are always erected on the judgment of engineers, or other men presumed to be competent, and rarely without some difference of opinion, and it is scarcely possible that any plan wholly unobjectionable to all could be adopted.

• • •

Defendants allege that both congress and the legislature of California have authorized the use of the navigable waters of the Sacramento and Feather Rivers for the flow and deposit of mining *debris*; and having so authorized their use, all the acts of the defendants complained of are lawful, and results of those acts, therefore, cannot be a nuisance, public or otherwise.

• • •

The conditions thus imposed upon California by the act of congress admitting her into the Union cannot be lawfully violated by obstructing, much less destroying, the navigation of her rivers and bays for purposes of having no relation to facilitate navigation or commerce. The power of congress to regulate commerce between the states would also, doubtless, enable it, by proper legislation, independent of these conditions imposed by the act of admission, to prevent the state from destroying or obstructing, or authorizing the destruction or obstruction of, the capacity for navigation of her navigable waters.

• • •

Woodruff's interests involved are by no means insignificant, no matter how much may have been said to belittle them. His block of stores, built on one of the most eligible business locations in Marysville, at a cost of at least somewhere between $40,000 and $60,000, his nearly 1,000 acres of farming land—among the best in the state—in Sutter county, called the Hock Farm, and his Eliza tract of over 700 acres on the opposite side of the river, in Yuba county and upon which a little settlement, embracing business houses and a public regular steamboat landing, once existed of which 125 acres in the aggregate on the two tracts are conceded to have already been destroyed certainly constitute an estate of no inconsiderable value.

• • •

The brief flood occasioned by the breaking of the English dam, in June last, afforded a striking illustration of what is liable hereafter to occur. This enormous deposit of *debris* in the Yuba, and near Marysville, and in the streams in the mountains above, is a continuing, ever-present, and, so long as hydraulic mining is carried on as now pursued it will ever continue to be, an alarming and ever-growing menace, a constantly augmenting nuisance, threatening further injuries to the property of complainant, as well as the lives and property of numerous other citizens similarly situated. Against the continuous and further augmentation of this nuisance the complainant must certainly be entitled to legal protection.

• • •

The supreme court of California has never recognized the validity of any custom to mine in such a manner as to destroy or injure the property of others, even in the district of diggings where the local customs and usages of miners are sanctioned by the statutes. But the California reports are full of cases where the principle has been enforced in the mines that every one must so use his own property as not to injure another.

• • •

After an examination of the great questions involved, as careful and thorough as we are capable of giving them, with a painfully anxious appreciation of the responsibilities resting upon us, and of the disastrous consequences to the defendants, we can come to no other conclusion than that complainant is entitled to a perpetual injunction. But as it is possible that some mode may be devised in the future for obviating the injuries, either one of those suggested or some other, and successfully carried out, so as to be both safe and effective, a clause will be inserted in the decree giving leave on any future occasion, when some such plan has been successfully executed, to apply to the court for modification or suspension of the injunction.

Let a decree be entered accordingly.

Source: "Woodruff v. North Bloomfield Gravel Mining Co. and Others." In The Federal Reporter 18 (1884), 756–757, 759–760, 763, 764, 765, 767, 770, 786, 792–793, 797, 802, 808–809.

EXCERPTS FROM "THE OCTOPUS: A STORY OF CALIFORNIA"

Frank Norris was a writer famous for muckraking accounts that criticized entrenched power. This excerpt comes from his novel castigating the Southern Pacific Railroad's power in California. It reveals, in rather overblown prose, the ways American culture frequently engendered nature (as feminine and passive) and economic activity (as male and aggressive). The passage is a highly symbolic meditation on the transformations California's Central Valley was experiencing at the time.

The day was fine. Since the first rain of the season, there had been no other. Now the sky was without a cloud, pale blue, delicate, luminous, scintillating with morning. The great brown earth turned a huge flank to it, exhaling with moisture of the early dew. The atmosphere, washed clean of dust and mist, was as translucent as crystal. Far off to the east, the hills on the other side of Broderson Creek stood out against the pallid saffron of the horizon as flat and as sharply outlined as if pasted on the sky. The campanile of the ancient Mission of San Juan seemed as fine as frostwork. All about between the horizons, the carpet of the land unrolled itself to infinity. But now it was no longer parched with heat, cracked and warped by a merciless sun, powdered with dust. The rain had done its work; not a cloud that was not swollen with fertility, not a fissure that did not exhale the sense of fecundity. One could not take a dozen steps upon the ranches without the brusque sensation that underfoot the land was alive, roused at last from its sleep, palpitating with the desire of reproduction. Deep down in the recesses of the soil the great heart throbbed once more, thrilling with passion, vibrating with desire, offering itself to the caress of the plow, insistent, eager, imperious. Dimly one felt the deep seated trouble of the earth, the uneasy agitation of its members, the hidden tumult of its womb, demanding to be made fruitful, to reproduce, to disengage the eternal renascent germ of life that stirred and struggled in its loins.

• • •

It was the long, stroking caress, vigorous male, powerful for which the earth seemed panting. The heroic embrace of a multitude of iron hands, gripping deep into the brown, warm flesh of the land that quivered responsive and passionate under this rude advance, so robust as to be almost an assault, so violent as to be veritably brutal. There under the sun and under the speckless sheen of the sky, the wooing of the Titan began, the vast primal passion, the two world forces, the elemental male and female, locked in a colossal embrace, at grapples in the throes of an infinite desire, at once terrible and divine, knowing no law, untamed, savage, natural, sublime.

    Source: Frank Norris. *The Octopus: A Story of California* (Garden City, NY: Doubleday & Company, 1901), 121–122, 125–126.

EXCERPTS FROM "THE HETCH HETCHY VALLEY"

The controversy over whether to build a dam in Yosemite National Park's Hetch Hetchy Valley was the first national conflict of its kind. A cofounder of the Sierra Club, John Muir, was the leading spokesperson for those who thought dams in national parks were unwise. Indeed, he saw it as sacrilegious; this passage, first published in the Sierra Club Bulletin *in 1908 reveals how Muir viewed such controversies in religious terms. Such language and perspectives still remain strong, showing that the legacy of Hetch Hetchy continues.*

It appears therefore that Hetch-Hetchy Valley, far from being a plain, common, rock-bound meadow, as many who have not seen it seem to suppose, is a grand landscape garden, one of Nature's rarest and most precious mountain mansions. As in Yosemite, the sublime rocks of its walls seem to the nature-lover to glow with life, whether leaning back in repose or standing erect in thoughtful attitudes, giving welcome to storms and calms alike. And how softly these mountain rocks are adorned, and how fine and reassuring the company they keep—their brows in the sky, their feet set in groves and gay emerald meadows, a thousand flowers leaning confidingly against their adamantine bosses, while birds, bees, and butterflies help the river and waterfalls to stir all the air into music—things frail and fleeting and types of permanence meeting here and blending, as if into this glorious mountain temple Nature had gathered here choices [sic] treasures, whether great or small, to draw her lovers into close confiding communion with her.

• • •

Garden- and park-making goes on everywhere with civilization, for everybody needs beauty as well as bread, places to play in and pray in, where Nature may heal and cheer and give strength to body and soul. This natural beauty-hunger is displayed in poor folks' window-gardens made up of a few geranium slips in broken cups, as well as in the costly lily gardens of the rich, the thousands of spacious city parks and botanical gardens, and in our magnificent National parks—the Yellowstone, Yosemite, Sequoia, etc.—Nature's own wonderlands, the admiration and joy of the world. Nevertheless, like everything else worth while, however sacred and precious and well-guarded, they have always been subject to attack, mostly by despoiling gainseekers—mischief-makers of every degree from Satan to supervisors, lumbermen, cattlemen, farmers, etc., eagerly trying to make everything dollarable, often thinly disguised in smiling philanthropy, calling pocket-filling plunder. "Utilization of beneficent natural resources, that man and beast may be fed and the dear Nation grow great." . . . Ever since the establishment of the Yosemite National Park by act of Congress, October 8, 1890, constant strife has been going on around its borders and I suppose this will go on as part of the universal battle between right and wrong, however its boundaries may be shorn or its wild beauty destroyed.

• • •

That any one would try to destroy such a place seemed impossible; but sad experience shows that there are people good enough and bad enough for anything. The proponents of the dam scheme bring forward a lot of bad arguments to prove that the only righteous thing for Hetch-Hetchy is its destruction. These arguments are curiously like those of the devil devised for the destruction of the first garden—so much of the very best Eden fruit going to waste; so much of the best Tuolumne water. Very few of their statements are even partly true, and all are misleading. Thus, Hetch Hetchy, they say, is a "low-lying meadow."

On the contrary, it is a high-lying natural landscape garden.

"It is a common minor feature, like thousands of others."

On the contrary, it is a very uncommon feature; after Yosemite, the rarest and in many ways the most important in the park.

"Damming and submerging it 175 feet deep would enhance its beauty by forming a crystal-clear lake."

Landscape gardens, places of recreation and worship, are never made beautiful by destroying and burying them. The beautiful lake, forsooth, should be only an eyesore, a dismal blot on the landscape, like many others to be seen in the Sierra. For, instead of keeping it at the same level all the year, allowing Nature to make new shores, it would, of course, be full only a month or two in the spring, when the snow is melting fast; then it would be gradually drained, exposing the slimy sides of the basin and shallower parts of the bottom, with the gathered drift and waste, death and decay of the upper basins, caught here instead of being swept on to decent natural burial along the banks of the river or in the sea. Thus the Hetch Hetchy dam-lake would be only a rough imitation of a natural lake for a few of the spring months, an open mountain sepulcher for the others.

"Hetch Hetchy water is the purest, wholly unpolluted, and forever unpollutable."

On the contrary, excepting that of the Merced below Yosemite, it is less pure than that of most of the other Sierra streams, because of the sewerage of camp grounds draining into it, especially of the Big Tuolumne Meadows campgrounds, where hundreds of tourists and mountaineers, with their animals, are encamped for months every summer, soon to be followed by thousands of travelers from all the world.

These temple destroyers, devotees of ravaging commercialism, seem to have a perfect contempt for Nature, and, instead of lifting their eyes to the mountains, lift them to dams and town skyscrapers.

Dam Hetch-Hetchy! As well dam for water-tanks the people's cathedrals and churches, for no holier temple has ever been consecrated by the heart of man.

Source: John Muir, "The Hetch Hetchy Valley," *Sierra Club Bulletin* 6 (January 1908), http://www.sierraclub.org/ca/hetchhetchy/hetch_hetchy_muir_scb _1908.html.

EXCERPTS FROM "HOW FIRE HELPS FORESTRY: THE PRACTICAL VS. THE FEDERAL GOVERNMENT'S THEORETICAL IDEAS"

Just past the turn of the twentieth century, a debate raged among foresters about whether fire could serve as a useful management tool or whether it should be excluded as much as possible from Western forests. The following excerpt stridently opposes the emerging federal fire paradigm of complete fire suppression. Unfortunately for Hoxie, the timing of publication coincided with the disastrous 1910 fires in northern Idaho and western Montana that effectively stymied further discussion about using fire as a management tool for several decades. This source also reveals the tension, never far from the surface, between public land policies and private land management.

Practical foresters contend and can demonstrate that from time immemorial fire has been the salvation and preservation of our California sugar and white pine forests. The white man found these forests to his liking five hundred and two thousand years after they sprang from the soil and it is admitted that the Indian for centuries, for his own convenience, no doubt, fired the forests at periods of about three years, it is stated; that is to say, he burned certain forest areas one year and certain other areas another, and so on, to the end that the burnings were about three years apart.

The practical invites the aid of fire as a servant, not as a master. It will surely be the master in a very short time unless the Federal Government changes its ways by eliminating the theoretical and grasping the practical.

• • •

I have said that fire always has been and always will be the salvation and preservation of our California sugar and white pine forests, and no doubt the forests of many other states. In a manner, however, fire running at will is master. This is not the practical aspect. It is the intention to deal herein with fire as a servant, whose coming is to be prepared for in advance. This preparation can be undertaken successfully in the summer months and the servant fire can be put to work in the fall months, or after the first rains when it would require aid rather than otherwise in its good work of destroying decayed logs and off-fall, the accumulation of a year or two or more; more than that, fire also destroys the destructive insects such as beetles, which are, it is declared by experts, by reason of the Federal Government's theoretical ideas of keeping fire out of the forests, becoming very much more destructive; their ravages have become very noticeable and in effect are the same as girdling.

Naturally the question is asked by the theorists: "How will the fire be kept from the standing live timber; how save that from the awful ravages of the fire deliberately turned loose in it?" The answer is: *In itself it won't burn*, as an all-wise nature has given it protection in the form of bark that is a nonconductor of heat; but that all-wise nature does not prevent a dead tree from falling or rolling against a live one and subjecting it to unusual and unnecessary hazard. The practical simply proposes to remove these extra hazards in advance of the servant fire. Should the extra hazard be a log, cut out a section of it immediately against the living tree and roll the cut out portion a safe distance from the particular tree endangered, or clear away limbs, as the case may be, and in this manner prepare for the servant fire; perchance, this may prepare for subsequent firing, of extra hazard in the shape of thickets of small trees.

It is claimed by the theorists that fire, even passing through forest areas previously prepared for it, will destroy young trees and the damage on this account would be irreparable. The claim will also be made by the theorists that the vegetable mold or forest carpet is subjected by this enemy fire to total destruction. The practical answer to this is that in the forests described an average from say five to fifteen trees (very much oftener than the former number) is all an acre will sustain. In nature it is simply the survival of the fittest, so that in mature forests the fittest control and, in effect, stunt and make dormant the younger growth by depriving it of the life-giving light and heat. And this younger growth will never make any considerable headway until the parent tree is removed, as can be readily demonstrated by cutting any of the trees subjected to the conditions described and counting the yearly rings.

• • •

The practical realizes that nothing but a miracle will keep fire out of the forest areas, for, to accomplish this, even lightning must be eliminated; further, it is realized that if the theoretical continues for a few years longer there will be no hope of saving these areas from useless unnecessary and enormous damage, as the accumulated fallen limbs and unusual and unnecessary hazard is many times greater in five or ten than in two or three years, for in the nature of things forest trees drop their lower limbs annually and others are broken by snow falling timber, or otherwise.

The practical says, "Let the fire at the proper time of year run at will," in the forest described, rather than not at all, as it can be clearly demonstrated that the class of forests named is greatly benefited rather than injured by this manner of treatment. The proof is abundant: magnificent virgin forests of the class described that have been subjected to fire at will, and not at selected favorable periods of the year, for the past five hundred to perhaps two thousand or more years.

• • •

Galen Clark, former guardian of Yosemite valley, who died recently, was a man who loved the wilds of nature so much that as early as the year 1857 he explored the country back of Wawona and is credited with being the first white man to see the magnificent Mariposa Big Tree Grove. This was only fifty-three years ago and for nineteen hundred and fifty years at least previous to this discovery these same magnificent trees had been subjected to fire *running at will*, many of such set by lightning sent by an all-wise Providence; these fires are evidenced by fire scars at the base or butts of some of these magnificent trees, such scars being the result of unusual hazard or such that the tree itself so damaged could not have supplied. The practical would have removed the cause a safe distance from the tree in advance on the servant fire. The miracle that is necessary to keep fire out of the forest has evidently not as yet occurred, as a very recent report of the honorable Secretary of Agriculture attributes to lightning nearly three hundred of the fires in the forest reserves in the United States during the year 1909. Why not by practical forestry keep the supply of inflammable matter on the forest cover or carpet so limited by timely burning as to deprive even the lightning fires of sufficient fuel to in any manner put them in the position of master? In short, remove the cause and in so doing remove the effect, as fires to the forests are as necessary as are crematories and cemeteries to our cities and towns; this is Nature's process for removing the dead of the forest family and for bettering conditions for the living.

The ideas embodied herein are an attempt at least toward the essence of the forestry subject; but the field is so great that many phases of it are not as fully covered as the writer would choose to have them. He is, however, heartily in accord in so far as practical forestry is concerned and he only hopes that proper hearing of this side of the question will be at least hastened by his efforts. He has no fear that those who have studied the subject for years, as he has, in the forests themselves, will not join their testimony to his earnest declaration that we must count on fire to help in practical forestry.

Source: George L. Hoxie, "How Fire Helps Forestry: The Practical vs. the Federal Government's Theoretical Ideas," *Sunset Magazine* 34 (August 1910): 145–146, 146–147, 148, 151.

EXCERPTS FROM "THE FIGHT FOR CONSERVATION"

Gifford Pinchot, the first chief of the U.S. Forest Service, ranked as the most prominent conservationist during the Progressive Era and claimed to have invented the term conservation. In this statement, taken from his 1910 book, The Fight for Conservation, *Pinchot outlines the three major principles he believed best characterized conservation as practiced by federal agencies. As the head of the USFS, he was in position to put these principles into practice.*

. . . The principles which govern the conservation movement, like all great and effective things, are simple and easily understood. . . .

The first great fact about conservation is that it stands for development. . . . Conservation does mean provision for the future, but it means also and first of all the recognition of the present generation to the fullest necessary use of all the resources with which this country is so abundantly blessed. Conservation demands the welfare of this generation first, and afterward the welfare of the generations to follow. . . .

In the second place conservation stands for the prevention of waste. . . . The first duty of the human race is to control the earth it lives upon. . . .

In addition to the principles of development and preservation of our resources there is a third principle. It is this: The natural resources must be developed and preserved for the benefit of the many, and not merely for the profit of a few. . . .

. . . The application of common-sense to any problem for the Nation's good will lead directly to national efficiency wherever applied. In other words, and that is the burden of the message, we are coming to see the logical and inevitable outcome that these principles, which arose in forestry and have their bloom in the conservation of natural resources, will have their fruit in the increase and promotion of national efficiency along other lines of national life. . . .

. . . So from every point of view conservation is a good thing for the American people.

Source: Gifford Pinchot, *The Fight for Conservation* (New York: Doubleday, Page and Company, 1910), 42, 44–45, 46, 50.

EXCERPTS FROM "THE ROMANCE OF A RIVER: A PAST, PRESENT AND FUTURE SURVEY OF IRRIGATION IN SOUTHERN IDAHO"

E. B. Darlington was the chief engineer for the Twin Falls North Side Land and Water Company in southern Idaho. This excerpt from Reclamation Record *in 1920 tells a typical celebratory story of reclamation. It is a history of easy, noble success, and Darlington expects it to extend into the future with careful guidance by professionals like himself. The reclamation ideal is evident as a duty to maximize water use to serve human economies. Notable by its absence is a discussion of environmental consequences or frustrations ubiquitous in irrigating the arid West.*

We are to-day at a period in Idaho's history where we can indulge in both a review of the past and a vision of the future. Fifty years ago Snake River ran uninterrupted to the sea. To-day over half its waters are diverted to the use of man. To-morrow we expect to see a complete utilization of this magnificent stream from Jackson Lake to Milner Dam.

Snake River and its tributaries are the arteries which carry the lifeblood of southern Idaho. The mighty torrents sent down from the Tetons, the Sawtooth

Mountains and the Owyhees vitalize 2,000,000 acres of our sun-kissed plains. Before many years have passed another million acres will be reclaimed from the desert and added to this emerald empire.

• • •

About 110 years ago the first white men came into the Snake River Valley. They were fur traders. One of the first camps established was near the present site of St. Anthony. Another party entered Idaho by way of the beautiful valley of the Teton. Upon reaching the Snake River, which at this point was placid and inviting, they embarked upon its broad bosom in canoes, not realizing the latent furies in the sparkling water. This first attempt of white men to put to their own uses the untamed Snake was disastrous. After many hazardous experiences they reached the wild waters where now stands the Milner Dam. There one of the canoes was wrecked and some of the party drowned, and travel by boat was abandoned.

It was not until 70 years later that the great river began to yield her beneficence to man. Pioneers from Utah about that time began to settle along its banks and upon its tributaries. In Mormon colonies they had learned how to make use of the life-giving water of the streams by spreading it over the thirsty soil. Nature's moisture deficiency was thus supplied and bountiful crops were the result. Of course the lines of least resistance were followed first. The earliest ditches were built by individuals. What a signal event it must have been in those early days to herald the first water from the river as it came down the newly constructed ditch upon some little farm carved out of the wilderness! Perhaps only those of us later comers who have been camped upon the desert and waited long hours for the arrival of the water wagon can appreciate the thrill there must have been in the hearts of those early families when the water first appeared.

As the land that could be served by individual ditches became occupied, small cooperative associations were formed for building ditches on a somewhat larger scale and to cover lands a little farther back from the river. These community ditches were not always successful. Operation of them sometimes worked out so that the man at the upper end got all the water and the man at the lower end all the upkeep. The next step was the organization of corporations to construct more pretentious works and to sell or rent water to the farmers under them. Dissatisfaction developed with this arrangement, as it was felt that the use of water was such a vital necessity there was danger in placing its control in the hands of interested officials. Most of these properties have been taken over by irrigation districts, by which form of organization the water users themselves own and control the irrigation system and water supply.

It was at about this stage of development that farseeing engineers and other men of vision . . . began to conceive the possibilities of redeeming vast areas upon the higher benches by means of monumental diversion works and far-reaching distributaries many miles in length. Great dams were to serve as ladders by which the water would climb to the mesas and there be diverted into canals, themselves resembling river channels in size and capacity. Project of such magnitude could only be undertaken under the protection of legislation safeguarding the investment of enormous amounts of capital. In 1894 the Carey Act was passed, but there was very little construction under it until the amendments of 1896 and 1901, providing that the cost of the works should be a lien upon the land. From 1901 to 1910 was the period of greatest activity under the Carey Act, and during that period practically all the great projects in Idaho were built. In 1902 the reclamation act was passed, and shortly afterwards work was undertaken on the Minidoka project and the Boise project. The development of these projects has been remarkable. Their agricultural development has been phenomenal. A short time ago reports were made public showing that the value of crops produced in the Minidoka project during 1919 exceeded the entire cost of the works. The North and South Side Twin Falls enterprises are also wonderfully prosperous.

The visions of the early seers have been far surpassed. What seems to be to many of their contemporaries to be impracticable dreams have been more than realized. Their splendid plans and designs have met every test. Aesthetics critics sometimes profess to believe that engineering works are a profanation of nature and a blasphemy upon the Almighty's scheme of things. How little is this view found in the truth! Is it not a better concept of these men of great vision and master designers to regard themselves as understudies of the Creator, delegated to bring forth upon the earth a better condition of life, a finer spirit of contentment, a higher state of development, and an advancement in human progress?

Thus has been brought about the wonderful reclamation that is now so manifest. Thus have been brought into use the latent agricultural resources of this great valley. What a wonderful experience it has been to watch the marvelous development by irrigation in the last 20 years. We who have been privileged to take a part in this development have rejoiced many times that we are engaged on a task that carried with it such inspirations and that produced such conspicuous results. Those of us who have dwelt for months in the sagebrush, who have worked long hours into the night with the weird wail of the coyote in our ears and the wind whipping our tents to shreds, who have baked in the sun, chilled in the blizzard, and choked in the dust, can now take keenest delight in the fair fields of alfalfa, the waving grain, the splendid buildings, the thriving stock, and the canals flowing full where once our lines of stakes were driven.

But what does the future hold in store for the Snake River Valley? We do not propose now to rest on our oars. The time is ripe for broader and greater visions and for harder and better efforts. We must join hands to procure the maximum use of the great river. Every year several million acre-feet run unused to the ocean. This water must be impounded and conserved for use at the time of greatest requirement. . . . Let us visualize the consequences of making this surplus water available for irrigation.

• • •

New development cannot be taken recklessly, however. We have won through to our present status in spite of many serious mistakes that have been made in the past. These will be warning signs for danger in the future. There must be a better settlement pattern than we have had hereto before. The conquest of new land is a bigger and harder proposition than we have realized. Men without adequate capital to finance their farming operations must have long-time credits at moderate rates of interest. They should have the opportunity to provide themselves with comfortable homes, good livestock and the right kind of equipment. Many new settlers need expert advice. Plans are being formulated on the North Side Twin Falls project to inaugurate a department of farm engineering, to furnish building plans, lay out service ditches, locate tap boxes and weirs, plan field layouts, experiment as to the best spacing of catch ditches and corrugations, make soil tests, plan rotation schedules and render every reasonable service to help get the new man satisfactory [sic] established, and to eliminate for him so far as possible the problems which in the past have only been solved after years of experience.

There is a song with the line "If the waters could speak as they roll along" that seems applicable to the Snake River. The waters of this Nile of the West would, indeed, speak of marvelous things—of great adventures and great accomplishments, of tragedy sometimes, but more often of happiness and prosperity—sometimes of tremendous damage done and again of great wealth produced. The potentialities either way are almost beyond imagination.

As citizens of the Snake River Valley, we have a great duty upon us—to see that the romantic river is an agency only of beneficence and that in the development of its full utilization these incalculable benefits shall be for the greatest good of the greatest number.

Source: E. B. Darlington, "The Romance of a River: A Past, Present and Future Survey of Irrigation in Southern Idaho," *Reclamation Record* 11 (March 1920): 122–123.

FRANKLIN D. ROOSEVELT'S EXTEMPORANEOUS REMARKS AT THE SITE OF THE BONNEVILLE DAM, OREGON (3 AUGUST 1934)

President Franklin D. Roosevelt toured the Far West early in his presidency, stopping at important construction sites, including the Bonneville Dam along the Columbia River. This speech reveals Roosevelt's conservation values of capturing natural power for economic development. Like regional boosters, Roosevelt envisioned a Northwest expanding with power, economic growth, and new families populating rural spaces.

Governor Meier, My Friends of Oregon and Washington:

There is an old saying that "seeing is believing" and that is why I came here today.

Until today I have never been familiar with more than the lower course of the Columbia River, but as far back as 1920 I had the privilege of coming out through these States—through all of the great Northwest—and I conceived at the time the very firm belief that this wonderful valley of the Columbia was one of the greatest assets, not alone of the Northwest, but of the United States of America. Back there, fourteen years ago, I determined that if I ever had the rank or opportunity to do something for the development of this great River Basin and for the territory that surrounds it, I would do my best to put this great project through.

Yes, "seeing is believing." Over a year ago, when we first established the principle of commencing great public works projects for every part of the Union, I became firmly convinced that the Federal Government ought immediately to undertake the construction of the Bonneville Dam and the Grand Coulee Dam, and so we got started. General Martin reminded me, as we were driving out here, that it was only on the 26th day of September last year—ten months ago only—that the definite allocation of money for the Bonneville project was made by me at the White House, and I think we have gone a long way in less than a year.

It has been my conception, my dream, that while most of us are alive we would see great sea-going vessels come up the Columbia River as far as the Dalles, and it was only this morning that the Secretary of War told me of a new survey that is being made by the Army Engineers. From that survey I hope it will be found to be, in part of wisdom, to enlarge these locks so that ocean-going ships can pass up as far as the Dalles. And, when we get that done and moving, I hope that we can also make navigation possible from the Dalles up, so we may have barge transportation into the wheat country.

I am reminded a good deal of another river, with a problem somewhat similar—a river on which I was born and brought up—the Hudson. It was only a comparatively few years ago—within the past ten years—that through an action of the Federal Government the channel of the Hudson River was so deepened that Albany, 140 miles from the sea, was made a seaport. You have a very similar case on the Columbia. In the same way, in the State of New York, above Albany, you meet the rapids and falls of the Mohawk. It was over a hundred years ago that Dewitt Clinton, a Governor of New York, built what was called "Clinton's Ditch," the Erie Canal, and carried the possibility of navigation by barge from the sea to the Great Lakes. And so I believe that the day will come on the Columbia when we shall not only extend sea-going navigation far back into the continent but, at the end of sea navigation, we shall be able to extend barge transportation still further back far north into the State of Washington and far into the State of Idaho. That is a dream, my friends, but not an idle dream, and today we have evidence of what man can do to improve the conditions of mankind.

There is another reason for the expenditure of money in very large amounts on the Columbia. In fact there are a good many reasons. While we are improving navigation we are creating power, more power, and I always believe in the old saying of "more power to you." I do not believe that you can have enough power for a long time to come, and the power we shall develop here is going to be power which for all time is going to be controlled by the Government.

Two years ago, when I was in Portland, I laid down the principle of the need for Government yardsticks so that the people of this country will know whether they are paying the proper price for electricity of all kinds. The Government can create yardsticks. At that time one had already been started on the Colorado River. Since then two other yardsticks have been undertaken, one in the Tennessee Valley, one here on the Columbia River, and the fourth, the St. Lawrence, is going to be started.

In this Northwestern section of our land, we still have the opening of opportunity for a vastly increased population. There are many sections of the country, as you know, where conditions are crowded. There are many sections of the country where land has run out or has been put to the wrong kind of use. America is growing. There are many people who want to go to a section of the country where they will have a better chance for themselves and their children, and there are a great many people who have children and need room for growing families. As a Roosevelt I am thinking about growing families.

Out here you have not just space, you have space that can be used by human beings. You have a wonderful land—a land of opportunity—a land already peopled by Americans who know whither America is bound. You have people who are thinking about advantages for mankind, good education, and, above all, the chance for security, the chance to lead their own lives without wondering what is going to happen to them tomorrow. They are thinking about security for old age, security against the ills and accidents that come to people and, above all, security to earn their own living.

Today I have seen a picture I knew before only in blueprint form. So far as topography goes, it conforms to the blueprints, and the chief engineer of this project tells me nothing stands in the way of its being completed on time, on schedule and according to plan.

Within three years, I hope the Bonneville Dam will be an actual fact and, as a fact, it will from then on militate very greatly to the benefit of the lives, not only of the people of Oregon and Washington but of the whole United States.

I know you good people are heart and soul behind this project and I think most of you are heart and soul behind what your government is trying to do to help the people of the United States. I wish I might stay here and survey everything in detail but, as you know, I have been on a long voyage and the sailor man does not stay put very long in one place.

I have been so much interested during this wonderful drive here that I have delayed things all along the road. That is why I am an hour late. Now I have to go to the train.

I want to tell you from the bottom of my heart what a privilege it is to come here and see this great work at first hand. May it go on with God's blessing and with your blessings.

Source: Samuel I. Rosenman, *The Public Papers and Addresses of Franklin D. Roosevelt: The Advance of Recovery and Reform 1934* (New York: Random House, 1938), 352–354.

EXCERPTS FROM "UDALL V. FEDERAL POWER COMMISSION"

This majority opinion written in 1967 by Justice William O. Douglas effectively stopped the construction of a dam at High Mountain Sheep on the Snake River. Although the main legal question hinged on whether the dam would be a federal or private one, Douglas used several laws that encouraged broader thinking about resource use, including recreational values. This decision helped reverse the trend of authorizing virtually any dam a public or private entity proposed. A committed conservationist, Douglas also used the word "ecology" for the first time in a Supreme Court opinion.

We do know that on the Snake-Columbia waterway between High Mountain Sheep and the ocean, eight hydroelectric dams have been built and another authorized. These are federal projects; and if another dam is to be built, the question whether it should be under federal auspices looms large. Timed releases of stored water at High Mountain Sheep may affect navigability; they may affect hydroelectric production of the downstream dams when the river level is too low for the generators to be operated at maximum capacity; they may affect irrigation; and they may protect salmon runs when the water downstream is too hot or insufficiently oxygenated. Federal versus private or municipal control may conceivably make a vast difference in the functioning of the vast river complex.

Beyond that is the question whether any dam should be constructed.

• • •

The objective of protecting "recreational purposes" means more than that the reservoir created by the dam will be the best one possible or practical from a recreational viewpoint. There are already eight lower dams on this Columbia River system and a ninth one authorized; and if the Secretary is right in fearing that this additional dam would destroy the waterway as spawning grounds for anadromous fish (salmon and steelhead) or seriously impair that function, the project is put in an entirely different light. The importance of salmon and steelhead in our outdoor life as well as in commerce is so great that there certainly comes a time when their destruction might necessitate a halt in so-called "improvement" or "development" of waterways. The destruction of anadromous fish in our western waters is so notorious that we cannot believe that Congress through the present Act authorized their ultimate demise.

• • •

Mr. Justice Holmes once wrote that "A river is more than an amenity, it is a treasure." . . . That dictum is relevant here for the Commission . . . must take into consideration not only hydroelectric power, navigation, and flood control, but also the "recreational purposes" served by the river. . . .

Fishing is obviously one recreational use of the river and it also has vast commercial implications. . . . The Commission, to be sure, did not wholly neglect this phase of the problem. In its report it adverted to the anadromous fish problem, stating that it was "highly controversial" and was not "clearly resolved on record." The reservoir is "the most important hazard" both to up-stream migrants and downstream migrants. Upstream migrants can be handled quite effectively by fish ladders. But those traveling downstream must go through the turbines; and their mortality is high. Moreover, Chinook salmon are "basically river fish and do not appear to adapt to the different conditions presented by a reservoir." The ecology of a river is different from the ecology of a reservoir built behind a dam. What the full effect on salmon will be is not known.

• • •

The need to destroy the river as a waterway, the desirability of its demise, the choices available to satisfy future demands for energy . . . are all relevant to a decision . . . but they were largely untouched by the Commission.

Source: Udall v. Federal Power Commission, 387 U.S. 428, 434–436, 437–438, 439–440, 450 (1967).

EXCERPTS FROM VARIOUS PIECES OF FEDERAL LEGISLATION

A series of federal laws passed in the 1960s and 1970s attempted to ameliorate the cumulative effect of industrial life in the United States. These laws asserted federal power into environmental questions in unprecedented ways. They included measures at preserving some areas free of economic development, promoted research and regulation of industrial pollution, and opened decision making to public input. These excerpts suggest the purposes and means of this new legislation.

Multiple-Use Sustained Yield Act (1960)

An act to authorize and direct that the national forests be managed under principles of multiple use and to produce a sustained yield of products and services, and for other purposes.

Be it enacted by the Senate and House of Representatives of the United States of America in Congress assembled, That it is the policy of the Congress that the national forests are established and shall be administered for outdoor recreation, range, timber, watershed, and wildlife and fish purposes.

• • •

Section 2

The Secretary of Agriculture is authorized and directed to develop and administer the renewable surface resources of the national forests for multiple use and sustained yield of the several products and services obtained therefrom. In the administration of the national forests due consideration shall be given to the relative values of the various resources in particular areas. The establishment and maintenance of areas of wilderness are consistent with the purposes and provisions of this Act.

• • •

Section 4

As used in this Act, the following terms shall have the following meanings:

(a) "Multiple use" means: The management of all the various renewable surface resources of the national forests so that they are utilized in the combination that will best meet the needs of the American people; making the most judicious use of the land for some or all of these resources or related services over areas large enough to provide sufficient latitude for periodic adjustments in use to conform to changing needs and conditions; that some land will be used for less than all of the resources; and harmonious and coordinated management of the various resources, each with the other, without impairment of the productivity of the land, with consideration being given to the relative values of the various resources, and not necessarily the combination of uses that will give the greatest dollar return or the greatest unit output.

(b) "Sustained yield of the several products and services" means the achievement and maintenance in perpetuity of a high-level annual or regular periodic output of the various renewable resources of the national forests without impairment of the productivity of the land.

Source: Multiple Use Sustained Yield Act of 1960, 16 U.S.C. §§ 528-531, June 12, 1960. http://www.fs.fed.us/emc/nfma/includes/musya60.pdf.

Wilderness Act (1964)

Section 2

(a) In order to assure that an increasing population, accompanied by expanding settlement and growing mechanization, does not occupy and modify, all areas within the United States and its possessions, leaving no lands designated for preservation and protection in their natural condition, it is hereby declared to be the policy of the Congress to secure for the American people of present and future generations the benefits of an enduring resource of wilderness. For this purpose there is hereby established a National Wilderness Preservation System to be composed of federally owned areas designated by Congress as "wilderness areas," and these shall be administered for the use and enjoyment of the American people in such manner as will leave them unimpaired for future use and enjoyment as wilderness, and so as to provide for the protection of these areas, the preservation of their wilderness character, and for the gathering and dissemination of information regarding their use and enjoyment as wilderness; and no Federal lands shall be designated as "wilderness areas" except as provided for in this Act or by a subsequent Act.

• • •

(c) A wilderness, in contrast with those areas where man and his own works dominate the landscape, is hereby recognized as an area where the earth and its community of life are untrammeled by man, where man himself is a visitor who does not remain. An area of wilderness is further defined to mean in this Act an area of undeveloped Federal land retaining its primeval character and influence, without permanent improvements or human habitation, which is protected and managed so as to preserve its natural conditions and which (1) generally appears to have been affected primarily by the forces of nature, with the imprint of man's work substantially unnoticeable; (2) has outstanding opportunities for solitude or a primitive and unconfined type of recreation; (3) has at least five thousand acres of land or is of sufficient size as to make practicable its preservation and use in an unimpaired condition; and (4) may also contain ecological, geological, or other features of scientific, educational, scenic, or historical value.

Source: Public Law 88-577. http://www.fsa.usda.gov/dafp/cepd/epb/statutes/ WildernessAct.pdf.

Wild and Scenic Rivers Act (1968)

Congressional declaration of policy

(b) It is hereby declared to be the policy of the United States that certain selected rivers of the Nation which, with their immediate environments, possess outstandingly remarkable scenic, recreational, geologic, fish and wildlife, historic, cultural, or other similar values, shall be preserved in free-flowing condition, and that they and their immediate environments shall be protected for the benefit and enjoyment of present and future generations. The Congress declares that the established national policy of dam and other construction at appropriate sections of the rivers of the United States needs to be complemented by a policy that would preserve other selected rivers or sections thereof in their free-flowing condition to protect the water quality of such rivers and to fulfill other vital national conservation purposes.

• • •

Management Direction

Section 10

(a) Each component of the national wild and scenic rivers system shall be administered in such manner as to protect and enhance the values which caused it to be included in said system without, insofar as is consistent therewith, limiting other uses that do not substantially interfere with public use and enjoyment of these values. In such administration primary emphasis shall be given to protecting its aesthetic, scenic, historic, archaeologic, and scientific features. Management plans for any such component may establish varying degrees of intensity for its protection and development, based on the special attributes of the area.

Source: Wild and Scenic Rivers Act, 16 U.S.C. §§ 1271-1287, October 2, 1968. http://www.nps.gov/rivers/wsract.html.

National Environmental Policy Act (1969)

Section 2

The purposes of this Act are: To declare a national policy which will encourage productive and enjoyable harmony between man and his environment; to promote efforts which will prevent or eliminate damage to the environment and biosphere and stimulate the health and welfare of man; to enrich the understanding of the ecological systems and natural resources important to the Nation; and to establish a Council on Environmental Quality.

• • •

Section 101

(a) The Congress, recognizing the profound impact of man's activity on the interrelations of all components of the natural environment, particularly the profound influences of population growth, high-density urbanization, industrial expansion, resource exploitation, and new and expanding technological advances and recognizing further the critical importance of restoring and maintaining environmental quality to the overall welfare and development of man, declares that it is the continuing policy of the Federal Government, in cooperation with

State and local governments, and other concerned public and private organizations, to use all practicable means and measures, including financial and technical assistance, in a manner calculated to foster and promote the general welfare, to create and maintain conditions under which man and nature can exist in productive harmony, and fulfill the social, economic, and other requirements of present and future generations of Americans.

(b) In order to carry out the policy set forth in this Act, it is the continuing responsibility of the Federal Government to use all practicable means, consistent with other essential considerations of national policy, to improve and coordinate Federal plans, functions, programs, and resources to the end that the Nation may –

1. fulfill the responsibilities of each generation as trustee of the environment for succeeding generations;

2. assure for all Americans safe, healthful, productive, and aesthetically and culturally pleasing surroundings;

3. attain the widest range of beneficial uses of the environment without degradation, risk to health or safety, or other undesirable and unintended consequences;

4. preserve important historic, cultural, and natural aspects of our national heritage, and maintain, wherever possible, an environment which supports diversity, and variety of individual choice;

5. achieve a balance between population and resource use which will permit high standards of living and a wide sharing of life's amenities; and

6. enhance the quality of renewable resources and approach the maximum attainable recycling of depletable resources.

(c) The Congress recognizes that each person should enjoy a healthful environment and that each person has a responsibility to contribute to the preservation and enhancement of the environment.

Section 102

The Congress authorizes and directs that, to the fullest extent possible: (1) the policies, regulations, and public laws of the United States shall be interpreted and administered in accordance with the policies set forth in this Act, and (2) all agencies of the Federal Government shall—

(A) utilize a systematic, interdisciplinary approach which will insure the integrated use of the natural and social sciences and the environmental design arts in planning and in decisionmaking which may have an impact on man's environment;

(B) identify and develop methods and procedures, in consultation with the Council on Environmental Quality established by title II of this Act, which will insure that presently unquantified environmental amenities and values may be given appropriate consideration in decisionmaking along with economic and technical considerations;

(C) include in every recommendation or report on proposals for legislation and other major Federal actions significantly affecting the quality of the human environment, a detailed statement by the responsible official on—

(i) the environmental impact of the proposed action,

(ii) any adverse environmental effects which cannot be avoided should the proposal be implemented,

(iii) alternatives to the proposed action,

(iv) the relationship between local short-term uses of man's environment and the maintenance and enhancement of long-term productivity, and

(v) any irreversible and irretrievable commitments of resources which would be involved in the proposed action should it be implemented.

Prior to making any detailed statement, the responsible Federal official shall consult with and obtain the comments of any Federal agency which has jurisdiction by law or special expertise with respect to any environmental impact involved. Copies of such statement and the comments and views of the appropriate Federal, State, and local agencies, which are authorized to develop and enforce environmental standards, shall be made available to the President, the Council on Environmental Quality and to the public.

National Environmental Policy Act of 1969 (NEPA)

Source: 42 U.S.C. §§ 4321-4347, January 1, 1970. http://ceq.eh.doe.gov/nepa/regs/nepa/nepaeqia.htm.

Clean Air Act (1970)

Section 101

(a) The Congress finds

(1) that the predominant part of the Nation's population is located in its rapidly expanding metropolitan and other urban areas, which generally cross the boundary lines of local jurisdictions and often extend into two or more States;

(2) that the growth in the amount and complexity of air pollution brought about by urbanization, industrial development, and the increasing use of motor vehicles, has resulted in mounting dangers to the public health and welfare, including injury to agricultural crops and livestock, damage to and the deterioration of property, and hazards to air and ground transportation;

(3) that air pollution prevention (that is, the reduction or elimination, through any measures, of the amount of pollutants produced or created at the source) and air pollution control at its source is the primary responsibility of States and local governments; and

(4) that Federal financial assistance and leadership is essential for the development of cooperative Federal, State, regional, and local programs to prevent and control air pollution.

(b) The purposes of this title are -

(1) to protect and enhance the quality of the Nation's air resources so as to promote the public health and welfare and the productive capacity of its population;

(2) to initiate and accelerate a national research and development program to achieve the prevention and control of air pollution;

(3) to provide technical and financial assistance to State and local governments in connection with the development and execution of their air pollution prevention and control programs; and

(4) to encourage and assist the development and operation of regional air pollution prevention and control programs.

(c) Pollution Prevention.- A primary goal of this Act is to encourage or otherwise promote reasonable Federal, State, and local governmental actions, consistent with the provisions of this Act, for pollution prevention.

Source: Clean Air Act (1970), 42 U.S.C. §§7401. http://www.epa.gov/oar/caa/caa.txt.

Endangered Species Act (1973)

Section 2

(a) FINDINGS. The Congress finds and declares that

(1) various species of fish, wildlife, and plants in the United States have been rendered extinct as a consequence of economic growth and development untempered by adequate concern and conservation;

(2) other species of fish, wildlife, and plants have been so depleted in numbers that they are in danger of or threatened with extinction;

(3) these species of fish, wildlife, and plants are of aesthetic, ecological, educational, historical, recreational, and scientific value to the Nation and its people;

(4) the United States has pledged itself as a sovereign state in the international community to conserve to the extent practicable the various species of fish or wildlife and plants facing extinction. . . .

(5) encouraging the States and other interested parties, through Federal financial assistance and a system of incentives, to develop and maintain conservation programs which meet national and international standards is a key to meeting the Nation's international commitments and to better safeguarding, for the benefit of all citizens, the Nation's heritage in fish, wildlife, and plants.

(b) PURPOSES. The purposes of this Act are to provide a means whereby the ecosystems upon which endangered species and threatened species depend may be conserved, to provide a program for the conservation of such endangered species and threatened species, and to take such steps as may be appropriate to achieve the purposes of the treaties and conventions set forth in subsection (a) of this section.

(c) POLICY.

(1) It is further declared to be the policy of Congress that all Federal departments and agencies shall seek to conserve endangered species and threatened species and shall utilize their authorities in furtherance of the purposes of this Act.

(2) It is further declared to be the policy of Congress that Federal agencies shall cooperate with State and local agencies to resolve water resource issues in concert with conservation of endangered species.

Source: Endangered Species Act (1973) 7 U.S.C. 136; 16 U.S.C. 460 et seq. http://www.fws.gov/Endangered/esa.html.

EXCERPTS FROM PRESIDENT BILL CLINTON'S "REMARKS ON OPENING THE FOREST CONFERENCE" IN PORTLAND, OREGON (2 APRIL 1993)

The conflict concerning logging in old-growth forests and endangered species reached a boiling point in the early 1990s. Early in Bill Clinton's presidency he convened a meeting, dubbed a "Forest Summit," that was designed to find common ground. This excerpt is drawn from President Clinton's opening remarks in April 1993. The statement reveals Clinton's desire to find a solution that can please everyone, something that seemed almost inconceivable at the time.

Good morning. I want to thank every one of you who are in the room today and also all of those who are outside—and there are certainly many who have come here—for caring enough to be here. We're here to discuss issues whose seriousness demands that we respect each other's concerns, each other's experiences, and each other's views. Together we can move beyond confrontation to build a consensus on a balanced policy to preserve jobs and to protect our environment.

• • •

We're all here to listen and to learn from you. We're here to discuss issues about which people feel strongly, believe deeply, and often disagree vehemently. That's because the issues are important and are related and intrinsic to the very existence of the people who live here in the Pacific Northwest.

We're discussing how people earn their livelihoods. We're discussing the air, the water, the forests that are important to your lives. And we're addressing the values that are at the core of those lives. From the trailblazers and the pioneers to the trapper and the hunters, the loggers and the mill workers, the people of the Northwest have earned their livings from the land and have lived in awe of the power, the majesty, and the beauty of the forests, the rivers, and the streams.

Coming from a State, as I do, that was also settled by pioneers and which is still 53 percent timberland—we have an important timber industry and people who appreciate the beauty and the intrinsic value of our woodlands—I've often felt at home here in the Northwest. I'll never forget the people I've met here over the last year-and-a-half whose lives have been touched by the issues that we're here to discuss. I remember the timber industry workers with whom I spoke at a town hall meeting in Seattle last July who invited me to come to their communities and learn about their problems.

I remember the families from the timber industry whom I met last September in Max Groesbeck's backyard in Eugene, Oregon. I was moved beyond words by the stories that people told me there and by their determination to fight for their communities and their companies and their families.

I was also inspired by Frank Henderson, who had lost his job as a timber worker and gone through retraining to learn thermoplastic welding and now owns a plastics welding business of his own. He was a guest of mine at the Inaugural, and I'm glad to have him here with us today.

And I remember Elizabeth Bailey of Hayfork, California. She's 11 years old and she was one of the girls and boys who visited me at the White House a few Saturdays ago to participate in our televised townhall meeting for children. Her parents, Willie and Nadine Bailey, have had to close their timber business because, in the past, politics seemed to matter more than people or the environment. And I'm glad that Nadine Bailey, a dedicated spokesperson for loggers, is also here with us today.

As I've spoken with people who work in the timber industry I've been impressed by their love of the land. As one worker told me at our meeting in the Groesbecks' backyard, "I care about Oregon a lot, the beauty of the country."

We're fortunate to have people with us today who bring not only a variety of experiences but a variety of views to the questions before the conference: How can we achieve a balanced and comprehensive policy that recognizes the importance of the forests and timber to the economy and jobs of this region? And how can we preserve our precious old-growth forests which are part of our national heritage and that, once destroyed, can never be replaced?

For too long, the National Government has done more to confuse the issues than to clarify them. In the absence of real leadership, at least six different Federal Agencies have hooked their horses to different sides of the cart, and then they've wondered why the cart wouldn't move forward. To make things worse, the rhetoric from Washington has often exaggerated and exacerbated the tensions between those who speak about the economy and those who speak about the environment.

Not surprisingly, these issues have very often ended up in court while the economy, the environment, and the people have all suffered. That's why it's so important that the people here today are meeting in a conference room, not a courtroom. Whatever your views, everyone who will speak today comes from the Northwest and will have to live with the results of whatever decisions we all make.

We're here to begin a process that will help ensure that you will be able to work together in your communities, for the good of your businesses, your jobs, and your natural environment. The process we begin today will not be easy. Its outcome cannot possibly make everyone happy. Perhaps it won't make anyone completely happy. But the worst thing we can do is nothing. As we begin this process, the most important thing we can do is to admit, all of us to each other, that there are no simple or easy answers.

This is not about choosing between jobs and the environment but about recognizing the importance of both and recognizing that virtually everyone here and everyone in this region cares about both. After all, nobody appreciates the natural environment more than the working people who depend upon it for fishing, for boating, for teaching their children to respect the land, the rivers, and the forests. And most environmentalists are working people and business people themselves, and understand that only an economically secure America can have the strength and confidence necessary to preserve our land, our water and our forests, as you can see in how badly they're despoiled in nations that are not economically secure.

A healthy economy and a healthy environment are not at odds with each other. They are essential to each other. Here in the Northwest, as in my own home State, people understand that healthy forests are important for a healthy forest-based economy; understand that if we destroy our old growth forest, we'll lose jobs in salmon fishing and tourism and, eventually, in the timber industry as well. We'll destroy recreational opportunities in hunting and fishing for all and eventually make our communities less attractive.

We all understand these things. Let's not be afraid to acknowledge them and to recognize the simple but powerful truth that we come here today less as adversaries than as neighbors and coworkers. Let's confront problems, not people.

Today I ask all of you to speak from your hearts, and I ask you to listen and strive to understand the stories of your neighbors. We're all here because we want a healthy economic environment and a healthy natural environment, because we want to end the divisions here in the Northwest and the deadlock in Washington.

If we commit today to move forward together, we can arrive at a balanced solution and put the stalemate behind us. Together, we can make a new start.

Thank you very much.

Source: Public Papers of the Presidents of the United States. http://www .gpoaccess.gov/pubpapers/search.html.

EXCERPTS FROM "AN INTERVIEW WITH JUDI BARI"

Judi Bari was an Earth First! organizer in California's redwood forests. She was a unique environmentalist in that she was also a labor organizer. This combination allowed her to attack timber corporations for their mistreatment of workers and the woods. Although her labor organizing did not succeed as much as her radical environmental organizing, Bari attempted a broader-based community activism than many environmentalists. Her car was bombed for her efforts. This short excerpt from a 1993 interview demonstrates the interconnectedness of her causes and its potential radicalism.

The timber workers have something to tell us that we don't know. They're out in the woods all day. They know exactly what's going on. Just giving them that respect of being intelligent, that made a lot of difference in their openness toward us. I began to meet workers by blockading them. We'd shut down some logging operation, and there they'd be, the perfect captive audience. When we'd start talking, because the way I was talking to them indicated that I had similar experiences and sympathies and understood what their life conditions were like, they were really interested in talking to me. I began to get to know some of the issues in the workplaces and some of the working conditions. I began to advocate for those positions from whatever public forum I got from being an

EF!er. I began to advocate for the conditions of the workers because the unions have been essentially busted in our area and there really isn't anybody advocating for them.

• • •

I don't believe that it's possible to save the Earth under capitalism because I think that capitalism is based on the exploitation of the Earth, just like it's based on the exploitation of workers. But I don't believe that traditional Marxist socialism is the answer either. Marx speaks only of redistributing the spoils of raping the Earth more equitably among the class of humans. He doesn't address the relationship of the society to the Earth, and I think that is one of the principal contradictions. We need to find a new way to live on the Earth without destroying the Earth or exploiting lower classes. It needs to be socially just and it needs to be biocentric. We are calling this Revolutionary Ecology.

Source: Douglas Bevington, "Earth First! in Northern California: An Interview with Judi Bari," in *The Struggle for Ecological Democracy: Environmental Justice Movements in the United States*, ed. Daniel Faber, 254–255, 270 (New York: The Guilford Press, 1998).

IMPORTANT PEOPLE, EVENTS, AND CONCEPTS

2,4,5-T (2,4,5-TRICHLOROPHENOXYACETIC ACID)   This chemical combined with 2,4-D to create Agent Orange, the toxic substance used extensively in the Vietnam War as a defoliant. Forest officials also applied 2,4,5-T to forests to battle pests. It reportedly resulted in birth defects and miscarriages among nearby residents.

ACRE-FOOT   An acre-foot is the volume of water that covers one acre to a depth of one foot. It is 1233.5 cubic meters, or 43,560 cubic feet.

ANADROMOUS   Anadromous fish are born in fresh water and then migrate to the ocean where they reach adulthood. They return to fresh water to spawn. This life cycle requires significant physiological transformations. Anadromous fish include various species of Pacific salmon *(Oncorhynchus)* and steelhead trout.

ANDRUS, CECIL DALE (1931–PRESENT)   A four-term Idaho governor (1970–1977 and 1987–1994) and secretary of the interior in Jimmy Carter's administration, Cecil Andrus earned a reputation for being environmentally responsive. He was first elected governor on a platform opposing a proposed open-pit mine in Idaho's White Cloud Mountains. As secretary of the interior, Andrus discontinued many federal construction projects, especially dams, in an effort to preserve natural resources and save money. His policy decisions often generated opposition among many westerners.

ANTHROPOCENTRISM   Anthropocentrism posits that humans are the center of the ethical world. This worldview is frequently used to justify numerous legal, governmental, and economic decisions concerning the environment. Holding to this ideology, policymakers and others make decisions using human values as the benchmark and without considering environmental ramifications.

ANTHROPOGENIC FIRE   Anthropogenic fires are set intentionally. These fires generally create a pattern, or fire regime, that develops over time. The purposes of anthropogenic fire regimes vary. Diverse groups have used anthropogenic fire as a tool for hunting, to promote various plant species, to remove brush that impeded travel, to reduce insect populations, or to enhance soil for agricultural purposes. In the twentieth century, an ideology of fire suppression has severely reduced anthropogenic fires.

ASSOCIATION OF FOREST SERVICE EMPLOYEES FOR ENVIRONMENTAL ETHICS (AFSEEE)   The Association of Forest Service Employees for Environmental Ethics comprises current and former U.S. Forest Service employees, citizens, activists, and other government employees. A nonprofit organization, AFSEEE emerged in the 1980s as a response to political demands on the Forest Service to increase timber harvests without taking the ecological complexities of the forests into account. The group's mission is to form a system of forest management that is both ecologically and economically sustainable, as well as being ethical. By forming this organization, employees of the Forest Service promoted a less traditional approach to forestry and essentially became an internal watchdog for Forest Service policies that could damage the integrity of forest ecosystems.

AYUNTAMIENTO   The *ayuntamiento* was the Spanish equivalent to a town council. This body was charged with making political decisions in the best interests of the community. A significant responsibility for the *ayuntamiento* was to provide access to water. Spanish law accorded water as a community resource, and no individual or group enjoyed unique privileges to that resource. The *ayuntamiento* used its power from the crown to distribute water in the best interests of the entire community.

BAIRD, SPENCER FULLERTON (1823–1887)   Spencer Fullerton Baird began his career in medicine, but he later became a noted naturalist and zoologist. Appointed as a professor of natural history at Dickinson College in 1846, Baird combined the fields of natural history and zoology. In 1850, Baird served as assistant secretary to Joseph Henry, the head of the Smithsonian. After Henry's death in 1878, Baird was promoted to secretary of the Smithsonian. He became particularly renowned for his work with fish species in both the Atlantic and Pacific oceans, establishing a marine laboratory in Massachusetts. Finally, he conducted work that was central to establishing hatcheries for Pacific salmon in the Far West.

BARI, JUDI (1949–1997)   Judi Bari became a noted environmental, labor, and social justice leader. As a representative of Earth First!, she organized nonviolent protests against corporate logging in California's redwood forests. Almost uniquely, Bari attempted to find common ground among timber workers and radical environmentalists. On 24 May 1990, she was involved in a car bombing that is still surrounded by controversy. She died of breast cancer on 2 March 1997.

BIG BLOWUP (1910 FIRES)   A series of forest fires, now referred to as the Big Blowup, that raged across more than 5 million acres of timberland in Idaho and Montana during August 1910. Those fires helped establish the Forest Service's policy of total fire suppression. In addition, this catastrophic blaze shaped future federal fire suppression policy by allowing Congressional authorization of federal emergency funds for firefighting.

BIOACCUMULATION   Bioaccumulation gathers substances into the tissues of a living organism in increasingly concentrated forms as it moves up through the food web. For instance, scientists and researchers found that fish and other organisms acquired various chemicals and toxic materials from pesticides, fertilizers, and chemical wastes in elevated levels. After accumulating in other organisms, the toxic substances may pass to higher predators, including humans, creating health concerns.

BIOCENTRISM (OR ECOCENTRISM)   Biocentrism argues that all life forms are equally valuable and that the whole is greater than the individual parts. It opposes anthropocentrism. Some environmentalists have used this view in their contention that human decision making must account for ecological precepts and must value all aspects of nature. Radical environmentalists, such as Earth First!, have helped popularize this concept, an outgrowth of deep ecology philosophy.

BIOREGIONALISM   Bioregionalism is a philosophy that espouses living within local ecosystems and their productive capacities in a sustainable way. Advocates define bioregionalism by watersheds, vegetation cover, microclimate, species makeup, and topography. Bioregionalism promotes deeper knowledge of local lands to allow a place-specific strategy of land use.

BIOTA   Biota refers to all the living life forms in a given area. The area can be as large as a forest or as small as a single fallen tree.

BONNEVILLE DAM   The construction of the Bonneville Dam was completed in 1938 as a way to harness the hydroelectric power of the Columbia River, the first dam of its kind built across the Columbia. Before the dam was constructed, numerous American Indian tribes protested because it would interfere with the anadromous fish runs, which constituted a traditional resource. As part of the New Deal legislation, the Bonneville Dam succeeded in its larger goal of providing cheaper power to the region. Critiques of the dam resulted in more attention to salmon runs along the river, and a submerged pipeline was added to the dam to assist in salmon spawning runs.

BOULDER CANYON PROJECT (HOOVER DAM)   On 25 June 1929, the Boulder Canyon Project authorized a federal dam along the Colorado River. The desire to control flooding, generate hydroelectricity, and provide irrigation water precipitated the dam's construction. The dam was named after Herbert Hoover to honor his long involvement in the project. It stands more than 700 feet high; is 1,244 feet across at the top, 660 feet thick at the base, and 45 feet thick at the top; and weighs 6.6 million tons. It was built to have a power-generating capacity of 2.8 million kilowatts and could store up to two years' average flow from the Colorado River in Lake Mead.

BROWER, DAVID ROSS (1912–2000)   David Brower was among the most important conservationists in the twentieth century. A leading contributor to the Sierra Club from the 1930s through the 1960s, Brower became the club's first executive director in 1952 and transformed the club into a very well organized conservation association. Under Brower's leadership, the membership grew and had reached 70,000 members when Brower stepped down in 1969. He successfully led campaigns to block the construction of several dams and prevented extensive tree-cutting. Brower's knack for advertising extended the club's influence outside California for the first time in its history. In 1969 Brower left the club after an internal dispute over a nuclear power plant in Diablo Canyon in California. His militancy in this particular conflict distanced him from the club's more cautious members. After leaving, Brower formed the Friends of Earth and the John Muir Institute.

BROWNFIELD   Brownfields are areas of vacant land contaminated by hazardous substances. In 1995, when the Brownfields Law passed, the Environmental Protection Agency received federal financial assistance to help clean up these areas.

BUNKER HILL MINING COMPANY   From small mining claims established in 1885 along the Coeur d'Alene River in northern Idaho, the Bunker Hill Mining

Company (shortened to the Bunker Hill Company in 1956) formed and produced silver, lead, and zinc until its closure in 1981. The company's work created significant pollutants as by-products to the mining and smelting process, distributing the wastes in the area's rivers, soil, air, and groundwater. After the Environmental Protection Agency and environmental legislation strengthened (and metal prices collapsed by the early 1980s), the Bunker Hill Company closed. The area surrounding the company is a 21-square-mile Superfund site, the largest in the nation.

BURBANK, LUTHER (1849–1926) Luther Burbank was a famed horticulturist from California whose plant-breeding experiments brought him world fame. His work improved the quality of plants and increased the world's food supply. Eventually, Burbank introduced more than 800 new varieties of plants, including more than 200 varieties of fruits, many vegetables, nuts and grains, and hundreds of ornamental flowers.

BUREAU OF LAND MANAGEMENT (BLM) The Bureau of Land Management is a federal agency charged with supervising public lands, especially those reserved for grazing purposes. Created in 1946, this agency united the formerly separate federal General Land Office and U.S. Grazing Service. At first the role of the BLM was ambiguous and complicated because of conflicting laws; however, the 1976 Federal Land Policy and Management Act clarified the BLM's roles and responsibilities. The Federal Land Policy and Management Act acknowledged the inherent value of the remaining public lands and stated that these lands would remain public. Currently, the BLM is responsible for balancing multiple uses for land among public land users, private groups or individuals, and traditional user groups.

CAREY ACT (1894) Originally sponsored by Senator Joseph Maull Carey from Wyoming, this legislation allocated up to 1 million acres of public lands to desert states to develop irrigation and promote agricultural settlement. The act responded to the failure of private companies to provide irrigation to western settlers. The Carey Act, though, was not viewed as a success because few states could adequately fund or develop the irrigation projects. One of the few exceptions to this was Idaho, which accounts for more than 60 percent of all lands irrigated under the Carey Act. The 1902 Newlands Act largely replaced the Carey Act.

CARRYING CAPACITY Carrying capacity is the number of living creatures a specified area can sustain for an indefinite time. It is determined by such factors as water and vegetation. If an area exceeds its carrying capacity, the environment will sustain damage.

CENTRAL VALLEY PROJECT   Begun in 1935 by the U.S. Bureau of Reclamation, this comprehensive water project provided a long-term plan for water use in the Sacramento River basin of California. Dams and canals were constructed to transport water throughout the valley. The Central Valley Project was a multi-purpose project with purposes including flood control, improved navigation, hydroelectric power, irrigation, and municipal and industrial water supplies. The Central Valley Project is also concerned with protecting the Sacramento–San Joaquin River Delta from seawater encroachment and nominally with protecting fish and wildlife.

CHAPARRAL   Chaparral is a type of shrub environment found most often in semiarid regions. It is a relatively imprecise term commonly used for the sclerophyllous brushlands that appear in some form in virtually all Mediterranean climates, such as Southern California. Chaparral typically grow in dense bunches, making them almost impassable by larger animals. Chaparral inhibits most timber growth, but their extensive root system prevents erosion. Chaparral is often pyrophytic; that is, it depends on fire.

CHURCH, FRANK (1924–1984)   While serving as a Democratic senator for Idaho from 1957–1981, Church championed various environmental causes. Most notably, he sponsored the national Wilderness Act in 1964 and the Wild and Scenic Rivers Act in 1968, while also being instrumental in establishing the Hells Canyon and the Sawtooth Wilderness national recreation areas. After his death in 1984, Congress renamed Idaho's River of No Return Wilderness to the Frank Church River of No Return Wilderness to honor his conservation efforts.

CIVILIAN CONSERVATION CORPS (CCC)   Founded in 1933 as part of the New Deal, the Civilian Conservation Corps provided relief to the unemployed. Originally named the Emergency Conservation Work, it was changed to the CCC in 1937. CCC workers were young men between the ages of seventeen and twenty-four years. Over nine years, the CCC hired more than 2 million men at the rate of $30 per month ($22 went directly to the worker's family). The CCC planted more than 2.25 billion trees, built 6 million small and large dams, and constructed countless wildlife shelters. In addition, the workers stocked rivers and lakes with fish and improved beaches and campgrounds for recreation. The CCC also provided labor for the conservation work of other agencies, such as the U.S. Forest Service and the Soil Conservation Service.

CLEAN AIR ACT (1963) In response to air pollution, this legislation set the national policy and created national standards for air pollution control. It was the follow-up amendment to the 1955 Act to Provide Research and Technical Assistance Relating to Air Pollution. The 1955 act granted funds to investigate air pollution issues. With the 1963 act, government agencies at the federal, state, and local levels enjoyed more power to investigate and enforce air pollution violations. Amendments to the Clean Air Act in 1990 increased the authority of states to control and standardize oxide emissions as well as any airborne deposits from surface mining. The amendments also created the Risk Assessment and Management Commission, which was charged with investigating any body that could jeopardize human health by exposure to potentially hazardous substances.

CLEAN WATER RESTORATION ACT (1966) The Clean Water Restoration Act increased the government's role in ameliorating water pollution. The act enlarged state appropriations for local waste-treatment facilities. Additionally, it created a restoration program for rivers and a research fund for water pollution solutions.

COASTAL ZONE MANAGEMENT ACT (1972) Designed to improve land-use planning for ocean-adjacent counties, as well as for any counties that border any of the Great Lakes, the Coastal Zone Management Act created broad guidelines to regulate competing users of beaches, wetlands, and marshlands. It attempted to prevent resource conflicts over wildlife protection, recreation areas, pollution, and transportation. The law grew out of a concern that coastal degradation would result if governments ignored the environmental impact of such projects. The act gave the secretary of commerce the power to provide grants to states to develop programs to safeguard future uses of these coastal areas.

COMMODITY/COMMODIFICATION Commodities are things of commercial value that are traded. The commodification of land, resources, or animals extracts nature from its ecological value and assigns it only market value. For example, a tree becomes lumber to sell to a growing town and an animal's fur becomes a coat.

DAWES ACT *See* General Allotment Act of 1887.

DEEP ECOLOGY Deep ecology is a philosophical viewpoint that takes a more holistic view of nature and places humans within the interconnectedness of nature instead of above it. It is basically a biocentric, or ecocentric, worldview that establishes the moral equality of all things. It combines spirituality, healing practices, and environmentalism. Deep ecology developed as an outgrowth of the countercultural movement of the 1960s and 1970s. Arne Naess, a Norwegian philosopher, coined the term, and William Devall and George Sessions popularized it in the United States. Deep ecology furnished a philosophical foundation to some radical environmentalism by the 1980s.

DDT (DICHLORO-DIPHENYL-TRICHLOROETHANE)   For many years one of the most widely used pesticides in the United States, DDT was first synthesized in 1874. Its effectiveness as an insecticide was only discovered in 1939. During World War II, the United States produced large quantities of DDT to control vector-borne diseases, such as typhus and malaria. After 1945, domestic use of DDT on farms and forests became widespread. During the next thirty years, approximately 1.35 billion pounds of DDT was used domestically. Although scientists voiced warnings as early as the mid-1940s, it was Rachel Carson's 1962 book, *Silent Spring*, that stimulated widespread public concern over use of the chemical. After 1959, DDT usage in the United States declined greatly, dropping from a peak of approximately 80 million pounds in that year to just under 12 million pounds in the early 1970s.

DESERT LAND ACT (1877)   This 1877 legislation was largely modeled after the earlier Homestead Act. It sanctioned the sale of 640-acre lots at twenty-five cents an acre for areas that were deemed without timber or minerals or uncultivable without irrigation. The low price was meant to encourage settlement in arid regions of the American West. To gain the land, a person had to live on the land and irrigate it without further federal assistance. The act mostly failed because few individuals could afford to irrigate the land. Subsequent legislation, such as the 1894 Carey Act and the 1902 Newlands Act, sought to fix this problem by providing federal aid for irrigation.

DISTURBANCE ECOLOGY   This particular subfield of ecology examines natural events and their effects on ecological systems. These disturbances include fires, tornadoes, insects, and diseases. Disturbance ecology recognizes the dynamic nature of natural systems and seeks to comprehend how ecosystems respond to such disturbances. It can also incorporate human-induced changes, something traditional ecology ignored for a long time.

DOUGLAS, WILLIAM O. (1898–1980)   As an associate justice of the U.S. Supreme Court from 1939 to 1975, William O. Douglas earned a reputation as a staunch supporter of the environment. Growing up in Washington State, Douglas spent time hiking the Cascade Mountains. That experience influenced his later judicial and extrajudicial career. He organized numerous protests, specifically against proposed roadways through wilderness areas. His books, including *A Wilderness Bill of Rights* (1965) and *My Wilderness: The Pacific West* (1960), expressed his views of nature, and judicial decisions such as *Udall v. Federal Power Commission* (1967) changed the way the Supreme Court considered environmental issues. After his death, a wilderness area in the Cascade Mountains honored his work as the William O. Douglas Wilderness Area.

EARTH FIRST!   Formed in the 1980s, Earth First! was the nation's best-known radical environmental group. From its roots in the Southwest, Earth First! grew rapidly in response to forest destruction in the Northwest. This group occasionally used sabotage techniques (called "ecotage"), such as damaging bulldozers by pouring sugar into their gas tanks and spiking trees. While their methods have elicited sharp controversy and legal sanctions, Earth First! has brought attention to some of the environmental impacts of corporations and extractive industries, especially to issues such as old-growth forest protection in the Far West.

ECOCENTRISM   *See* Biocentrism.

ECOLOGY   Ecology studies the relationships between living things and their environments. It has become the predominant model for managing forests, rivers, and other parts of the landscape. By using ecology, government agencies and land managers have tried to understand more comprehensively how their projects affect living and nonliving entities within their habitats.

ECOSYSTEM   An ecosystem encompasses not only the living and nonliving aspects of a defined place but also the interactions among them. Because of the interconnectedness of an ecosystem, all parts in some manner rely on each other and are affected by each other. Thus, if one aspect is changed, the ramifications will be felt by all the other components of the system. An ecosystem was traditionally viewed as self-containing and self-sustaining. Some attribute the ecosystem concept to noted ecologist Arthur Tansley, who first used the term in an academic paper in 1935, while others credit it to Raymond Lindeman, who used the term in 1942. But in 1953 Eugene P. Odum made it widespread when he applied the concept in his book *Fundamentals of Ecology* to explore connections on a large scale among various systems such as watersheds, plants, animals, humans, and climate as opposed to simply exploring each system separately. Until this time most scientists studied individual systems rather than the interactions of multiple factors, especially humans.

ECOSYSTEM MANAGEMENT   *See* New Perspectives.

ECOTAGE   Combining the prefix "eco" with the word "sabotage," ecotage refers to acts of vandalism usually aimed at industries or companies that radical environmentalists believe harm the environment. Sometimes referred to as monkeywrenching or, more recently, ecoterrorism, ecotage is a practice used by radical environmentalist groups such as Earth First! or the Earth Liberation Front. Acts of ecotage range from spiking trees to uprooting genetically altered vegetables to burning down office buildings.

EL NIÑO–SOUTHERN OSCILLATION (ENSO)   El Niño–Southern Oscillation (ENSO) is an ocean and atmospheric condition originating in the western and central Pacific Ocean. It influences weather and the environment for North and South America. Typically, the Pacific Ocean has warm water along continental coastlines in Central and North America. ENSO occurs every few years when the warmest water moves from the western Pacific further south and east to the coastlines of both American continents. This transfer of warm water affects the atmosphere, often creating thunderstorms. After thunderstorms, the moisture and wind move into the upper atmosphere thus changing the jet streams. The jet streams, in turn, shape storms and weather across the Americas. Because it changes temperatures and other conditions, ENSO can deeply affect fish populations and climate patterns.

ENDANGERED SPECIES CONSERVATION ACT (1973)   The Endangered Species Conservation Act guides federal policies for assessing endangered species and directing protection for endangered and threatened species. A fundamental basis of this law is its principle that species ought to enjoy the right to continued existence. Passed in 1973, this act has come under a great deal of criticism for placing wildlife, fish, and plants ahead of economic development, especially as environmentalists have used it to save broad patches of habitat. Environmentalists have successfully used this law to stop or slow economic activity such as logging and irrigating to protect habitat for the northern spotted owl and varieties of Pacific salmon. Its administration is generally slow, costly, and controversial. Nevertheless, it remains popular among broad segments of the American populace.

ENVIRONMENTAL IMPACT STATEMENTS   The 1969 National Environmental Policy Act required all federal agencies to submit environmental impact statements summarizing possible effects of federal projects. By forcing agencies to consider ecological impacts, the law introduced a new level of planning for various projects. In addition, an environmental impact statement is opened for a period of public comment, giving the law a democratic element that allows the public to challenge proposed plans.

ENVIRONMENTAL JUSTICE/ENVIRONMENTAL JUSTICE MOVEMENT (EJM)   The environmental justice movement began in the 1960s as a response to the growing awareness of environmental damage in regions typically inhabited by less powerful and represented segments of U.S. society. This damage was traditionally inflicted by companies, corporations, or industries who thought they could do so because the groups they were affecting had little or no legal recourse or simply would not take the time to protest. The term "environmental justice" refers to the belief that groups needed a collective voice to be heard in decisions concerning their neighborhoods. Common issues are locating garbage dumps or toxic waste sites in poor neighborhoods. The EJM tends to focus on the grassroots and offers an alternative to both mainstream and radical environmentalism. Moreover, EJM leaders are often women, unlike other environmental organizations. This term is closely tied to "environmental racism."

ENVIRONMENTAL PROTECTION AGENCY (EPA)   The U.S. government established the EPA in 1970, largely in response to public sentiment for environmental regulations. Its mission focuses on preserving human health and maintaining a cleaner and safer environment. The EPA primarily serves as the federal government's environmental research and enforcement tool. The executive agency works to implement national standards for various environmental laws and programs (such as the Clean Air Act), ensures that states and industries comply with those standards, and researches environmental problems.

ENVIRONMENTAL RACISM   Environmental racism refers to the bias that members of ethnic and racial minorities face from both companies and the federal government in environmental matters. Since the 1960s, more attention has been given to this form of racism as studies have shown that these groups face more health and safety issues from environmental damage and poor environmental practices. Further, environmental laws and policies have been shown to protect whites more than other groups.

FEDERAL AIR POLLUTION CONTROL ACT (1955)   The Federal Air Pollution Control Act was the first piece of federal legislation to recognize air pollution as a problem and concern for human health. It allocated money for research, but did little to actually prevent or curb air pollution other than to recognize that it is harmful to humans, livestock, agriculture, and the environment. It also lacked any enforcement mechanism.

FEDERAL LAND POLICY AND MANAGEMENT ACT (FLPMA) (1976)   The Federal Land Policy and Management Act (FLPMA) clarified the federal policy for multiple use of any public lands under the Bureau of Land Management's jurisdiction, some 450 million acres. This land was not to be sold for private use, although it could be used for grazing. The act required the Bureau of Land Management to manage public lands using multiple-use and sustained-yield strategies to ensure resources for both the present and the future, establishing policies that protect, develop, and enhance public lands for economic, recreational, and aesthetic benefits.

FEDERAL POWER COMMISSION (FPC)   Established in 1920 by the Federal Water Power Act, the Federal Power Commission served as an independent agency licensing hydroelectric projects on the nation's waterways. It evaluated water development plans and regulated electric utilities between states. The FPC was the agency responsible for approving proposed hydroelectric dams. In 1977, the commission was merged with the Federal Energy Regulatory Commission.

FEDERAL WATER POLLUTION ACT AMENDMENT OF 1972   Designed to both limit sewage discharge and establish standards for water quality, the Federal Water Pollution Act Amendment of 1972 sought to eliminate all water pollution discharges by 1985. The act created the National Water Quality Commission to evaluate the development of programs devoted to water quality.

FIRE REGIME   Fire regimes result from patterns of burning. Regimes can be influenced or caused by anthropogenic burning, but they also exist naturally within ecosystems. They depend on a combination of the physical, climatic, and biological environment within the ecosystem, as well as human practices.

FLOOD CONTROL ACT (1917)   Officially called the Ransdell-Humphreys Act, this legislation gave the secretary of war the authority to spend up to $10 million each year to prevent flooding of both the Mississippi and Sacramento rivers. The failure of the act was apparent when, in 1928, the Mississippi River overflowed to disastrous results. Subsequent Flood Control Acts over the course of the twentieth century attempted to invest in engineering to prevent floods.

FOREST ECOSYSTEM MANAGEMENT ASSESSMENT TEAM (FEMAT)   In 1993, President Bill Clinton instructed the Forest Ecosystem Management Assessment Team to create a federal management plan that recognized the need to sustain the old-growth forests and economy in the Pacific Northwest. Members of the team came from several government resource management agencies, including the Forest Service, Fish and Wildlife Service, National Marine Fisheries Service, National Park Service, Bureau of Land Management, and Environmental Protection Agency. The team ultimately produced a report in 1993 that detailed ten options for federal forest management in the region. With a few changes, Option 9 became the legal management plan in 1994.

FORSMAN, ERIC   A bulk of the credit for starting the northern-spotted-owl debate goes to Eric Forsman. In 1972, as a graduate student at Oregon State University, he began studying the owl. His research revealed that the owl's population was declining. Forsman also discovered that the owl required large tracts of old-growth forests for adequate habitat. After the Forest Service turned away his efforts to have owl habitat preserved, the northern spotted owl debate heated up and later turned into a national debate.

FRANKLIN, JERRY (1936–PRESENT)   Jerry Franklin is a leading expert on sustainable forest management whose strategies center on science. His work stresses the complexities of forest ecosystems where he argues against clear-cutting and encourages leaving forest debris and numerous trees behind when logging. His approach is usually called New Forestry and has been seen as controversial by traditional timber managers who find that it prevents much harvesting and by radical environmentalists who do not like that the approach allows logging under certain circumstances. In 1959, Franklin became a research forester for the Forest Service, and later moved on to become a professor at Oregon State University and then at the University of Washington. Franklin's work in the 1980s and 1990s put him squarely within the debates over old-growth forests and the northern spotted owl.

FRIENDS OF THE THREE SISTERS   In 1954, members of a Eugene, Oregon–based outdoor group called the Obsidians organized a committee called the Friends of the Three Sisters to challenge a proposal that would have allowed logging in the Three Sisters Wilderness Area in central Oregon. The members of the committee worked to develop valid scientific arguments against the proposed logging and even organized hiking trips into the wilderness area, hoping to influence government officials and the public. They lost their campaign but continued efforts for further preservation while inspiring other local groups to organize on behalf of local wilderness areas.

GENERAL ALLOTMENT ACT OF 1887 (DAWES ACT)   Sponsored by Senator Henry Dawes of Massachusetts, the General Allotment Act authorized the president to break up communally held Indian reservations, to distribute the lands to individual Indians, and to sell the "surplus" land to non-Indians. The federal government held titles to these individual allotments in trust for 25 years, at which point the Indians received full title to it and U.S. citizenship. The nominal goal of the Dawes Act was to turn American Indians into small farmers, but its main effect was to vastly reduce Indian land holdings. Indeed, the act diminished Native lands from 138 million acres when the act was passed to 52 million acres when it was repealed in 1934.

GENERAL MINING LAW (1872)   The General Mining Law was designed to protect miners, specifically for patent rights. It gave placer miners the right to explore and purchase mining rights in the public domain with no benefit to taxpayers. Sponsored by Senator William Morris Stewart of Nevada and signed into law by President Ulysses S. Grant, this law was specifically intended for western states, effectively allowed for free mining in the West, and still guides mining policy in the West by allowing mining companies to purchase land at 1872 prices. In addition, the law does not contain any environmental requirements allowing miners and mining companies to extract resources and leave behind wastes, laying the burden of cleanup on the public's shoulders. Nonprofit groups and environmentally minded government officials still work to change this archaic law.

GRAND COULEE DAM   Begun in 1933 and completed in 1941, the Grand Coulee Dam is located on the Columbia River in Washington State. At the time, it was the largest dam in the United States. The Grand Coulee Dam was built for three different purposes: irrigation, hydroelectric power, and flood control. While irrigation may have been a leading factor in the initial construction, the timing on the completion of the dam shifted the focus, and irrigation came much later. Electricity became much more important as war industries grew in the Northwest to help with World War II. Currently, the Grand Coulee Dam supplies more than 6.5 million kilowatts of power to the region. The Grand Coulee Dam did not accommodate fish passage, however, and thus effectively wiped out the salmon population upstream.

GREEN RUN INCIDENT   On 2 December 1949 officials at the Hanford Nuclear Reservation intentionally released about 8,000 curies of radioactive iodine into the air without notifying or evacuating the public downwind or nearby. Compared with the far more famous Three Mile Island meltdown, 500 times more radioactive material was released from Hanford in the Green Run incident, yet residents nearby Three Mile Island were informed and evacuated, whereas residents of Hanford were not.

HANFORD, WASHINGTON   Hanford was the site of a nuclear facility in south-central Washington that went online with its first reactor in 1944 as part of the Manhattan Project. The site produced plutonium for nuclear weapons and performed energy research and development. With the end of the Cold War in 1991, weapon development ceased. The area houses abundant radioactive wastes that officials must attempt to clean up at the cost of billions of dollars. Chemical wastes leaked into the groundwater and Columbia River, spreading to other Northwest residents.

HETCH HETCHY VALLEY CONTROVERSY   Whether to build a dam in Yosemite National Park's Hetch Hetchy Valley developed into a prominent national controversy. Just past the turn of the twentieth century, San Francisco needed more water for its growing population and turned to this valley with the Tuolumne River as the city's choice for a municipally controlled dam site. Conservationists disagreed about the merits of the program. Historians have focused on the split within the movement between those represented by Gifford Pinchot who desired careful development of resources and John Muir who wanted sacred places like national parks to be preserved for tourists. Both sides tried to persuade President Theodore Roosevelt and other politicians. Finally, after national campaigns, President Woodrow Wilson supported Congress, which in 1913 passed a bill that allowed for the dam to be built within the valley. Although it is usually depicted as a battle between conservation and preservation, at the time, the controversy largely focused on the fact that San Francisco wanted a public water supply to counter monopolistic private water suppliers.

HODEL, DONALD PAUL (1935–PRESENT)   Donald Hodel, a Republican, served as both secretary of energy and, later, secretary of the interior under Ronald Reagan in the 1980s. He received his law degree from the University of Oregon Law School in 1960 and served as a top administrator in Bonneville Power Administration, the federal agency that headed development of public power in the Northwest. This background prepared him for his later political positions. Hodel believed nuclear power was the most efficient means to supply electricity to the nation, a position that led to much environmentalist resentment. Upon becoming secretary of the interior, Hodel announced the policy that any unused dirt roads or footpaths would be considered right-of-ways. This so-called Hodel policy threatened the overall management policy for federal lands, as it obstructed future allocation of wilderness areas. When Bruce Babbitt became secretary of interior for Bill Clinton, he rescinded the Hodel policy.

HOMESTEAD ACT (1862)   President Abraham Lincoln signed the Homestead Act in 1862 as part of the Republican Party's goal of promoting Western settlement. The law provided settlers with 160 acres of the public domain. In exchange for the land, settlers had only to pay an $18 filing fee, build a house on the land, and live there for five continuous years. To be eligible to receive this land, the settler had to be the head of a household and at least twenty-one years old. Immigrants, former slaves, and small farmers took advantage of this act, eventually claiming more than 270 million acres of public land. Much of the land, however, quickly found its way into corporate hands.

HOOVER DAM   *See* Boulder Canyon Project.

INDICATOR SPECIES   Scientists and land managers use indicator species to determine the biological quality of large ecosystems. They are typically used as an early warning of pollutants or other changes in an ecosystem. Indicator species are a practical tool for managers who cannot monitor all life within large ecosystems. By designating species that are central to functioning ecosystems, they can observe larger patterns of environmental health. The northern spotted owl has become the most famous and controversial indicator species, used for West Coast old-growth forests.

INTEGRATED PEST MANAGEMENT   Integrated pest management (IPM) constitutes an environmentally sensitive approach to pest management that does not rely on chemicals. IPM programs use information on the life cycles of pests and their interaction with the environment, such as natural predators. This information, in combination with available pest control methods, is used to manage pest damage by economical means and with minimal hazards to people, property, and the environment.

INTENSIVE FORESTRY   Intensive, or high yield, forestry has been seen by many as a way to produce more timber on less land to meet the growing demand for wood and paper products. Through a combination of nursery management, site preparation, weed control, fertilization, and other forestry techniques, foresters intensively manage forests with chemicals and technology. Advocates see it as a way to reserve some lands just for timber production, allowing wilderness areas to be preserved for noncommercial uses. Critics find intensive forestry an overly optimistic approach that does not adequately understand or value the biological functions of forests.

INTERMIX FIRE (URBAN-WILDLAND INTERFACE FIRE) The intermix fire on the border between suburban and rural or wild areas has become an archetypal fire problem in recent decades. These intermix fires are caused as urban areas expand into combustible wild or rural habitat. Many catastrophic fires occur along these interface areas because of years of fire suppression coupled with expanding, urban development. These fires cause massive property damage and produce high-profile attention; they are among the most costly fires.

INVASIVE SPECIES Invasive species have been introduced, either intentionally or not, to a region they did not originally inhabit. Invasive species often came with newly arriving human populations. Often, these new species threaten native species and become classified as pests or weeds that pose a distinct danger to the preexisting ecosystems.

IRRIGATION DISTRICT Irrigation districts were formed as a result of the 1887 Wright Act in California. A community with a two-thirds majority could form an irrigation district that functioned as a mini-legislature with the ability to tax and condemn property. The districts would raise funds to construct the first public irrigation projects in California, and the district held the water rights. Irrigation districts sought to support small landholdings to compete with large industrial farms. They increased specialty crop farming and slightly decreased average farm sizes. However, legal challenges and costly expenses caused most districts to fail.

IRRUPTION Irruption is a rapid increase in animal or plant populations. While scientists cannot always determine causes of irruptions, there are often human disturbances or other environmental changes that affect food supplies and contribute to irruptions. Typically, the species that experiences the irruption follows it with a population decrease, as food supplies or other resources give out in response to the increase.

KEYSTONE SPECIES Keystone species are integral to the survival of the other species within a given ecosystem. Thus, the removal of a keystone species dramatically alters the entire composition of an ecosystem. For example, removing a predator can allow other species to increase, sometimes exceeding an ecosystem's carrying capacity. Similarly, eliminating a keystone species that is prey can decrease other species' ability to survive.

KLAMATH RECLAMATION PROJECT   Developed by the Bureau of Reclamation, the Klamath Reclamation Project is a large irrigation system consisting of several dams that remove water from Upper Klamath Lake and the Klamath River in Oregon and transport it to the Klamath Basin turning it into a giant agricultural region. The project began in 1906 and construction finished in the 1920s. The project drained lakes and wetlands, eliminating important waterfowl habitat on the Pacific flyway. In recent years, concern over that declining habitat as well as the project's contribution to declining salmon has produced substantial political conflict. Federal agencies have had to shut off irrigation water for farmers to protect endangered fish, an action that has adversely affected farmers and pleased environmentalists.

LAND AND WATER CONSERVATION FUND ACT (1965)   In 1962, President John F. Kennedy called for a fund to pay for land acquisition and facility management for recreation areas. Established in 1965, the Land and Water Conservation Fund Act was funded by admission fees for federal areas, public land sales, and taxes on fuel for motorboats. States receive 60 percent of the funds and the federal government uses the remainder.

LUX V. HAGGIN (1886)   This 1886 decision established water rights in California, creating the so-called California Doctrine. The court case pitted two large landowners against each other and sought to resolve whether prior appropriation or riparian water rights would have precedence. The California Supreme Court decided that riparian rights were inherent in private lands. However, prior appropriation rights superseded riparian rights if the water user claimed the water rights before the riparian owner. Thus, timing mattered significantly.

MAGNUSON, WARREN GRANT (1905–1989)   Warren G. Magnuson of Washington served in the House of Representatives beginning in 1936, moving to the Senate in 1944. He became well known for, among other things, his activism in the arena of fish protection and environmentalism. He sponsored both the Safe Drinking Water Act of 1972 as well as the 1976 Toxic Substances Control Act. The Magnuson Fishery Management and Conservation Act of 1976 is his most well-known legacy, which protected coastal and anadromous species through an established conservation zone for fish management. It also instituted Regional Fishery Management Councils to supervise fishery management plans and maximize annual yields.

MARINE MAMMAL PROTECTION ACT (1972)   The Marine Mammal Protection Act was enacted to prevent extinction of marine mammals by prohibiting unsustainable hunting. The act applied to U.S. citizens as well as anyone within U.S. territorial waters. Furthermore, the law prohibited any importation of mammals or their by-products. The act protects any mammal classified as an endangered species from hunting except those used by subsistence hunting by Native American tribes.

MAXIMUM SUSTAINED YIELD Maximum sustained yield is a tool used by managers and economists. Theoretically, it establishes the highest yield that can be continuously taken annually from a stock under typical environmental conditions without significantly affecting the reproduction process. The maximum sustained yield has been used to determine allowable fish harvests or timber sales. It has been a notoriously poor measure of ecological reality.

MCCALL, TOM (1913–1983) Tom McCall became the governor of Oregon in 1966 and was instrumental in creating several state environmental programs. Before becoming governor, McCall was a noted journalist whose documentary, *Pollution in Paradise*, about pollution in the Willamette River, encouraged remedial action. During his tenure until 1975, the Department of Environmental Quality formed to develop plans to curb pollution, and the Land Conservation and Development Commission was established to manage the state's revolutionary land-use planning. McCall also urged the passage of the Bottle Bill, which provided deposits on drink containers in further efforts to reduce pollution and promote recycling. Finally, he presided over the state when the legislature asserted state ownership of the remaining beaches of Oregon to prevent overdevelopment. Although McCall received much of the credit for these strong environmental measures, he built on the foundation of his predecessor, Bob Staub.

MCKAY, JAMES DOUGLAS (1893–1959) Douglas McKay from Oregon became secretary of interior in 1952. Among other things, he proposed dam development along the Snake River near Hells Canyon. Critical conservationists called him "Give Away McKay" for his penchant for promoting state and private development of public resources.

MCNARY, CHARLES LINZA Charles McNary, a Republican senator from Oregon from 1918 to 1944, was influential in creating various forest legislation. He was instrumental in passing the Clarke-McNary Reforestation Act of 1924. This act required cooperation between the states and the secretary of agriculture to both procure tree and plant seeds for forest and to restock cutover or burned areas of forest. He also cosponsored the Sweeney-McNary Forest Research Act of 1928, which extended federal forestry research programs. In 1928, he helped pass the Woodruff-McNary Act, legislation that authorized federal authorities to create new national forests.

MISSION 66   After World War II, the National Park Service promoted visitation through Mission 66. Designed to improve the deterioration in park conditions and facilities, Mission 66 sought to improve the infrastructure to support the large influx of tourists. The project began in 1956 and ended by meeting its completion goal in 1966, thus, giving it the name Mission 66. During that period the National Park Service spent more than $1 billion on improvements to roads, hotels, and other buildings and resources. Planners for the project also developed the idea of the visitor center and built about 100 new visitor centers during this decade.

MONO LAKE COMMITTEE   The Mono Lake Committee started in 1978 as a response to the degradation of the Mono Basin ecosystem in northern California as a result of the city of Los Angeles continuing to transport and use the lake's water. Since its inception, members of this nonprofit organization have worked to protect and restore the ecosystem in the Mono Basin. They have focused in particular on raising the water level. In 1994, the work from this organization paid off and helped produce Decision 1631 from the California State Water Resources Control Board that set permanent stream flows within the basin and a water level of 6,392 feet for Mono Lake. These efforts will not bring the lake back to its original level but will significantly protect and restore a portion of the area damaged by Los Angeles's water imperialism.

MOTHERS OF EAST LOS ANGELES   Composed primarily of Mexican American women, Mothers of East Los Angeles originally formed in 1985 to oppose the placement of a state prison in East Los Angeles near the multicultural Boyle Heights neighborhood. This grassroots effort succeeded, and the group continued to oppose development projects that would harm the environment and quality of life for residents in East Los Angeles and adjacent communities, such as siting toxic waste incinerators in the neighborhood. It is an excellent example of part of the environmental justice movement.

MUIR, JOHN (1838–1914)   John Muir is known as one of the founding American conservationists. His books include *A Thousand-Mile Walk to the Gulf*, in which he declared his devotion to the natural world. *The Mountains of California* and *My First Summer in the Sierra* are considered classics of California history and nature writing. In 1892, he helped found the Sierra Club and became its first president. Muir's writing and later lobbying, such as for Hetch Hetchy Valley and national parks generally, brought national attention to many places deserving protection. He viewed nature, especially seeming wild places, as sanctuaries of God's creation and urged their salvation as an almost religious cause.

MULHOLLAND, WILLIAM (1855–1935)   William Mulholland was an Irish immigrant who arrived in Los Angeles in 1877. After working for a private water company, he became superintendent of the Los Angeles Water Company in 1902 after the city asserted municipal ownership of the water supply. Mullholland was instrumental in gaining control over Owens Valley water and overseeing the construction of the aqueduct to Los Angeles. His efforts at water planning for Los Angeles were imperialistic, successful, and widely supported in the metropolis. He supervised the construction of the Saint Francis Dam from 1924–1926. The dam ultimately failed and killed more than 400 people, a fate that devastated Mulholland personally and professionally. In response, he was replaced as head of the water company in 1929. Perhaps no other individual is more responsible for the shape of Los Angeles than Mulholland.

MULTIPLE-USE SUSTAINED-YIELD ACT OF 1960 (MUSY)   The Multiple-Use Sustained-Yield Act establishes Forest Service policy as managing national forests for outdoor recreation, timber, rangelands, watersheds, and wildlife and fish habitats. MUSY marked the first time the Forest Service intentionally put outdoor recreation alongside the traditional uses of the national forests and resulted in some areas being set aside for recreational use instead of profit. However, the Forest Service largely used the vagueness of the act to continue to produce timber at high levels.

MUTUAL WATER COMPANY   Mutual water companies were cooperatives owned privately by local residents for the purpose of providing cheap water supplies to members. First established in California, these companies challenged prevailing water rights doctrines and practices in the state.

NATIONAL AMBIENT AIR QUALITY STANDARDS (NAAQS)   Required by the Clean Air Act, National Ambient Air Quality Standards are levels the Environmental Protection Agency sets and enforces for various pollutants from an assortment of sources to protect public health and the environment. The Clean Air Act established two types of these standards. Primary standards set restrictions to safeguard public health with particular emphasis on children, the elderly, and those with special health problems. Secondary standards set restrictions to safeguard public well being, including limits to guard against vision impairment and damages to livestock and agriculture.

NATIONAL ENVIRONMENTAL POLICY ACT (NEPA)   The National Environmental Policy Act was signed 1 January 1970 by President Richard Nixon in response to the environmental debates of the 1960s. The act provided a national policy to guide human interactions with the environment, encouraged efforts that would effectively end or prevent environmental damage, and furthered understanding of both ecological systems and natural resources. Among other things, NEPA created the Council on Environmental Quality, an advisory board to the president that was designed to sift through the activities of various government agencies and provide cogent policy advice. Furthermore, and perhaps most revolutionary, NEPA required the preparation of environmental impact statements for any federal agency that has a proposed development project and required federal agencies to provide summaries of how their projects could potentially affect the environment with suggestions as to how to minimize these effects. After the publication of an environmental impact statement, a public comment period opens during which citizens and organizations may argue for or against projects. It is arguably the most important federal environmental policy.

NATIONAL FOREST MANAGEMENT ACT OF 1976 (NFMA)   The National Forest Management Act required the secretary of agriculture to evaluate and manage national forests using an ecologically based approach and abiding by sustained-yield and multiple-use standards. The act defined sustained-yield as being nondeclining and required that wildlife populations be properly managed, giving significant gains to environmental advocates. However, the act also calls for proper management of goods and services produced from the nation's forests, keeping traditional forestry values intact. The law in large part was inspired by controversies over clear-cutting in national forests.

NATIONAL PARK SERVICE (1916) (NPS)   The Army had long administered national parks, but they had deteriorated into a dilapidated state. The National Park Service was created in response to these conditions and in reponse to the Hetch Hetchy controversy. After the lobbying of Secretary of the Interior Franklin K. Lane and Stephen T. Mather and as part of the Progressive reform era, Congress decided a specific government agency was needed to oversee the parks. Mather became the first director in 1916. His belief that beautiful scenery should be the true mark of a nation's success guided the service's strategies. The NPS was one of the vehicles that allowed Americans to see the importance and uniqueness of their natural environment. The NPS largely focused on the West and the majority of federally protected lands were there. The early national parks offered beautiful landscapes with few other economic alternatives. Throughout the twentieth century, though, the NPS contended with the unintended side effects of creating tourist-friendly park regions, such as overuse, pollution, and litter. As the landscape became more recognized as a national treasure, the role of the NPS expanded to protect the land more. Currently, NPS oversees 355 sites, including national monuments that possess natural, historical, or cultural value.

NATIONAL TRAILS SYSTEM ACT (1968)   Congress created the National Trails System Act to promote outdoor recreation near urban areas and support the maintenance of hiking and horseback trails for recreational use. The act designated three levels of trails. A National Scenic Trail can only be designated as such by Congress because it typically crosses state lines and is several hundred miles long. Scenic Trails maintain an absolute prohibition of any type of motorized vehicle or mountain bike. National Recreation Trails allow some types of motorized vehicles and mountain bikes. Side and Connecting Trails are much shorter than the other two classes and serve to provide access points to the other two. Any of these trails may pass over any federal or nonfederal land; however, in the case of both National Recreation Trails and Side and Connecting Trails, agreements with any affected landowners must be reached before construction can begin.

NEUBERGER, RICHARD (1912–1960)   Richard Neuberger served in the U.S. Senate as an Oregon Democrat from 1955 until his death in 1960. Before serving in Congress he was a nationally noted journalist and served in the Oregon state legislature. Throughout his political career, he supported liberal causes and environmental measures. Most notably, he pushed for the conservation of land in the Klamath Basin and protection of unique dunes on the Oregon coast that later became the Oregon Dunes National Recreation Area. As a conservationist, though, he believed technical solutions could mitigate problems caused by dams, which he tirelessly supported on the Columbia and Snake rivers.

NEW FORESTRY   New Forestry is a forest management strategy that aims to balance the needs of industry, wildlife, and communities sustainably. It calls for a focus on ecosystem structure and function when devising cutting and management plans. This new approach is oriented toward maintaining biodiversity and soil productivity with an emphasis on recognizing the complexities of the forest ecosystem.

NEWLANDS (OR RECLAMATION) ACT (1902)   The Newlands Act was designed to improve upon the 1894 Carey Act, which excluded federal financial assistance for irrigation projects in arid or semiarid regions. This 1902 legislation created the U.S. Reclamation Service (later called the U.S. Bureau of Reclamation) to oversee reclamation and irrigation projects funded from public land sales in Western states. More than 3 million acres received irrigation within the first four years of the act's inception. Annual water fees for users augmented the funds generated by public land sales. The purpose was to create irrigation projects in arid regions, ensuring the future repayment from settlers. This success in creating many dams and irrigation works also failed because most farmers could never repay the federal investment, as had been the original goal. Many western regions still rely on the government to at least partially help maintain these large-scale irrigation projects.

1910 FIRES   *See* Big Blowup.

NEW PERSPECTIVES, OR ECOSYSTEM MANAGEMENT   New Perspectives became a guiding Forest Service program in 1992. It was an experimental program to test different land management practices. Also called ecosystem management, the program responded to demands to balance traditional production of timber and other goods with sustaining a healthy ecosystem. Occasionally vague, the policy has been left open to interpretation, but essentially New Perspectives tries to manage forest ecosystems from ecological principles by incorporating goals based in science as well as political and economic values. New Perspectives helped include broader ecosystem management into official Forest Service policy, showing the influence of environmentalists in challenging the Forest Service's traditional priorities.

NORTHERN SPOTTED OWL (STRIX OCCIDENTALIS CAURINA)   The northern spotted owl symbolized the forest management debate starting in the 1980s and continuing today. The owl prefers a habitat consisting of large tracts of old-growth forests in the Far West. Environmentalists pushed for legal protection for the owl through provisions in the Endangered Species Act to protect those forests. The debate quickly became a media event pitting the smallish, brown owl and environmentalists against timber communities and jobs. In 1990, federal officials listed the owl as threatened, thereby protecting its habitat.

OPTION 9   In 1994, Option 9 from the Forest Ecosystem Management Assessment Team report became, with few adjustments, the official federal plan for Northwest forests. The plan opened one-fifth of old-growth forests to logging and allowed the remaining portion to be open for thinning. Roadless areas received no protection, watersheds received limited protection, and the logging production levels were lower than anticipated. As such, the plan disappointed virtually all sides of the forest debate except politicians. Environmentalists believed it did not push protections far enough, and timber interests believed it did not open forests for enough timber production.

ORGANIC ACT OF 1897 The Organic Act tried to establish priorities for the newly created federal forest reserves. It emphasized a constant supply of timber to meet the nation's needs. It guided the Forest Service until the National Forest Management Act replaced it in 1976.

ORGANIC FARMING Organic farming produces food and other agricultural products without most synthetic chemicals common in fertilizers and pesticides used in conventional agriculture. Organic farmers use nonchemical means to prevent pests and stimulate plant growth. It arose from the reform spirit of the 1960s but has become dominated by some large firms, especially in California, that do not embrace all aspects of that reform vision.

OUTDOOR RECREATION RESOURCES REVIEW COMMISSION (1958–1962) Established to compile, review, and study all data concerning outdoor recreation needs and prospects. The commission recommended roles that federal, state, and regional governments, as well as nongovernmental businesses, should provide for outdoor recreation. The commission's findings resulted in the establishment of the Bureau of Outdoor Recreation, and later the President's Council on Recreation and Natural Beauty (1966–1968).

PACIFIC GAS & ELECTRIC COMPANY (PG&E) The Pacific Gas & Electric Company, which was formed in 1905, served as the utility that provided much of California, especially the north, with electricity and natural gas. The company operates natural gas power plants, hydroelectric plants, and a nuclear power plant. Much of the energy required to generate electricity for its customers comes from fossil fuels, which has caused health problems for residents living near some of the company's facilities and multiple legal battles. PG&E has also faced public opposition to plans to build a nuclear power plant at Bodega Bay. Most famously, residents of the town of Hinkley, California, claimed that PG&E polluted drinking water with chromium resulting in a variety of serious health problems and won a settlement of more than $300 million after *Anderson v. Pacific Gas and Electric Co.* (1993). This issue was dramatized in the feature film *Erin Brockovich.*

PACIFIC FLYWAY   The Pacific Flyway is one of four major North American flyways, or migration routes, for birds. It encompasses the western Arctic and the Pacific Coast and Rocky Mountain regions of Canada, the United States, and Mexico. The flyway continues south to merge with other flyways. Various migratory birds and waterfowl use these routes to travel between their breeding and wintering grounds. The draining of wetlands in the Pacific Flyway along the West Coast has compromised its ability to support the continued existence of these birds.

PACIFIC RIM OF FIRE   Located around the edge of the Pacific Ocean, the Pacific Rim (or Ring) of Fire is a ring of volcanoes. The area is a U-shaped arc, stretching north along the western coasts of both South and North America and then west and south along the seaboards of Asia and Australia. Collisions of various tectonic plates result in the concentration of volcanoes, as well as intense seismic activity. Volcanoes in the Cascade Mountains, including Mount Saint Helens and Mount Lassen, make up part of the Pacific Rim of Fire.

PEMMICAN   Pemmican is a type of American Indian food prepared from dried meat strips, which are then pounded into a paste and mixed with fat and berries. Pemmican stored well, allowing for easy trade and preservation of food for later consumption. Although often associated with bison meat, Northwest tribes prepared pemmican with salmon.

PEOPLE V. TRUCKEE LUMBER (1897)   This 1897 court case pitted the state of California against the Truckee Lumber Company. The company dumped wastes from a sawmill into the Truckee River harming fish that passed through the river and spawned there, as well as other aquatic life. The California Supreme Court held that fish were the property of the people of the state of California and the wastes disturbed the rights the people held to their property. Therefore, the people received a favorable ruling ordering the Truckee Lumber Company to discontinue polluting the river. It was a revolutionary decision that allowed state power to be used to limit economic activities to protect wildlife.

PERIPHERAL CANAL   In the 1960s, the Sacramento–San Joaquin Delta in California experienced continual high salinity problems that made the water virtually useless for agriculture and industry. The Central Valley Project hoped to cure some of the salinity ills in the delta, but the salinity troubles continued. Local officials, as well as reclamation engineers, proposed the Peripheral Canal to directly move fresh Sacramento River water south without letting it mix with delta water to alleviate salinity problems. The Peripheral Canal soon became an important political issue as residents of northern California viewed it as just another imperialistic move of Southern California. Changing environmental values led residents to raise numerous ecological issues, but the bill to create the canal still passed in 1980. Eventually, voters squashed the plan in 1982 through a referendum for the repeal of the previously passed bill.

PESTICIDES   A pesticide is any substance or mixture of substances intended to prevent, destroy, or repel any pest. The term "pesticide" applies to herbicides, fungicides, and various other substances used to control pests. Pesticides are also any substance or mixture intended for use as a plant regulator, defoliant, or desiccant.

PETROFARMING   Petrofarming is the practice of farming by using and relying on pesticides, fertilizers, and other petroleum-based chemicals to increase production. This practice often results in higher yields, but it harms the land and water by promoting continuous and intensive farming and not allowing the land time to rest and recover. Petrofarming was central to the agricultural complex that emerged after World War II.

PRIOR APPROPRIATION   Prior appropriation related to water rights in the West. It gives water rights to those who use it on a first-come, first-served basis. Regardless of whether or not they own land that has water running though it, water users can claim water rights under prior appropriation so long as they continue to use it. Prior appropriation contrasts with the riparian rights used in other regions of the United States, which gives property owners rights to use, undiminished and unharmed, water flowing through or adjacent to their property. Ultimately, the West adapted both prior appropriation and riparian rights.

RAY, DIXY LEE (1914–1994)   Dixy Lee Ray, a conservative Democrat, served as Washington state's first female governor from 1976 to 1980. Trained as a marine biologist, Ray chaired the Atomic Energy Commission from 1972 to 1975 under presidents Richard Nixon and Gerald Ford. She came to believe environmental problems had been overstated in the popular media and by environmental groups. To combat this problem, she wrote two books on environmental alarmism, *Trashing the Planet* (1990) and *Environmental Overkill* (1993).

REDWOOD SUMMER   Redwood Summer occurred in 1990 in northern California's redwood forests. Environmental activists from across the country and members of Earth First! and other organizations gathered and initiated a series of protests over the logging of ancient redwoods. With Earth First! organizing the event, the summer was meant to bring a mass movement to the woods, not unlike the civil rights campaigns in the 1960s. Several confrontational episodes pitted environmentalists against timber companies and loggers. Redwood Summer also saw an increase in tree-sitting as a strategy for environmentalists. The efforts did little to thwart logging, but the media coverage brought national attention to the cause.

RIPARIAN RIGHTS   Riparian rights govern water rights in most of the United States, except in Western states. These rights are granted based on the physical location of ownership. An owner whose land runs along a river or a stream has the right to use water as long as that use does not alter or diminish the waterway in any way that would affect other riparian rights holders downstream. In the West, riparian rights coexist, with adaptations, with prior appropriation rights.

ROADLESS AREA REVIEW AND EVALUATION (1972) (RARE)   In 1972, the Forest Service designed the Roadless Area Review and Evaluation to determine which roadless areas within national forests deserved designation as wilderness and which should be opened as commercial areas. The agency conducted a second review during the Carter administration because of widespread disagreement. RARE II, as the second version was known, classified all wilderness areas into three categories: areas under the National Wilderness Preservation System, areas to be used for off-road vehicles, and areas to be further studied.

SAVE-THE-REDWOODS LEAGUE   Established in 1918, the Save-the-Redwoods League worked to protect ancient redwood forests in California so that they might endure forever. The nonprofit organization worked to obtain or purchase tracts of redwoods, contributed to the creation of state and national parks, and sought to influence legislation sympathetic to its cause. The League occasionally found itself at odds with other environmental groups who were less willing to work with timber companies.

SEATTLE AUDUBON V. EVANS (1991)   *Seattle Audubon v. Evans* was one of the more prominent legal cases that pitted the northern spotted owl and environmentalists against timber industries. In a 1991 ruling by Federal District Judge William Dwyer, this case resulted in a temporary stoppage of 135 planned timber sales in Pacific Northwest forests because the judge found that federal agencies had violated wildlife laws. The suit slowed old-growth logging and precipitated moves toward reform.

SIERRA CLUB   Founded in 1892, the Sierra Club started as a product of John Muir's and others' interest in the Sierra Nevada. Early on, the club sponsored hikes and then involved itself more politically with attempts to save Hetch Hetchy Valley from a proposed dam. After World War II, under the leadership of David Brower, membership soon exploded, and the Sierra Club became perhaps the most powerful environmental organization in the nation. It is now firmly ensconced in the mainstream and campaigns for conservation and environmental protection across the globe.

SOIL CONSERVATION ACT (1935)   The Soil Conservation Act, part of New Deal reforms, grew out of a five-year soil erosion study undertaken by the Forest Service. In the wake of the Dust Bowl tragedy, the study found that soil erosion posed a serious threat to agriculture and required long-term solutions at the federal level. The legislation created the Soil Conservation Service, which used Civilian Conservation Corps workers to assist in soil conservation work during the Depression. The Soil Conservation Service directed research into areas such as methods to control wind and water erosion. Later renamed the Natural Resources Conservation Service, conservationists in this agency began working with individuals and communities to find workable solutions to soil erosion.

STATE WATER PROJECT   California's State Water Project called for transporting water from the Sacramento River system to the San Joaquin Valley then moving it over the Tehachapi Mountains into Southern California. First proposed in 1919 by Lt. Robert B. Marshall of the U.S. Geological Survey, the state project was partially designed to get around acreage restrictions that were part of federal reclamation efforts. Eventually, voters approved a $170 million bond act in December 1933. But in the midst of the Great Depression, revenue bonds were unmarketable so no funding could be found in the state, so the federal government took over and called it the Central Valley Project. The plan reemerged in the 1940s and Governor Edmund "Pat" Brown moved it forward in the 1950s. It finally was completed in 1972 and gave agricultural users subsidized water.

STONE, LIVINGSTON (1836–1912)   A noted fish preservationist, Livingston Stone was a founding member and secretary of the American Fisheries Society. He worked to preserve fish populations, specifically salmon, through artificial propagation. In the mid-1870s, Stone built the first federal fish culture station at California's McCloud River. Although enthusiastic, Stone grasped salmon biology incompletely and misidentified Pacific salmon as members of the same genus as Atlantic salmon.

SUCCESSION   Succession is the process by which one ecological community gradually replaces another. This process begins when a new species, either animal or plant, overtakes the area inhabited by a pioneer species. Ecologists once believed succession led theoretically to a climax community when the new species and old species no longer changed but remained in a self-replicating community. This view has been challenged sharply by disturbance ecology. Succession demonstrates that ecological communities are constantly changing.

SUPERFUND   Created as a response to the Love Canal incident, in which a community in upstate New York experienced serious health problems because of buried toxic waste, Superfund was the common name for the Comprehensive Environmental Response, Compensation, and Liability Act signed into law by President Jimmy Carter in 1980. Superfund was designed to financially support hazardous waste cleanups. Unfortunately, although the initial appropriations for Superfund totaled more than $10 billion, this proved too little for the 1,200 sites that needed to be cleaned. Criticized for being ineffective and underfunded, within its first thirteen years, Superfund had only completed 150 site cleanups. In the Pacific Northwest, Hanford Nuclear Reservation and the Bunker Hill mining complex in northern Idaho represent two major Superfund sites where cleanup work continues.

SUSTAINED-YIELD   Sustained-yield forestry attempts to perpetually harvest the same amount of timber from season to season as forests produce. Ideally, sustained-yield forestry will allow no declines in output and will support timber communities in perpetuity. National forest managers put the idea in place as an attempt to curb clear-cutting and maintain reasonable harvests in the future. Too often, forest managers have pushed up their predictions for forest production to meet economic and political goals that move past sustained-yield.

TAYLOR GRAZING ACT (1934)   As part of the New Deal, the Taylor Grazing Act was designed to protect all public rangelands. The act rescinded the 1862 Homestead Act by withdrawing the remaining land in the public domain from settlement. These tracts became grazing districts and are currently administered by the Bureau of Land Management.

10:00 A.M. POLICY   In the 1930s the U.S. Forest Service launched a fire suppression policy that called for extinguishing any wildfires by 10 a.m. the day following its detection. If that was not achieved, the goal was to control it by 10 a.m. the following day, and so on. This policy symbolized the Forest

Service's blanket policy that all fires deserved suppression. It put more men in the backcountry to fight fires, including smokejumpers.

TRADITIONAL ENVIRONMENTAL (OR ECOLOGICAL) KNOWLEDGE   Typically associated with indigenous groups, traditional environmental knowledge is knowledge passed on orally to each generation; it revolves around human relationships with specific environments, usually long inhabited by individual groups.

UDALL, STEWART (1920–PRESENT)   Stewart Udall served as a U.S. Representative from Arizona from 1955–1961 and secretary of the interior from 1961–1969, during a particularly active period for implementing environmental reforms. He recognized the need for environmental conservation and federal management of public lands. He also helped expand the National Park System significantly during his tenure as secretary of the interior, adding several new parks, monuments, recreation areas, historic sites, and wildlife refuges. Udall continues his conservation work today as an author and environmentalist.

UNITED FARM WORKERS' ORGANIZING COMMITTEE (UFWOC)   The United Farm Workers' Organizing Committee was a labor union formed in 1966 when the National Farm Workers Association and the Agricultural Workers Organizing Committee merged. Led by Cesar Chavez, the union used mostly nonviolent means to achieve better working conditions and improve wages. A key complaint of the UFWOC was workers' exposure to pesticides. Their most famous campaign came when they launched a grape boycott. Assisted by Americans sympathetic to their cause who boycotted grapes in stores across the nation, the UFWOC won a significant contract with California grape growers in 1969 after a boycott lasting several years.

U.S. ARMY CORPS OF ENGINEERS   The U.S. Army Corps of Engineers consists of civilian and military personnel who primarily work on engineering and environmental projects across the nation. The Corps' existence dates back to 1802 when Congress established a Corps of Engineers, but its most recognizable work occurred in the middle part of the twentieth century, particularly after World War II. During this period, the Corps, along with other agencies such as the U.S. Bureau of Reclamation, engaged in a competition of sorts to construct countless dams.

U.S. BUREAU OF RECLAMATION   Created by President Theodore Roosevelt in 1902 as the U.S. Reclamation Service, this federal agency was responsible for the supervision and administration of federal reclamation and irrigation projects. The Bureau of Reclamation constructed numerous dams, power plants, and canals to help settle and develop the West.

U.S. DEPARTMENT OF AGRICULTURE (USDA)   President Abraham Lincoln established the Department of Agriculture in 1862 to supervise policy for the nation's agrarian economy. Now, the USDA oversees food safety guidelines, leads federal programs to combat hunger in America, promotes agriculture at home and abroad, and assists rural Americans with their financial and cultural needs. In addition to this, since 1905 the department has operated as the steward for the nation's national forests and rangelands through its most powerful agency, the Forest Service. The Natural Resources Conservation Service also falls under the authority of the USDA .

U.S. DEPARTMENT OF THE INTERIOR   The Department of the Interior was created in 1849 within the executive branch of the federal government. This agency has been largely responsible for land preservation and development. The department supervised such issues as homesteading, American Indian concerns, and land patents. In the late nineteenth century, the Department of the Interior increased its authority and jurisdiction, as western emigration escalated. Establishing the transcontinental railroad and maintaining diplomatic relations between non-Indians and Indians were two of the larger issues facing the department. The later creation of national parks and forest reserves during the Progressive Era also increased the jurisdiction of the department, although the forest reserves were later moved to the Department of Agriculture. In the twentieth century, the Department of the Interior has focused on a variety of conservation and preservation programs for the federal government. Most notably, the National Park Service is within this department.

U.S. FISH COMMISSION   In 1871, Congress created the commission to study the reasons for a decline in fisheries off the U.S. coast. For the next 30 years the commission built hatcheries and investigated fish and aquatic life along the U.S. coast and in nearby rivers. Practically, research declined as the commission focused overwhelmingly on artificially propagating fish. By 1902, the commission reorganized into the U.S. Bureau of Fisheries and later became part of the U.S. Fish and Wildlife Service in 1939.

U.S. FISH AND WILDLIFE SERVICE   Officially created in 1940, the U.S. Fish and Wildlife Service actually incorporated governmental agencies dating back eight decades, including the Bureau of Fisheries and the Bureau of Biological Survey.

The purpose of the service was to supervise and manage all fish and wildlife resources within the nation. Conserving and protecting fish and wildlife and their natural habitats have become the agency's main responsibilities. After the Endangered Species Act in 1973, these responsibilities expanded to include developing recovery plans for endangered species.

U.S. FOREST SERVICE (USFS)   Incorporated in 1905, the Forest Service presided over national forests throughout the nation. It recognized that the federal government had to take a more active role in the forest management. The larger goal of the USFS has been to provide continuous forest resources for both public enjoyment and private gain. Throughout the twentieth century, the USFS has expanded its role to combine multiple uses within the nation's forests, including recreation and timber needs.

URBAN-WILDLAND INTERFACE FIRE   *See* Intermix fire.

WEIR   A weir is a fencelike structure used to catch fish in slower-moving streams and along banks of rivers. It blocks fish from swimming upstream or traps them where fishers can harvest the fish in great numbers. American Indians used weirs for millennia to harvest salmon in great numbers.

WILD AND SCENIC RIVER ACT (1968)   The Wild and Scenic River Act protected specific rivers from dam construction and commercial development. It set up three different classifications for protected rivers: wild rivers, scenic rivers, and recreational rivers. The act also served to define specific restrictions for these rivers, as well as protecting a segment of land on either side of the river. The strips of land average 320 acres for every mile of river. The act was designed to conserve the natural resources within the rivers for both environmental and recreational purposes. It was a response to the rapid dam-building of the post–World War II era.

WILDERNESS   While the term is ambiguous and constantly shifting, wilderness can be widely defined as any area that is not shaped or largely inhabited by people. The Wilderness Act of 1964 further defined wilderness as an area "untrammeled by man, where man himself is a visitor."

WILDERNESS ACT (1964)   The Wilderness Act made it the official policy of the federal government to preserve the wilderness resources of the nation. The act defined wilderness and established the National Wilderness Preservation System. Originally, the act established 9.1 million acres of federally protected wilderness in national forests. Large tracts of roadless areas where most economic developments are prohibited characterize wilderness areas. The National Wilderness Preservation System has expanded significantly since the 1960s, but placing lands within the system has become politically contentious.

WRIGHT ACT (1887)   Passed one year after the California Supreme Court handed down the *Lux v. Haggin* decision, the Wright Act created irrigation districts in areas that had either 50 people or a majority of landowners. The Wright Act effectively ended the monopoly large landowners held over water rights. Within three decades, more than fifty irrigation districts were established.

YOSEMITE NATIONAL PARK ACT (1890)   The Yosemite National Park Act turned this area of California over to federal managers. In 1864, President Abraham Lincoln signed a bill that effectively preserved the area within the public trust under California jurisdiction. This was a historic act, as it was the first time the federal government had protected a region simply for scenic reasons. The 1890 act reasserted federal control but kept scenic preservation the guiding principle.

ZANJA MADRE   The *zanja madre*, or main trench, was the principal irrigation ditch emanating from a colonial Spanish town's water source, typically a river or reservoir. From that trench would be built several smaller irrigation trenches that would feed into local commons or farms.

ZANJERO   During the colonial period in Spanish California, a *zanjero* was a special administrator of water rights appointed by the local *ayuntamiento*. The *zanjero* oversaw irrigation systems and regulated the amount of water given to certain types of uses. The *zanjero* retained a lot of power in the local community. His power, however, came only with the consent of the *ayuntamiento*, who could relieve the *zanjero* if he did not distribute the water fairly and according to the public good.

CHRONOLOGY

200 million years BP (before present) The North American continent meets with the Pacific Ocean floor, resulting in familiar coastal landmarks.

100 million BP The Okanogan microcontinent merges with the North American continent.

50 million BP The North Cascade microcontinent joins with the North American continent.

17–11 million BP The Columbia Plateau is created by basalt lava flows from volcanic activity.

25,000–11,000 BP Beringia, the land bridge between Siberia and Alaska, remains open.

18,000 BP The climate cycle shifts and summers warm, signaling the end of the last major glaciation.

15,000 BP Lake Bonneville breaks free and carves out riverbeds such as the Snake River Plain.

14,000 BP Human population increases throughout the Western Hemisphere.

14,000–11,000 BP Receding glaciers create recognizable landscapes in the region.

13,000–11,000 BP Megafauna, such as the mammoth and mastodon, die off in North America in what is known as the Pleistocene extinctions.

12,000 BP Glacial Lake Missoula repeatedly drains in massive floods, creating the channeled scablands.

7,900 BP Mount Mazama begins a series of eruptions that affect much of the Northwest and ultimately creates Crater Lake.

4,000–2,000 BP Generalized resource specialization occurs among Coastal Native societies.

2,500 BP Northwest tribes develop means to store salmon for extended periods. They also begin to cultivate specialized crops, such as tobacco, which subsequently leads to sedentary settlements within the region.

1,500 BP Native hunting pushes pinnipeds to offshore island rookeries.

1492 CE (common era) The first sustained contact between Europe and North America occurs with the arrival of Christopher Columbus.

1542 Spanish explorer Juan Rodríguez Cabrillo lands at San Diego Bay.

1580 Sir Francis Drake's western coast voyage of North America occurs.

Early 1700s The horse is reintroduced to the Northwest and incorporated into the cultures of many tribes.

1741 Russian fur traders establish permanent operations in the Alaskan Aleutian Islands.

1750s onward Disease follows trade routes across the continent and into the West.

1769 The Spanish "Sacred Expedition" extensively explores California and brings with it permanent European settlement on the West Coast. The expedition introduces European crops and livestock to the region. Missions begin in California.

1774 The Spanish Juan Pérez expedition explores the Northwest Coast, making landfall at 54° 40'.

1775–1783 A smallpox epidemic of continental proportions reaches the Far West.

1776 James Cook sets out on his third world voyage, eventually reaching Nootka Sound in 1778.

1777 The first recorded epidemic is reported at the mission Santa Clara in Spanish California.

1778 Fur trade to China is initiated when James Cook's crew receives profits of 1,800 percent.

1780s Fur trade in the northwestern environments begins to show effects on animal populations and ecosystems.

1781 Yumans kill Fray Francisco Tomás Hermenegildo Garcés in response to Euro-American cattle destroying Yuman crops.

1784 Captain Cook's journals are published, initiating greater interest in the Northwest Coast.

1785 An Indian uprising takes place at Mission San Gabriel, precipitated in part by resource scarcity.

1786 The first evidence of localized extinctions is found in sea otter populations off Northwest Coast.

1790s The fur market is largely saturated.

1792–1794 George Vancouver follows up Cook's voyage with an extensive exploration of the northwestern coastline.

1792 Robert Gray, an American merchant, explores the Columbia River by boat.

1793 The Alexander Mackenzie expedition crosses the continent and establishes the presence of the North West Company in the Pacific Northwest.

1803 The Louisiana Purchase secured from France doubles the physical size of the United States.

1805–1806 The Lewis and Clark Expedition crosses through the Pacific Northwest collecting scientific and cultural data.

1806 Pneumonia, diphtheria, and measles are evident in Spanish missions.

1810s Russia harvests on average several thousand sea otters yearly from their holdings in North America.

1810 Neophytes (converted Indians) construct dams to facilitate irrigation at Spanish missions.

1812 Russia establishes Fort Ross north of San Francisco, California.

1818 The United States and Great Britain establish a treaty that settles on a joint occupation of Oregon Country.

1819 The Adams-Onis Treaty, a United States and Spanish agreement, signed away any claim of the Spanish to Oregon and northward.

1820 The sea otter population nears extinction because of overhunting by trappers.

1820s Smallpox, cholera, and scarlet fever become regular diseases in western settlements.

1820s Agriculture intensifies at Fort Vancouver by order of the Hudson's Bay Company.

1821 The North West Company and the Hudson's Bay Company merge and form a single British presence in the Pacific Northwest.

1821–1848 The newly independent Mexican government inherits control of the Spanish mission system and secularizes each mission in order to parcel out mission lands to Mexican citizens.

1822–1823 Hudson's Bay Company trades 10,000 beaver and otter pelts.

1823 Spanish missionaries and soldiers establish twenty-one missions in California by 1823.

1824 Hudson's Bay Company introduces seventeen cattle to its post at Fort Vancouver.

1824 Hudson's Bay Company traders meet American traders in southeastern Idaho, initiating an international and corporate competition for access to fur resources in the Northwest. Hudson's Bay Company begins a fur desert policy.

1825 A flood shifts the channel of the Los Angeles River by 20 miles.

1826–1848 Six million hides are shipped from California in lucrative hide and tallow trade.

1828–1830 Severe drought in California devastates ranchos.

1830s American missionaries and middle class farm migrants begin to arrive in the Oregon Territory.

1834 Missionary Jason Lee arrives in Willamette Valley.

1836 The Whitman and Spaulding missionary parties arrive in inland Northwest.

1839 Hudson's Bay Company forms the Puget's Sound Agricultural Company to diversify its business.

1840s Invasive grasses and weeds dominate many pastures and fields in California.

1842 The Oregon overland trail system is established, bringing a wider migration of American settlers.

1843 American officials learn of gold in California.

1845 Cayuse destroy irrigation works at the Whitman mission.

1846 A treaty between the United States and Great Britain finally establishes the 49th parallel as the border between the United States and British Canada.

1846–1847 A measles outbreak strikes Northwest populations.

1848 The United States acquires much of the Southwest, including California, from Mexico.

1848 James Marshall discovers gold at Sutter's Mill in the Sierra Nevada foothills.

1849 More than 80,000 migrants arrive in California to strike it rich in the gold rush, earning the name the "forty-niners."

1849 The U.S. Department of the Interior is established.

1850 California enters the United States of America.

1850 The Oregon Land Donation Act is passed.

1850s The U.S. government establishes a series of treaties with Native American tribes to secure land rights in the Northwest and California.

1852 "Mother of the Forest"—a giant sequoia tree 300 feet high, 92 feet in circumference, and about 2,500 years old—is cut down for display in carnival sideshows. The tree was in Calaveras Grove, part of what will become Yosemite National Park.

1852 The last healthy salmon-spawning run is reported for the Sacramento River.

1858 Colonel George Wright massacres about 700 Indian horses.

1859 The state of Oregon enters the United States of America.

1860 San Francisco–based journalist Thomas Starr King writes an eight-article series on Yosemite for the *Boston Evening Transcript*.

1860s Hydraulic mining arrives in the Northwest as an efficient, yet devastating method of extracting minerals.

1861–1862 Floods damage Sacramento because of the fallout of hydraulic mining alterations to river flows.

1862 The U.S. Department of Agriculture is established.

1862 The Homestead Act is passed, which entitles any head of household to 160 acres of land after five years of settlement.

1862 The Federal Morrill Land Grant Act grants each state 30,000 acres of publicly held lands to establish colleges of agriculture and mechanical skills.

1862–1872 A series of land grants is given to railroads to promote expansion.

1864 Senator John Conness of California introduces a bill to protect the Yosemite Valley as a state park in California.

1864 Hume, Hapgood, and Company forms the first salmon canning company in California.

1865 The Boise River loses salmon runs because of stream blockages from mining.

1868 Sacramento builds a canal to divert floodwaters around the city.

1870 One-third of California's commercial timber is already exhausted.

1870 The U.S. Fish Commission is founded.

1870s Mine sediments work their way to the San Francisco Bay.

1870s Technological improvements (e.g., fish wheels and fish traps) increase fish harvests in the Northwest and California.

1870s Local newspapers in the Northwest report wheat prices in distant markets.

1870s–1880s Railroad lines arrive in the Northwest and begin to link regional economies.

1871 The Redwood Lumber Association, a cartel, is founded in California to stabilize timber market.

1872 Yellowstone National Park becomes the first national park.

1872 California law requires cattle and sheep ranges to be enclosed.

1872 The U.S. Fish Commission establishes a salmon hatchery on McCloud River.

1873 The Timber Culture Act is passed.

1874 Los Angeles claims pueblo water rights granting the city ownership of all Los Angeles River water.

1876 The Southern Pacific Railroad arrives in Los Angeles.

1877 The Desert Land Act passes to encourage settlement of arid lands in the western United States.

1877 An El Niño season arrives in Oregon and surrounding areas.

1878 The Free Timber Act encourages settlement in the West.

1878 The Timber and Stone Act passes to encourage settlement of land in the western United States that is incapable of sustained agricultural production.

1878 John Wesley Powell's *Report on Arid Lands* describes the condition of the lands west of the 100th meridian as arid and in great need of irrigation if farming is to survive.

1879 The U.S. Geological Survey is created.

1880 The mining rush is renewed in Idaho territory, this time led by industrial miners.

1880 A U.S. census reveals that 92 percent of fishers in the West are immigrants, one-third of whom are Chinese.

1881 The Division of Forestry is established as an agency of the Department of Agriculture.

1882 John Dolbeer's donkey engine revolutionizes timber extraction by eliminating the reliance on animal power.

1883 The territorial legislature of Washington enacts a law making it a misdemeanor for farmers to let Canadian thistles go to seed on their land.

1883–1884 Mineral discoveries on the Coeur d'Alene River in Idaho attract 10,000 miners to the region.

1884 *Woodruff v. North Bloomfield Gravel Mining Co. and Others* is decided in the Ninth U.S. Circuit Court, effectively ending hydraulic mining in California goldfields after years of legal contests.

1886 *Lux v. Haggin* establishes the California Doctrine of water rights, combining elements of prior appropriation and riparian rights.

1887 The Hatch Act establishes agricultural experiment stations.

1887 The Wright Act is passed in California; it allowed communities with a two-third majority to form irrigation districts.

1887 The Dawes Act seeks to realign Native American tribes to an agriculturally based subsistence, destroy communal land holdings, and sell "surplus" land to Euro-Americans.

1889 Washington enters the United States of America.

1889–1890 A severe winter in the Northwest causes stock losses of 60 to 90 percent.

1890 Yosemite becomes a national park, along with the Sequioa and General Grant national parks.

1890 Idaho enters the United States of America.

1890 The U.S. Census proclaims the end of the frontier.

1891 The Forest Reserve Act is passed. The president now has the power to establish forest reserves on public lands.

1892 The Sierra Club is founded.

1892 River blockages stop salmon migration on Bruneau and Grande Ronde rivers.

1893 A depression strikes the United States, bankrupting many mining, railroad, and timber companies.

1894 The Carey Act passes Congress, giving desert states a million acres of public domain to use to promote irrigation.

1895 *Vernon Irrigation Co. v. Los Angeles* confirms the city's pueblo water right to all of the Los Angeles River's water.

1897 The Organic Act is passed by Congress setting guidelines for managing the forests of the United States.

1897 *People v. Truckee Lumber Company* is decided by the California Supreme Court and gives the state power to regulate industrial water uses that damaged fish habitat.

1898 Gifford Pinchot is named head of Forestry Division.

1899 Mount Rainer National Park is created.

1899 The River and Harbor Act forbids pollution of navigable waterways.

1899 California has 1.5 million irrigated acres.

1900 In response to the devastating population decrease of industrially hunted birds, the Lacey Act passed to protect wildlife.

1900–1904 The Los Angeles population doubles from 100,000 to 200,000 in four years.

1901 The Right of Way Act is passed, allowing the secretary of the interior to permit rights of way through publicly held lands in California.

1902 The Reclamation (or Newlands) Act establishes the Reclamation Service in the Department of the Interior to promote federal irrigation projects.

1902 Crater Lake National Park is created.

1905 Forest reserves are transferred from Department of the Interior to the Department of Agriculture. They are renamed national forests and managed by the new U.S. Forest Service.

1905 The Los Angeles Board of Water Commissioners announces "Titanic Project to Give City a River."

1906 The Antiquities Act reserves areas of scientific or historical importance as national monuments.

1906 The San Francisco earthquake and fire devastated the city.

1907 Between 1,000 and 1,200 truckloads of gravel and sand are taken daily from the Los Angeles River.

1908 The Winters Doctrine confirms western Indian tribal water rights.

1908 Mount Lassen National Monument is created.

1908 Heyburn State Park in Idaho is created as the first state park in the Northwest.

1908–1913 Construction of Los Angeles aqueduct takes place.

1909 The Mount Olympus National Monument is created.

1910 The Big Blowup, a major fire complex in northern Idaho and western Montana, burns 5 million acres.

1910 The Los Angeles city council passes an ordinance to stop garbage dumping in Los Angeles River bed.

1911 The American Game Protective and Propagation Association is founded.

1912 The Los Angeles River bed receives twenty-seven truckloads of garbage daily, despite laws prohibiting this dumping.

1913 The Hetch Hetchy Dam is approved.

1913–1914 Major floods cause extensive damage in Southern California.

1914 The Los Angeles County Flood Control District is created.

1915 The Celilo Canal is finished, making shipping along the Columbia River more predictable.

1916 The National Park Service Act is passed, creating the National Park Service.

1918 The Save-the-Redwoods League is founded.

1919 California has 4.2 million irrigated acres.

1920 The Mineral Leasing Act limits mining on federal lands.

1920 The Federal Water Power Act authorizes the Federal Power Commission to issue licenses for hydroelectric improvements.

1920 Idaho can boast of 13,000 miles of irrigation ditches and canals through 18,000 farms that water 2 million acres.

1920s Californians lead the nation in gasoline dependency.

1920s Californians' automobile ownership grows 451 percent while the population grows only 155 percent.

1923 The Snake River Committee of Nine is formed to arbitrate between river users.

1927 All Owens Valley banks close.

1927 Los Angeles hires the Olmsted Brothers and Harland Bartholomew and Associates to study the region's parks, ultimately creating *Parks, Playgrounds and Beaches for the Los Angeles Region,* a report that encourages development of more public space in the city.

1928 On March 12 the St. Francis Dam gives way in Los Angeles, killing more than 500 people.

1928 The California legislature bans oil drilling from piers.

1930 The Idaho Primitive Area is created.

1930s The Great Depression affects virtually all industries.

1933 The Civilian Conservation Corps is established as part of Franklin Roosevelt's New Deal.

1933 The Tennessee Valley Authority is established.

1933 The Tillamook Fire along the Oregon coast consumes a quarter-million acres of prime forest.

1934 The Taylor Grazing Act allows for federal regulation of the public domain.

1935 The U.S. Forest Service institutes the 10:00 a.m. fire policy.

1935 The Wilderness Society is founded.

1935 The Federal government authorizes the Central Valley Project.

1936 The National Wildlife Federation is founded.

1936 A referendum in California passes that allows oil companies to slant drill for oil.

1937 The Federal Aid in Wildlife Restoration Act grants money to states to fund wildlife protection.

1938 The Bonneville Dam is completed.

1938 The Olympic National Park is created.

1938 The Willamette Valley Project passes Congress to reconfigure the river for flood control, navigability, irrigation, and hydroelectricity.

1938 Los Angeles experiences the largest flood on record.

1940 The U.S. Fish and Wildlife Service consolidates federal conservation and protection programs focused on fish and wildlife.

1941 Aqueducts begin diverting water from streams feeding Mono Lake into the Los Angeles water supply.

1941–1945 The United States involvement in World War II, brings many industries out of the depression and encourages full production.

1942 The Boulder Canyon Project's All-American Canal brings Colorado River water to the Imperial Valley in California.

1942 Los Angeles completes a 242-mile aqueduct from the Colorado River.

1942 The Grand Coulee Dam is completed.

1946 The U.S. Bureau of Land Management is established.

1946 The U.S. Fish and Wildlife Service unsuccessfully proposes moratorium on dam-building until they can study how the dams affect salmon.

1947 *United States v. California* declares that the United States holds title to offshore oil deposits.

1947 California passes the first state air pollution act.

1948 The Federal Water Pollution Control Law is enacted.

1948 The Columbia River experiences a major flood.

1949 The Atomic Energy Commission intentionally releases a large amount of toxic iodine-131 into the atmosphere from Hanford, Washington. This becomes known as Green Run.

1950s Integrated pest management enjoys a brief period of enthusiasm.

1950s More than a thousand people a week relocate to Southern California.

1950 Dr. Arie Haagen-Smit identifies causes of smog in Los Angeles.

1954 Heavy smog conditions shut down industry and schools in Los Angeles for most of October.

1954 The U.S. Forest Service proposes reclassifying portions of the Three Sisters area of the Central Cascades sparking an important wilderness debate.

1955 On November 29 an accident kills several researchers at the experimental breeder reactor No. 1 at Arco, Idaho.

1955 The Federal Air Pollution Control Act is passed recognizing air pollution as a national problem.

1956 The Water Pollution Control Act provides financial support for water treatment plants.

1958 Leading conservationists stage a protest hike in Washington's Olympic National Park to draw attention to a proposed road along the Pacific coast.

1959 California becomes the first to impose automotive emissions standards.

1960 The Multiple Use-Sustained Yield Act is passed.

1961 The Save the San Francisco Bay Association is formed.

1962 Rachel Carson publishes *Silent Spring*.

1963 The Clean Air Act opens hearings and legal proceedings.

1964 The National Wilderness Preservation System is established by the Wilderness Act.

1964 The Bay Conservation and Development Commission is created.

1965 The Water Quality Control Act is passed.

1965 Lyndon B. Johnson hosts the White House Conference on Natural Beauty.

1966 The National Historic Preservation Act is passed.

1967 *Udall v. FPC* effectively stops dam construction in Hells Canyon at High Mountain Sheep.

1967 Automobiles in Southern California pump 20 million pounds of carbon monoxide into the air a day.

1968 The Wild and Scenic Rivers Act protects free-flowing, undammed rivers.

1968 The American Smelting and Refining Company proposes an open-pit molybdenum mine in Idaho's White Clouds.

1968 Redwoods National Park is established.

1969 An oil spill in Santa Barbara, California, demonstrates the problems of pollution.

1969 Greenpeace is founded.

1969 David Brower is forced out of the Sierra Club.

1969 The National Environmental Policy Act is passed and signed on 1 January 1970; it required federal agencies to prepare environmental impact statements for federal projects affecting the quality of the environment.

1970 The Resource Recovery Act (Solid Waste Disposal Act) is passed.

1970 The first Earth Day is celebrated on April 22.

1970 The Environmental Protection Agency is founded.

1970 The Clean Air Act is passed.

1970 The United Farm Workers' Organizing Committee gains a contract with farmers, based in part on complaints about pesticides-related health issues.

1970 California counts nearly 100 nuclear power plants within the state.

1970s California farmers reportedly spend $500 million for 100 million pounds of pesticides.

1972 DDT is banned for domestic use in the United States.

1972 The Federal Water Pollution Control Act (Clean Water Act) is passed.

1972 The Federal Environmental Pesticide Control Act is enacted.

1972 The Ocean Dumping Act is passed.

1972 The State Water Project is completed in California.

1972 Congress creates the Sawtooth National Recreation Area in Idaho.

1973 The Endangered Species Act is passed.

1973 The California Certified Organic Farmers becomes the first certification agency in the state.

1976 On June 5 Idaho's Grand Teton Dam fails, causing 14 deaths and millions of dollars worth of damage.

1976 The Toxic Substances Control Act is approved.

1976 The National Forest Management Act is passed.

1976 The EPA and Clean Air Act end the manufacturing of automobiles that run on leaded gasoline.

1976 The California Nuclear Regulatory Commission orders nuclear plant closed because of concerns over faults.

1977 The Clean Air Act amendments are passed.

1977 The Surface Mining Control and Reclamation Act is enacted.

1978 The Endangered American Wilderness Act designates lands to be protected in the western states to increase habitats and watersheds, as well as scenic and historical locations.

1978 The National Energy Act is passed.

1978 The Redwood National Park is enlarged.

1978 The Mono Lake Committee is formed.

1979 A reactor at the Three-Mile Island nuclear power station suffers a partial core meltdown because of a cooling malfunction.

1980 The "Superfund" is established by the Comprehensive Environmental, Response, Compensation and Liability Act.

1980 Mount Saint Helens erupts.

1980 The Peripheral Canal Bill passes the California legislature and is later overturned by statewide referendum in 1982.

1981 Earth First! is founded, and by 1988 it has roughly 12,000 members.

1983 Earth First! protestors block a bulldozer on a logging road on Bald Mountain in the Siskiyou National Forest.

1983 A former smelter site in Kellogg, Idaho, is named a Superfund site, the largest in the nation.

1984 The Mothers of East Los Angeles is formed.

1987 George Alexander, a California millworker, is seriously injured after his bandsaw shatters from a spiked tree.

1988 Americans are alarmed about drought conditions and the "greenhouse effect."

1989 The North American Wetlands Conservation Act is passed.

1989 The Association of Forest Service Employees for Environmental Ethics is organized.

1990 The northern spotted owl is added to the Endangered Species List, sparking controversy in the Pacific Northwest.

1990 A car bomb maims environmental activist Judi Bari in Oakland, California.

1990 Redwood Summer brings activists to the northern California coast to protest logging practices.

1990–1993 In California, 4,500 buildings burn from wildfires.

1991 The decision in *Seattle Audubon v. Evans* temporarily halts 135 timber sales in Pacific Northwest forests.

1992 The U.S. Forest Service begins the New Perspectives program for forest management.

1993 The Northwest Forest Conference, or Timber Summit, convenes in Portland to resolve old-growth logging and endangered species issues.

1993 Jack Ward Thomas is appointed to head the U.S. Forest Service, the first biologist to hold the position.

1993 Settlement with Los Angeles decreases the city's diversions from Mono Lake by one-third.

1994 The California Desert Protection Act sets aside millions of California desert wilderness.

1994 The U.S. Forest Service adopts Option 9 of the Forest Ecosystem Management Assessment Team's report with slight modifications as its forest management plan, creating the Northwest Forest Plan.

1994 The Northridge earthquake devastates Southern California.

1997 Julia Butterfly Hill climbs into a 55-meter-tall (180 foot) California coast redwood tree where she remains for a year and a half protesting logging.

1997 The United States is one of the signatories of the Kyoto Protocols in Japan. The protocols call for greenhouse emissions to be reduced below the 1990 standards by 2012. The U.S. Senate later refuses to ratify the treaty.

1999 Groundbreaking starts at Oakland's Fruitvale Transit Village.

2000 President Bill Clinton creates the Hanford Reach National Monument, encompassing the last free-flowing part of the Columbia River that flows past the Hanford Nuclear Reservation.

2002 Because of agricultural withdrawals, 33,000 salmon die in low water levels in Klamath River.

2002 California passes a law requiring a limit on carbon dioxide emissions for all automobiles sold in the state. It is the first law in the nation that requires automakers to limit emissions of greenhouse gases.

2002 The EPA adopts California emissions standards for off-road recreation vehicles.

2003 The Healthy Forests Restoration Act becomes law. It creates a new plan to reduce the threat of wildfires and improve forest health through fuel reduction practices.

2004 Automakers sue the state of California over laws requiring them to limit emissions of greenhouse gases in automobiles.

2004 The Kyoto Protocols take effect after Russia ratifies the agreement; the United States does not ratify it.

2006 Conservationists and farmers agree on a plan to restore fish and wildlife in the San Joaquin River in California.

2006 Ventura County's (CA) watershed managers receive a $5 million grant to prepare the Matilija Dam for removal from the Ventura River watershed.

2006 Congress passes the Northern California Coastal Wild Heritage Wilderness Act. The act would classify more than 270,000 acres in Northern California as wilderness.

REFERENCES AND FURTHER READING

This bibliographic essay does not exhaustively account for all books consulted or cited in this book. Instead, it suggests the most salient books for the major themes explored. In addition, the essay highlights where there are scholarly debates. In short, this essay is a place to begin for an orientation of the region's environmental history.

No environmental history has covered the entire region focused on in this volume. However, a number of classic books of environmental history have investigated portions of this region. Moreover, some of the best new works in the field examine portions of the Pacific West. So, while there is no single overview of the Far American West's environmental history, there are the building blocks for it and some of the best work in the field concentrates here.

For nonspecialists, the prehuman landscape can be difficult to understand, so basic textbooks offer the best place to begin. Geological foundations are revealed in the Roadside Geology Guides (Missoula, MT: Mountain Press) by David Alt and Donald W. Hyndman; see *Roadside Geology of Oregon* (1978), *Roadside Geology of Washington* (1984), *Roadside Geology of Idaho* (1989), and *Roadside Geology of Northern and Central California* (2000). Other overviews include Elizabeth L. Orr and William N. Orr's *Geology of the Pacific Northwest* (New York: McGraw-Hill, 1996) and Robert M. Norris and Robert W. Webb's *Geology of California,* 2nd edition (New York: John Wiley and Sons, 1990). Volcanoes are the topic of Stephen L. Harris, *Fire and Ice: The Cascade Volcanoes,* revised edition (Seattle: The Mountaineers and Pacific Search Press, 1980). Arthur R. Kruckeberg's, *The Natural History of Puget Sound Country* (Seattle: University of Washington Press, 1991) offers a detailed overview of western Washington's ecology.

Scholars have vigorously debated Native peoples' impact on the environment. Archaeologists such as Brian M. Fagan in *Ancient North America: The Archaeology of a Continent,* 3rd edition (New York: Thames and Hudson, 2000) and Gary Haynes in *The Early Settlement of North America: The Clovis Era* (New York: Cambridge University Press, 2002) offer technical approaches to human arrival in North America. The essays in Paul S. Martin and Richard G.

Klein's edited volume *Quaternary Extinctions: A Prehistoric Revolution* (Tucson: University of Arizona Press, 1984) argue for human-induced extinctions of dozens of mammalian species approximately 12,000 years before the present. In *After the Ice Age: The Return of Life to Glaciated North America* (Chicago: The University of Chicago Press, 1991), ecologist E. C. Pielou has furnished an accessible scientific study that places North America in the context of colonizing or recolonizing plants, animals, and humans since the last ice age. A variety of topics are covered in *Wilderness and Political Ecology: Aboriginal Influences and the Original State of Nature*, edited by Charles E. and Randy T. Simmons (Salt Lake City: University of Utah Press, 2002). These essays, based largely on scientific data, tend to frame the impact of Native peoples' activities on the land, plant, and animal resources through models in evolutionary theory and minimize conservationist perspectives.

While the scientific scholarly tradition has tended to privilege material culture and scientific evidence, anthropologists have offered highly detailed and sophisticated studies of how American Indians shaped their surrounding environment with much greater attention to oral traditions and historical documentation. Thomas C. Blackburn and Kat Anderson edited a classic study in *Before the Wilderness: Environmental Management by Native Californians* (Menlo Park, CA: Ballena Press, 1993) with a wide representation of these approaches. Rodney Frey, collaborating with the Schitsu'umsh, produced *Landscape Traveled by Coyote and Crane: The Worlds of the Schitsu'umsh (Coeur d'Alene Indians)* (Seattle: University of Washington Press, 2001), a deeply sensitive cultural study of human-nature interactions in the inland Northwest. Similarly, Eugene S. Hunn, with James Selam and Family, has described mid-Columbia River human ecology in *Nch'i-Wána "The Big River": Mid-Columbia Indians and Their Land* (Seattle: University of Washington Press, 1990). Alan G. Marshall's, "Unusual Gardens: The Nez Perce and Wild Horticulture on the Eastern Columbia Plateau," found in *Northwest Lands, Northwest Peoples: Readings in Environmental History*, edited by Dale D. Goble and Paul W. Hirt (Seattle: University of Washington Press, 1999) presents a useful picture of the relationship between the Nez Perce, their resource base, and rising social complexity attributed to scarcity.

Another area of debate centers on indigenous use of fire. Scientists are apt to conclude that pre-Columbian anthropogenic fire was haphazard and infrequent, as found especially in Thomas R. Vale, editor, *Fire, Native Peoples, and the Natural Landscape* (Washington, DC: Island Press, 2002). Historians and anthropologists generally see anthropogenic fire as being more prevalent, as evidenced in the essays in Robert Boyd, editor, *Indians, Fire, and the Land in the Pacific Northwest* (Corvallis: Oregon State University Press, 1999), and especially the work of Stephen J. Pyne: *Fire in America: A Cultural History of Wildland and Rural Fire* (Princeton, NJ: Princeton University Press, 1982. Reprint, Seattle: University of Washington Press, 1997) and *Fire: A Brief History* (Seattle: University of Washington Press, 2001). The division of this fire debate generally reflects the split between humanistic and scientific approaches to the precontact landscapes.

The first significant impacts of contact between western Native peoples and Europeans came through the vector of disease. Alfred W. Crosby's classic article, "Virgin Soil Epidemics as a Factor in the Aboriginal Depopulation in America," (*William and Mary Quarterly* 33 [April 1976]: 289–299), popularized the idea that European diseases found a virgin soil among Native Americans and wreaked havoc on comparatively uninitiated immune systems. David S. Jones, in "Virgin Soils Revisited," (*William and Mary Quarterly* 60 [October 2003]: 703–742), has further examined this idea and emphasized the more complex epidemiological interactions in so-called virgin soil epidemics. In California, Sherburne F. Cook has best documented diseases at the time of contact in "The Impact of Disease" (in *Green Versus Gold: Sources in California's Environmental History*, edited by Carolyn Merchant [Washington, DC: Island Press, 1998]: 55–59). Robert Boyd had done the most thorough job of investigating disease among Northwest Coast Indians in *The Coming of the Spirit of Pestilence: Introduced Infectious Diseases and Population Decline among Northwest Coast Indians, 1774–1874* (Seattle: University of Washington Press, 1999). Elizabeth A. Fenn's lively *Pox Americana: The Great Smallpox Epidemic of 1775–82* (New York: Hill and Wang, 2003) documents a continental smallpox epidemic that devastated Northwest Natives and demonstrates the far-reaching connections throughout North America at the end of the eighteenth century. Diseases wreaked havoc with Native peoples' ability to maintain subsistence practices and contributed to rebellion as explained in Steven W. Hackel, "Sources of Rebellion: Indian Testimony and the Mission San Gabriel Uprising of 1785" (*Ethnohistory* 50 [Fall 2003]: 643–669).

Exchange of disease both accompanied and prepared the way for greater incursions of European and American trade and colonization. Alfred W. Crosby provided the classic explanation of European's ecological advantages in Neo-Europes (places in which Europeans and their descendants enjoyed a demographic takeover—North and South America, Australia, New Zealand) in *Ecological Imperialism: The Biological Expansion of Europe, 900–1900* (New York: Cambridge University Press, 1986). Recently, David Igler showed the connection between trade and disease in the eastern Pacific in "Diseased Goods: Global Exchanges in the Eastern Pacific Basin, 1770–1850" (*American Historical Review* 109 [June 2004]: 692–719).

The first economic relationship that affected the Pacific West after European contact was the fur trade. Several studies of the maritime and land-based trade provide essential context but little ecological information; see James R. Gibson, *Otter Skins, Boston Ships, and China Goods: The Maritime Fur Trade of the Northwest Coast, 1785–1841* (Seattle: University of Washington Press, 1992); Alexandra Harmon, *Indians in the Making: Ethnic Relations and Indian Identities Around Puget Sound* (Berkeley: University of California Press, 1998); and Elizabeth Vibert, *Traders' Tales: Narratives of Cultural Encounters in the Columbia Plateau, 1807–1846* (Norman: University of Oklahoma Press, 1997). Adele Ogden discusses some environmental ramifications of the sea otter trade in "Sea Otters Encounter Russians" (in *Green Versus Gold: Sources in California's Environmental History*, edited by Carolyn Merchant [Washington, DC: Island Press, 1998]: 89–98). An excellent study that demonstrates the intertwined imperial and economic motives that combined to decimate beavers and their habitat along the Snake River Plain is Jennifer Ott's "'Ruining' the Rivers in the Snake Country: The Hudson's Bay Company's Fur Desert Policy" (*Oregon Historical Quarterly* 104 [Summer 2003]: 166–195).

A series of place-based studies ably demonstrate the impact of initial Euro-American colonization on Native peoples and local landscapes. For California, the best overview is William Preston's "Serpent in the Garden: Environmental Change in Colonial California" in *Contested Eden: California before the Gold Rush* (edited by Ramón A. Gutiérrez and Richard J. Orsi [Berkeley: University of California Press, 1998]: 260–298). Two essays in *Green Versus Gold: Sources in California's Environmental History* demonstrate the ecological change of California rangelands because of livestock grazing; see Raymond F. Dasmann, "The Rangelands" (194–199) and Paul F. Starrs, "California's Grazed Ecosystems" (199–205).

For the Northwest, a number of studies explore the nineteenth century's era of adjustment. The first environmental history of the region is Richard White's case study of Island County, Washington, *Land Use, Environment, and Social Change: The Shaping of Island County, Washington* (Seattle: University of Washington Press, 1980. Reprint, Seattle: University of Washington, 1992); it remains highly influential for the region and field as a whole. Indeed, in many ways, the following case studies of other subregions largely recapitulate White's study: on the Willamette Valley, see Peter G. Boag, *Environment and Experience: Settlement Culture in Nineteenth-Century Oregon* (Berkeley: University of California Press, 1992); for the region west of the Cascades, see Robert Bunting, *The Pacific Raincoast: Environment and Culture in an American Eden, 1778–1900* (Lawrence: University Press of Kansas, 1997); for the Blue Mountains of northeastern Oregon, see Nancy Langston, *Forest Dreams, Forest*

Nightmares: The Paradox of Old Growth in the Inland West (Seattle: University of Washington, 1995); and for Oregon as a whole, see William G. Robbins, *Landscapes of Promise: The Oregon Story, 1800–1940* (Seattle: University of Washington Press, 1997). These studies together describe how European ideas and actions, especially those centered on establishing and growing the capitalist economy, reduced diverse ecosystems into increasingly monocultural landscapes. In turn, simplified environments frequently complicated plans Euro-Americans had for controlling nature for economic purposes.

Surveys of various economies that developed in the nineteenth century after initial colonization reveal these economic and environmental values. The previous case studies move beyond initial colonization to delineate economic development, focusing on agricultural and timber economies. Further studies of nineteenth-century agriculture include a classic historical geography of the inland Northwest in D. W. Meinig, *The Great Columbia Plain: A Historical Geography, 1805–1910* (1968. Reprint, Seattle: University of Washington Press, 1995). Andrew Duffin takes a more critical approach in "Remaking the Palouse: Farming, Capitalism, and Environmental Change, 1825–1914" (*Pacific Northwest Quarterly* 95 [Fall 2004]: 194–204). An excellent and innovative study of agriculture in southern Idaho is Mark Fiege's *Irrigated Eden: The Making of an Agricultural Landscape in the American West* (Seattle: University of Washington Press, 1999), which abandons the traditional declensionist narrative of environmental history in favor of seeing hybrid landscapes where it becomes nearly impossible to disentangle the human from the natural.

For California, especially, agricultural development depended on manipulating scarce water resources. The literature on reclamation is vast and full of controversy, although much of the debate centers on developments within the twentieth century. The best overview of California is Norris Hundley Jr.'s *The Great Thirst: Californians and Water: A History*, revised edition (Berkeley: University of California Press, 2001). Donald Worster offers a highly critical study of reclamation ideology, seeing state power combining with business interests to hurt small landowners and the West's ecology. See his *Rivers of Empire: Water, Aridity, and the Growth of the American West* (New York: Oxford University Press, 1985). Marc Reisner's journalistic account covers much of the same ground in *Cadillac Desert: The American West and Its Disappearing Water*, revised and updated (New York: Penguin Books, 1993). Fiege's *Irrigated Eden* and Hundley's *The Great Thirst* see less of a conspiracy of interests than Worster or Reisner, while simultaneously recognizing significant ecological transformations. Donald J. Pisani also sees less monolithic state power in *To Reclaim a Divided West: Water, Law, and Public Policy, 1848–1902* (Albuquerque: University of New Mexico Press, 1992), which emphasizes fragmentation and national concerns. He continues his story in *Water and American Government: The Reclamation Bureau, National Water Policy, and the West, 1902–1935* (Berkeley: University of California, 2002), but here he seems to find Worster slightly more persuasive. Ultimately, these studies tell more about engineers than farmers or other westerners.

Additional studies address ranching and grazing. The best overview is Terry G. Jordan's *North American Cattle-Ranching Frontiers: Origins, Diffusion, and Differentiation* (Albuquerque: University of New Mexico, 1993). In *Counting Sheep: From Open Range to Agribusiness on the Columbia Plateau* (Seattle: University of Washington Press, 1982), Alexander Campbell McGregor tells of the development of a large-scale sheep-raising operation. David Igler's *Industrial Cowboys: Miller & Lux and the Transformation of the Far West, 1850–1920* (Berkeley: University of California Press, 2001) uncovers the incredible scale of economic mobilization and integration in California's leading agricultural enterprise.

Mining did much to shape the Far West. Essays in *Green Versus Gold: Sources in California's Environmental History* depict the impact and importance of the California Gold Rush; see Randall Rohe, "Mining's Impact on the Land" (125–135) and Robert Kelley, "Mining on Trial" (120–125), which emphasizes the legal struggle to halt hydraulic mining. Most recently, Andrew C. Isenberg has produced an important study of mining in California. *Mining California: An Ecological History* (New York: Hill and Wang, 2005) suggests that the Gold Rush initiated a pattern of initial exploitation, followed rapidly by outside investment and brisk industrialization, resulting in far-reaching ecological damage. Isenberg argues that the pattern is also reproduced in timber and agricultural economies in California first and then throughout the West. Katherine G. Aiken's *Idaho's Bunker Hill: The Rise and Fall of a Great Mining Company, 1885–1981* (Norman: University of Oklahoma Press, 2005) explains the rise of corporate mining and its economic impacts in northern Idaho.

Fishing is the last industry that arose prominently in the nineteenth century. Two stellar studies examine the industry. Arthur F. McEvoy's *The Fisherman's Problem: Ecology and Law in the California Fisheries, 1850–1980* (New York: Cambridge University Press, 1986) examines the ways legal traditions have shaped fisheries exploitation in California and challenges traditional explanations of declining yields. In the Northwest, Joseph E. Taylor III argues that simple political decisions, disguised as science, have always shaped fisheries management and elided the complex social and ecological causes of salmon decline in *Making Salmon: An Environmental History of the Northwest Fisheries Crisis* (Seattle: University of Washington Press, 1999). Both McEvoy's and Taylor's work offer sophisticated explanations for complex environmental histories.

These various economic developments powerfully incorporated the Far West into a national and global economy while ransacking natural resources with remarkable rapidity. James Willard Hurst's legal treatise, *Law and the Conditions of Freedom in the Nineteenth-Century United States* (Madison: University of Wisconsin Press, 1956), helps contextualize the legal and economic foundation of this exploitation. The response to this industrialization of nature took various forms, often called conservation. The best overview of the institutional response and the unfortunate longevity of many of these programs is found in Charles F. Wilkinson, *Crossing the Next Meridian: Land, Water, and the Future of the West* (Washington, DC: Island Press, 1992). The classic statement of conservation, though, remains Samuel P. Hays, *Conservation and the Gospel of Efficiency: The Progressive Conservation Movement, 1890–1920* (1959; reprint, New York: Atheneum, 1975). Hays argued that federal managers attempted to rationalize resource exploitation to control it and make it as efficient as possible. Donald J. Pisani sees more diversity among conservationists; see "The Many Faces of Conservation: Natural Resources and the American State, 1900–1940" (in *Taking Stock: American Government in the Twentieth Century*, edited by Morton Keller and R. Shep Melnick [New York: Woodrow Wilson Center Press and Cambridge University Press, 1999]: 123–155). The work of Robert Gottlieb emphasizes urban and industrial roots to the conservation and environmental movements to find the movement far more inclusive in terms of race, class, and gender; see *Forcing the Spring: The Transformation of the American Environmental Movement* (Washington, DC: Island Press, 1993).

While the reclamation efforts were one attempt at regulating nature more efficiently, strong examples also exist in forestry. See Langston's *Forest Dreams, Forest Nightmares* for a model showing how conservation-minded foresters attempted reform with surprising results, which revealed that controlling nature would not proceed as smoothly as conservationists expected. Stephen J. Pyne's *Year of the Fires: The Story of the Great Fires of 1910* (New York: Viking, 2001) tells the engaging story of immense fires in Idaho in 1910 and the fire-control reforms toward total suppression that resulted in the aftermath. Richard A. Rajala presents a neo-Marxist view of forestry reform in *Clearcutting the Pacific Rain Forest: Production, Science, and Regulation* (Vancouver: UBC Press, 1998), seeing the timber industry turn to technology to control nature and workers and capturing state forestry reforms to blunt any true reform. Best at capturing the spirit of conservationists is Gifford Pinchot's *The Fight for Conservation* (1910; reprint, Seattle: University of Washington Press, 1967), a book that forcefully announced conservation as a crusade.

Although much Progressive Era conservation focused on traditional rural resource-extractive industries, the habit of transforming nature efficiently infected urban westerners. In Seattle, engineers straightened rivers, built locks to connect inland waterways, and seriously disrupted both the preexisting ecosystems and subsistence patterns; see Matthew W. Klingle, "Fluid Dynamics: Water, Power, and the Reengineering of Seattle's Duwamish River" (*Journal of the West* 44 [Summer 2005]: 22–29); Lisa Mighetto, "The Strange Fate of the Black River: How Urban Engineering Shaped Lake Washington and the Duwamish River Watershed," (*Journal of the West* 44 [Fall 2005]: 47–57); and Coll Thrush, "City of the Changers: Indigenous People and the Transformation of Seattle's Watersheds" (*Pacific Historical Review* 75 [February 2006]: 89–117). Robbins discusses similar transformations of Oregon's Willamette River *Landscapes of Promise.*

For California, the urban conservation stories have long centered on two episodes of urban imperialism with San Francisco and Los Angeles acquiring the distant watersheds to promote metropolitan growth. Robert Righter's recent study, *The Battle Over Hetch Hetchy: America's Most Controversial Dam and the Birth of Modern Environmentalism* (New York: Oxford University Press, 2005), recasts the familiar story of San Francisco's imperialistic takeover of Tuolumne River as a story of public versus private power and urban versus tourist use of the valley. Gray Brechin accounts for San Francisco's powerful political reach in *Imperial San Francisco: Urban Power, Earthly Ruin* (Berkeley: University of California Press, 1999). For Los Angeles, see Reisner's *Cadillac Desert,* while John Walton's *Western Times and Water Wars: State, Culture, and Rebellion in California* (Berkeley: University of California Press, 1992) focuses on the rebellious response in the Owens Valley. William Deverell's *Whitewashed Adobe: The Rise of Los Angeles and the Remaking of Its Mexican Past* (Berkeley: University of California, 2004) and Andrew Hurley's "Aqueducts and Drains: A Comparison of Water Imperialism and Urban Environmental Change in Mexico City and Los Angeles" (*Journal of the West* 44 [Summer 2005]: 12–21) show the social problems caused by urban hydraulic engineering. The irony of flooding in arid Los Angeles is well documented in Jared Orsi's "Flood Control Engineering in the Urban Ecosystem," (in *Land of Sunshine: An Environmental History of Metropolitan Los Angeles,* edited by William Deverell and Greg Hise [Pittsburgh: University of Pittsburgh Press, 2005], 135–151). Orsi's study also shows how experts believed they could engineer their way out of water problems. Other studies of water in Los Angeles can be found in Mike Davis, *Ecology of Fear: Los Angeles and the Imagination of Disaster* (New York: Metropolitan Books, 1998) and Blake Gumprecht, *The Los Angeles River: Its Life, Death, and Possible Rebirth* (Baltimore, MD: The Johns Hopkins University Press, 1999). Finally, Greg Hise and William Deverell republished a 1930 plan for urban landscape reform that would have developed the metropolis in more ecologically sensitive ways; see *Eden by Design: The 1930 Olmsted-Bartholomew Plan for the Los Angeles Region* (Berkeley: University of California Press, 2000).

The early twentieth century also found westerners increasingly interested in protecting nature for aesthetic and recreational purposes, sometimes called the preservationist movement. The classic two histories remain Roderick Nash, *Wilderness and the American Mind*, 3rd edition (New Haven: Yale University Press, 1982) and Alfred Runte, *National Parks: The American Experience*, 3rd edition (Lincoln: University of Nebraska Press, 1997). Both see the movement toward park preservation as part of an American exceptionalism that took pride of place in the West. New studies have revised their arguments. Mark David Spence in *Dispossessing the Wilderness: Indian Removal and the Making of the National Parks* (New York: Oxford University Press, 1999) shows how park preservation was intimately tied to removing American Indian inhabitants from the landscape. Paul S. Sutter has shown how a feeling against the rising commercialization of wilderness led to further drives for wilderness preservation in *Driven Wild: How the Fight Against Automobiles Launched the Wilderness Movement* (Seattle: University of Washington Press, 2002). David B. Louter examines the role of automobiles in national parks using Olympic National Park as a case study; see "Wilderness on Display: Shifting Ideals of Cars and National Parks" (*Journal of the West* 44 [Fall 2005]: 29–38). J. M. Neil's *To the White Clouds: Idaho's Conservation Saga, 1900–1970* (Pullman: Washington State University Press, 2005) documents efforts at recreational development in Idaho, effectively showing that local developments were as important as national frameworks.

By the mid-twentieth century, greater efforts at wilderness preservation, recreational development, and increased conflict were common. In Idaho, local politicians long tried to preserve the Sawtooth Mountains; see Sara E. Dant Ewert, "Peak Park Politics: The Struggle over the Sawtooths, from Borah to Church" (*Pacific Northwest Quarterly* 91 [Summer 2000]: 138–149). Kevin R. Marsh explores the efforts to preserve wilderness in the Cascade Mountains, showing the importance of specific locales in "'This Is Just the First Round': Designating Wilderness in the Central Oregon Cascades, 1950–1964" (*Oregon Historical Quarterly* 103 [Summer 2002]: 210–233). Adam M. Sowards details one significant conservationist's efforts on behalf of Northwestern landscapes in "William O. Douglas's Wilderness Politics: Public Protest and Committees of Correspondence in the Pacific Northwest" (*Western Historical Quarterly* 37 [Spring 2006]: 21–42). William G. Robbins explores various conflicts and modern efforts at preservation in *Landscapes of Conflict: The Oregon Story, 1940–2000* (Seattle: University of Washington

Press, 2000). For Southern California, the struggle to preserve beaches against oil exploration is illustrated by Sarah S. Elkind in "Black Gold and the Beach: Offshore Oil, Beaches and Federal Power in Southern California" (*Journal of the West* 44 [Winter 2005]: 8–17). Susan R. Schrepfer documents the decades-long battle to preserve redwoods in *The Fight to Save the Redwoods: A History of Environmental Reform, 1917–1978* (Madison: University of Wisconsin Press, 1983). From a national perspective, Richard West Sellars examines the often quixotic approach of the National Park Service to promote ecological integrity in *Preserving Nature in the National Parks: A History* (New Haven, CT: Yale University Press, 1997).

The middle of the twentieth century also saw intense efforts to exploit new energy resources with significant environmental results. An overview for California is James C. Williams, *Energy and the Making of Modern California* (Akron, OH: University of Akron Press, 1997). Richard White's extended meditation on energy in the Columbia River in which he argues for a near-merger of the organic and the machine is *The Organic Machine: The Remaking of the Columbia River* (New York: Hill and Wang, 1995). Protest over nuclear power is described in Thomas Wellock, "The Battle for Bodega Bay" (in *Green Versus Gold: Sources in California's Environmental History*, ed. Carolyn Merchant [Washington, DC: Island Press, 1998], 344–349), while the long-term ecological and health impacts of nuclear development are contained within Michele Stenehjem Gerber, *On the Home Front: The Cold War Legacy of the Hanford Nuclear Site*, 2nd edition (Lincoln: University of Nebraska Press, 2002).

The themes of economic exploitation with dire ecological results continued into the second half of the twentieth century. Accordingly, many studies introduced above continued the subject matter developed in an earlier period. For reclamation, see Hundley's *The Great Thirst* and Worster's *Rivers of Empire*; for forestry see Robbins's *Landscapes of Conflict* and Langston's *Forest Dreams, Forest Nightmares*; for fisheries, see McEvoy's *The Fisherman's Problem* and Taylor's *Making Salmon*.

Additionally, some trends intensified or changed. For modern agriculture, California offers an exemplar of agribusiness. A sharp critical analysis is found in Richard A. Walker's *The Conquest of Bread: 150 Years of Agribusiness in California* (New York: The New Press, 2004). Meanwhile, Julie Guthman explores the rise of organic farming and its takeover by agribusiness in the important book, *Agrarian Dreams: The Paradox of Organic Farming in California* (Berkeley: University of California Press, 2004). Robbins's *Landscapes of Conflict* covers Northwestern agricultural developments well, arguing that modern farms became increasingly large and chemical-dependent.

Forestry also evolved into increasingly exploitive trends as the post–World War II timber economy relied on new technologies and available public timber resources with devastating results. The best overview is Paul W. Hirt, *A Conspiracy of Optimism: Management of National Forests since World War Two* (Lincoln: University of Nebraska Press, 1994) in which he argues that the Forest Service maintained overly optimistic harvest goals knowing they were neither ecologically nor economically sustainable. Robbins's *Landscapes of Conflict* and Langston's *Forest Dreams, Forest Nightmares* offer good case studies in Oregon of similar problems. Postwar timber exploitation led to fragmented habitats and threatened species; thus, the Pacific West's forest history is tied to endangered species and environmentalists' controversies. William Dietrich's *The Final Forest: The Battle for the Last Great Trees of the Pacific Northwest* (New York: Penguin, 1992) offers an outstanding journalistic account that presents all sides sympathetically. In her account of the radical organization Earth First!, Susan Zakin explores those activists' efforts to protest public land logging; see *Coyotes and Town Dogs: Earth First! and the Environmental Movement* (Tucson: University of Arizona Press, 1993). An interview with Earth First! activist Judi Bari is found in Douglas Bevington, "Earth First! in Northern California: An Interview with Judi Bari" (in *The Struggle for Ecological Democracy: Environmental Justice Movements in the United States*, edited by Daniel Faber [New York: Guilford Press, 1998], 248–271). From another perspective, Alston Chase provides a sharp attack on environmentalists and ecologists who he sees as having compromised science for activism in his occasionally polemical book *In a Dark Wood: The Fight over Forests and the Rising Tyranny of Ecology* (New York: Houghton Mifflin, 1995).

Environmental racism and environmental justice are important topics that have received relatively scant attention for the Pacific West. Laura Pulido's *Environmentalism and Economic Justice: Two Chicano Struggles in the Southwest* (Tucson: University of Arizona Press, 1996) and Mary Pardo's "Mexican American Women Grassroots Community Activists: 'Mothers of East Los Angeles'" (in *A Sense of the American West: An Environmental History Anthology*, edited by James E. Sherow [Albuquerque: University of New Mexico Press, 1998]: 243–260) offer important studies of Mexican farmworkers' problems with ecological health and community organizing, respectively. Ellen Stroud has produced an important piece in "Troubled Waters in Ecotopia: Environmental Racism in Portland, Oregon" (*Radical History Review* 74 [Spring 1999]: 65–95). For context of pollution and industrial collusion, generally, see Devra Davis, *When Smoke Ran Like Water: Tales of Environmental Deception and the Battle Against Pollution* (New York: Basic Books, 2002).

Modern environmental problems require creative approaches to solve. Two examples that show the potential when multiple parties agree to work toward a solution are found in Nancy Langston, *Where Land and Water Meet: A Western Landscape Transformed* (Seattle: University of Washington Press, 2003) and William A. Shutkin, *The Land that Could Be: Environmentalism and Democracy in the Twenty-First Century* (Cambridge: Massachusetts Institute of Technology Press, 2000). Although neither case study is a perfect solution, each demonstrates possibilities of democratic involvement in environmental management.

Several resources are useful for environmental history chronologies or definitions. The most readily available overviews are Mark Grossman, *The ABC-CLIO Companion to the Environmental Movement* (Santa Barbara, CA: ABC-CLIO, 1994); Carolyn Merchant, *The Columbia Guide to American Environmental History* (New York: Columbia University Press, 2002); and Roderick Frazier Nash, editor, *American Environmentalism: Readings in Conservation History*, 3rd edition (New York: McGraw-Hill, 1990).

This bibliographic essay highlights the main works used to explore the central topics of this book. No doubt, many other important studies exist; they were not deliberately overlooked. Moreover, scholars are rapidly producing new environmental histories that push the field in new directions. Within just a few years, then, it is likely that this essay would be very different. That, of course, is the sign of a healthy, vibrant field.

INDEX

ABOUT THE AUTHOR

Adam M. Sowards, Ph.D., is assistant professor of history and director of the Institute for Pacific Northwest Studies at the University of Idaho in Moscow, Idaho, where he is also a faculty member in the interdisciplinary environmental science program. He earned his Ph.D. from Arizona State University in 2001. He has researched environmental history from various perspectives, including histories of forests, reclamation, and ranching and has written biographical studies of major figures in American conservation. Professor Sowards has published multiple articles on environmental and western U.S. history in such scholarly journals as *Western Historical Quarterly, Journal of the West, Idaho Yesterdays,* and *Journal of the Southwest.* He has also contributed to essays in *John Muir in Historical Perspective* (Sally M. Miller, ed., 1999, Peter Lang) and *The Human Tradition in the American West* (Benson Tong and Regan A. Lutz, eds., 2002, Scholarly Resources). He is currently completing a book on U.S. Supreme Court Justice William O. Douglas's environmental politics and beginning a study that investigates the historical relationship between urban areas and surrounding forests.

DATE DUE